PETROLEUM
DEVELOPMENT
GEOLOGY

PETROLEUM DEVELOPMENT GEOLOGY

Second Edition

Parke A. Dickey
Emeritus Professor of Geology
University of Tulsa

PennWell Publishing Company
Tulsa, Oklahoma

Copyright © 1979, 1981 by
PennWell Publishing Company
1421 South Sheridan Road/P. O. Box 1260
Tulsa, Oklahoma 74101

Library of Congress Cataloging in Publication Data

Dickey, Parke Atherton, 1909–
    Petroleum development geology.
    Bibliography.
    Includes index.
    1. Petroleum—Geology. I. Title
TN870.5.D52   1981   553.2′8   81-11943
ISBN 0-87814-174-X   AACR2

Printed in the United States of America

2   3   4   5   85   84   83   82

Contents

Preface

The geologist has contributed to the petroleum industry since its beginning. Until recently, primary emphasis has been in the development and application of methods for discovering new oil and gas fields. Most petroleum geologists therefore work in exploration departments, and once a field has been discovered its exploitation has largely been left to the drillers and petroleum engineers.

In recent years the need for geologists in oil- and gas-field development has become more obvious. Oil reservoirs are seldom uniformly porous and permeable rocks, folded into domes. They consist of a complex of permeable sands formed in narrow river channels and beaches, separated from each other both vertically and laterally by impermeable shale. The average oil reservoir resembles more a plate of spaghetti than a stack of pancakes. Carbonate reservoirs are still more complicated because secondary solution and fracturing have altered the primary porosity. The pattern of oil and gas wells ought not to be a uniform square grid at an arbitrarily determined spacing. Rather it should be designed to fit the geology of the reservoir, so that the maximum amount of oil and gas can be recovered with the minimum number of wells, in a time determined by economic considerations.

The application of enhanced recovery methods requires a detailed knowledge of permeability distribution in the reservoir; otherwise expensive recovery media flow off in unpredictable directions and recover little additional oil, or even none at all.

This book was written as a text for an advanced geology course at the University of Tulsa, and it has been used in short courses. It is intended primarily for geologists who are being assigned to development work. However, most of the techniques peculiar to petroleum geology are described, and it is hoped that it will be useful as a textbook for university courses in petroleum geology.

Some of the methods in this book are used by development geologists but are not taught in universities. New employees are normally sent to sit on wells, where they are supposed to learn about drilling and completion methods. This experience is indispensible to a practical understanding, but it should be supplemented with some sort of instruction on principles. That's where *Petroleum Development Geology* comes in.

The book should also be useful to engineers engaged in detailed reservoir studies involving mathematical simulation and enhanced recovery projects. These people need to know how the geological data they depend on was obtained, and they must visualize petroleum reservoirs as they exist in the real world.

Most of our methods are undergoing rapid improvement. By the time newer techniques can be incorporated into a text, they are obsolete. The emphasis here, therefore, is on principles rather than on operating practice. The latter can be learned best in the field and from service company personnel and brochures.

This work was designed primarily as a teaching aid, not a reference book. The references at the end of each chapter list some, but not all, of the authoritative papers on the subject. This is not a summary of other people's opinions as published in the literature but is based on my own experience and understanding. Some of the ideas and principles are new, introduced here for the first time. Many ideas which are obsolete or which I do not agree with are not mentioned. However, no broadly based text like this can contain much original material, and each chapter attempts to summarize the current thinking on that topic.

Preface to the Second Edition

The printing of a second edition of *Petroleum Development Geology* has provided the author an opportunity to improve the book. Typographical and other errors have been corrected, and many additions have been made. The chapter on carbonate rocks has been expanded, and a new chapter has been added on oil in carbonate reservoirs. The chapter on fluid behavior in reservoirs has been completely rewritten. A final chapter has been added on evaluating a discovery—where to drill the second and third wells.

Since the publication of the first edition three years ago, the application of geology to oil and gas field development has increased substantially. Published reports of enhanced-recovery projects indicate that it is now customary to precede the field development with a detailed geological study. Many companies have organized task forces of geologists, engineers, and geophysicists working together to plan development programs. This change of policy is very gratifying and will lead to increased production of oil and gas.

Parke A. Dickey
Owasso, Oklahoma
November 1981

1

The Objectives of Development Geology

1. DEVELOPMENT GEOLOGY

The development or production geologist applies his knowledge and skills to the development of an oil or gas field after the first successful exploratory well has been drilled. He must take the geophysical, geological, and engineering information obtained during the exploration period, reinterpret it with the results of the first well, and then plan a development program.

Information on the regional geology and the samples, cores, and wire-line logs from the first one or two wells make it possible to predict the size, shape and lateral patterns of the reservoir rock. Drillstem and formation interval tests provide data on the reservoir parameters such as porosity, permeability, fluid saturation, gas-oil ratio, and water-oil contact.

With this information a model can be constructed that will predict the reservoir performance and the total recoverable reserves. Such early estimates are naturally only approximate, but they are most needed at the start of a development program when decisions must be made. The first decision is whether to proceed with the development. The next step is to estimate the number of wells, their approximate location, and the production facilities which will be required. With estimates of daily oil and gas production, it is possible to draw up a table predicting the yearly cash flow from the project. This table is required to justify the investment of capital in the development.

The need for geological information on the oil reservoir is especially acute when an old field is being evaluated for enhanced recovery. When fluids are injected to recover the residual oil, they must sweep the reservoir effectively. The whole development plan, including choice of method, well spacing, pattern, and completion practice, must conform to the geology of the reservoir.

An oil or gas prospect is developed by extensive and very expensive studies of regional geology, including subsurface geology and geophysics. The exploratory wells may cost several million dollars to drill. It would be foolish to risk the loss of a prospect as a result of ignorance or carelessness on the part of the driller, the engineer, or the well-site geologist. Some formal training is very desirable for all three.

A good explorationist does not turn over his responsibility to an engineer when a rig is moved onto his prospect. He is the best-qualified person to plan the drilling program, including mud weights and casing points, as well as logging, testing, and stimulation. He is also the best-qualified person to make the economic forecasts and evaluations.

2. EXAMPLES OF APPLICATIONS OF DEVELOPMENT GEOLOGY

The function of petroleum development geology can best be explained by giving examples of its application to actual oil fields. The first, Brent, shows how it was possible to plan the whole development and estimate the reserves while the first four wells were being drilled. The second, Hawkins, is an example of an oil field for which a different production practice was designed, 35 years after its discovery. The third, Bradford, shows how detailed geological studies are used to plan secondary recovery operations.

Brent, British North Sea. One of the largest oil fields to be discovered in recent years is Brent, in the British North Sea (Kingston and Niko, 1975). It was discovered by Shell-Esso in 1971 with Shell UK as the operating company. Figure 1–1 is a seismic cross section made before the field was discovered (Bowen, 1975). It shows a thick relatively undisturbed sedimentary series to a depth of 2 to 3 seconds. Here a very prominent high amplitude reflector suggests an erosional surface, with steep east- and gentle west-dipping slopes. Below is a generally west-dipping sequence, broken by normal faults. The discovery well was drilled almost 1 km down-dip to the west of the crest of the buried high (Figure 1–2). It is in 142 m of water and lies 150 km northeast of Shetland. It was 370 km north of the nearest U.K. well control in the North Sea.

It was reasonable to suppose that the undisturbed sediments were Tertiary or Cretaceous, but the age of the strata below the unconformity was quite unknown—they could have been Precambrian. They turned out to be Jurassic. Below the unconformity was about 115 m of black and gray shales which lie unconformably on a section of Jurassic sands deposited in a fluvio-deltaic environment. The first well found an oil column of 60 m while the second well, drilled on the crest of the structure a year later (Figure 1–2), found 53 m of gas overlying 114 m of net oil sand. Both wells found water below the oil. The second well cored the entire productive interval, and three production tests were made. Many formation interval tests showed that the oil was under-saturated with gas at the oil-water contact, but the solution gas-oil ratio increased upward to the gas-oil contact. The second well was drilled more deeply and found the Lower Jurassic Statfjord sands below, which were water bearing at this location.

Because of the remote and expensive location, decisions regarding further development had to be made quickly. These involved the design of platform and production facilities, which in turn required a plan for the production method. Geological studies of depositional environment supplied predictions of the nature, extent, and heterogeneities of the sand bodies. The seismic data gave the extent of the field above the known water-oil contact. The reserves were estimated as 1.0 billion bbl (140,000,000 tons) of oil and 2.0 trillion cu ft (56 billion cu m) of gas. It seemed important to maintain a high oil production

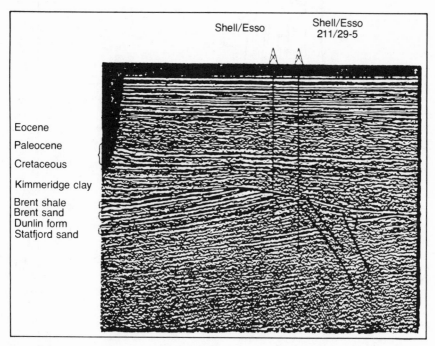

Fig. 1–1 Seismic line near Brent field *(after Bowen, courtesy Applied Science Publishers Ltd.)*

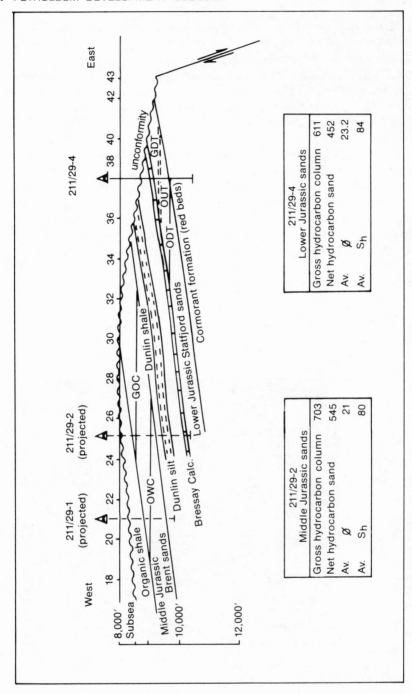

Fig. 1–2 Preliminary cross section, Brent field *(after Kingston and Niko, courtesy JPT)*

rate as soon and as long as possible to repay the high investment cost of development. The design of the first platform therefore planned production of 100,000 bbl/d (14,000 tons per day) injection of 200,000 bbl/d (31,000 cu m/d) of water, and all the produced gas.

A year later, in 1973, the third well confirmed the extension of the field to the north, and the fourth well, drilled on the east flank of the high, found the Lower Jurassic sands productive. This extended the pool to the east and added 0.3 billion bbl (43 million tons) of oil and 1.4 trillion cu ft (40 billion cu m) of gas but did not substantially change the development plans.

Later, more elaborate reservoir simulation studies were undertaken. Geological studies of the cores subdivided the reservoir into four depositional cycles, and it was concluded that they would behave as three separate reservoirs. The upper and lower show good lateral continuity, but the middle (lower deltaic plain) cycles 2 and 3 may have poor lateral continuity. An important uncertainty which remains is the strength of the natural water encroachment.

As a result of these models it was possible to predict the ultimate reserves, the optimum production rate, and the reservoir performance and to design the production facilities, after only 4 wells had been drilled, in a great multibillion-barrel field.

Prudhoe Bay, Alaska, is another giant field whose geology was worked out in great detail after only a few wells had been drilled. Based on these studies, the enormous expense of the transalaska pipeline was committed (Eckelman et al., 1975).

Hawkins, Texas. The Hawkins field in East Texas is an example of an old field whose production mechanism was changed as a result of the careful studies of the geology and past reservoir performance (King and Lee, 1976). The field was discovered in 1940 and produced more than 500 million bbl (71 million tons) through 1974. The producing sand is the Cretaceous Woodbine, which was extensively cored and studied geologically during the original development. Hawkins is in the same aquifer as East Texas, discovered in 1930, which had produced enormous amounts of oil and water, and this had lowered the pressure as far away as Hawkins. The coring revealed a layer of asphalt above the oil-water contact in the western segment of the field (Figure 1–3). This had been a pressure seal, so that the initial reservoir pressure was 1,985 psi (13,680 kPa), while the aquifer pressure was 1,830 psi (12,620 kPa). East of the main N-S fault the asphalt seal was less effective and the field operated under a strong water drive.

In the western segment, water penetrated the asphalt layer in the north, providing a water drive there. In the south, production was by

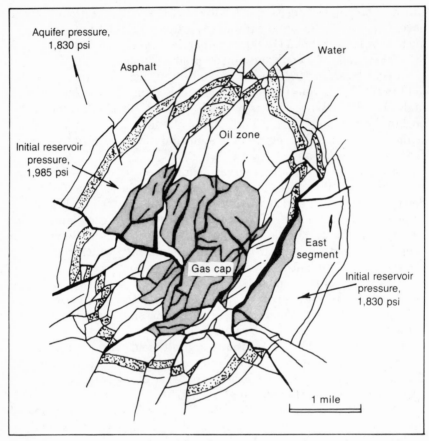

Fig. 1–3 Hawkins field showing location of asphalt layer and faults *(after King and Lee, courtesy JPT)*

gravity drainage, which caused the gas cap to migrate from north to south. Oil was invading the gas cap on the north. As a result of a study in 1974, it was decided to inject inert gas (products of combustion) into the gas cap to maintain the pressure and stop oil migration. Oil will be produced mainly by gravity drainage. It is estimated that an additional 189 million bbl (27 million tons) of oil can be recovered by this production practice.

Bradford, Pennsylvania. The Bradford field was discovered in 1871 and would have produced about 200 million bbl (28 million tons) of oil under primary solution gas drive. It was the first field into which water was deliberately injected in order to increase oil recovery (Fettke,

1950). The operators recognized that secondary recovery required an understanding of reservoir geology and fluid behavior, so the foundations of the modern science of reservoir behavior were laid in Pennsylvania in the 1930s. Maps drawn on top of the sand showed a sheet sand with an anticlinal structure. Careful studies of cable-tool samples showed that, instead, the pool consisted of a series of separate sand bodies which overlapped like shingles, all dipping to the west (Figure 1–4) (Wilson, 1950). Attempts were made to inject water at different rates into the different units.

Hydraulic fracturing increases the rate at which water can be injected but tends to open the natural joints so that water flows directly to the producing wells. The pattern is therefore arranged so that the lines of intake and producing wells are parallel to the natural fracture system. The water then pushes the oil in a direction at 90° to the artificially opened fractures.

Other interesting examples of plans for secondary and tertiary recovery based on geological interpretations of the size and shape of the individual sand bodies are Rangely, Colorado (Larson, 1974), Monahans, Texas (Dowling, 1975), and Louden, Illinois (Harris, 1975).

3. GEOLOGICAL FACTORS AFFECTING RESERVOIR BEHAVIOR

In the old days when oil fields produced under primary mechanisms, it was advantageous but not really essential to have an accurate picture of the geology of the reservoir. The oil came out of its own accord. Heterogeneities and other geological characteristics helped or hindered recovery, but there was not much the operator could do to take advantage of them.

During most of the period from 1940 to 1960, petroleum engineers tried to convince the operators that it was not necessary to drill wells on close spacings to recover the oil. One well to 40, 80, or even 160 acres (16, 32, or 64 hectares) would drain the reservoir, and much money would be saved by not drilling unnecessary wells. In the United States there usually are many different operators in the same oil pool. The companies who wanted to save money by producing the same amount of oil with fewer wells, say one to 40 acres, were unable to do so if other operators drilled more wells, say 4 wells on 40 acres (16 h). The oil would migrate toward the area with the closest well spacing.

When secondary recovery came along and fluids were injected into the reservoir, it became obvious to the more intelligent operators that they did not sweep evenly but bypassed over, under, or alongside the beds containing much of the residual oil. In secondary recovery, and

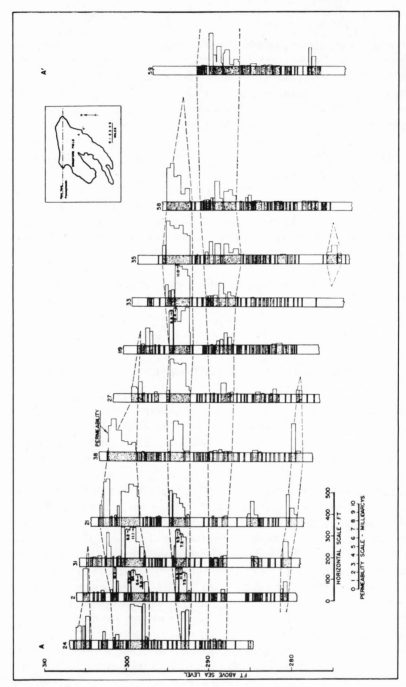

Fig. 1—4 Cross section showing individual sand reservoirs in Bradford field, Pennsylvania *(after Wilson, courtesy API)*

8

even more in tertiary recovery, a good sweep is essential to the success of the project. Generally sweep efficiency can be increased by shortening the well spacing. The principal reason why tertiary oil is much more expensive to produce than primary or secondary oil is that it requires many more wells.

Many of the enhanced-recovery methods gave good results in the laboratory but very poor results when they were tried out in the field, and the concept of *conformance* was developed. It was hoped that a mathematical expression could be derived to relate ideal behavior to what happened in the real and imperfect world of geology. This term revealed a serious fallacy in thinking. The method of producing the reservoir must *conform* to the geology; the geology cannot be expected to conform to the mathematical equations. When waterflooding or enhanced-recovery operations are contemplated, they must be planned to fit the geology of the reservoir. The type of enhanced-recovery method chosen, the well spacing and pattern, and the whole operation must be planned to conform to the local geology.

4. SIZE AND SHAPE OF A RESERVOIR

Continuity of strata. Sandstones are always stratified and are usually multiple, separated by thin partings of shale. The individual sand bodies represent single beaches, point bars, or channels that are piled alongside or on top of each other. Permeability is good within each element, but poor from one element to another. The permeability barriers are usually too thin to show on electric logs, although they can be clearly seen in cores.

In intensive operations, it is desirable to distinguish the individual sand bodies and determine their permeability separately. If wells are drilled closely enough, it is possible to delineate each individual bed. Then it is possible to make selective well completions and inject fluids into the different sand bodies under different pressures.

The more continuous sand bodies are laterally, the better the chance for enhanced recovery. Horizontal partings probably help keep the fluids moving parallel to the bedding. Unfortunately, they often separate layers of markedly different permeability. Heterogeneity of the reservoir is one of the chief causes of failure in enhanced recovery.

Carbonate reservoirs are difficult to delineate in this manner. In some upstanding reefs there is good vertical permeability. In such cases, gravity drainage, gas-cap drive, or solvent drives are favored. In the more usual case of fracture porosity, the direction of the fractures can sometimes be determined, but not their average spacing or actual position.

The size and shape of the aquifer determines whether the field will produce under water drive or dissolved gas drive. Some oil sands have great lateral extent. Among these are the Woodbine of the East Texas basin and the Wilcox of the Seminole area of Oklahoma. Oil pools in these sands produce under a sustained water drive. In many other places it is common for the water to advance into the reservoir a few thousand feet and then stop its advance. This suggests that, even in large aquifers, the water which invades the reservoir is coming from the volumetric expansion of the water in the aquifer, and there is no replenishment from the surface. On the other hand, oil is most commonly found in thin, lenticular reservoirs of near-shore depositional environments. This is probably because there were not enough barriers in the big thick sands to screen out and retain the oil. It is, therefore, extremely important to make an early guess as to the size of the aquifer in a newly discovered field. The whole later performance will depend on it.

Many carbonate reservoirs are in hydraulic connection with large aquifers. This is especially true of reefs, where there may be a large volume of water-filled porosity, while the oil occupies a small volume at the tops of the cupolas. If there is good vertical permeability, water will rise up, replacing the oil as it is withdrawn. On the other hand, some reefs are isolated and do not connect to large regions of porous carbonate.

Porosity types. There is a wide variety of sizes and shapes of pores. These are determined by the original grain size distribution and even more by the amount and nature of interstitial clays and secondary silica. Pores in sandstones look like smooth-walled corridors on the one hand, or irregular holes more or less filled with fluffy clay on the other. Presumably, the smooth, uniform pores give the best recovery, but very little attention has been given to this factor. Even more important is the state of wettability of the interior surfaces, but little attention has been given to this factor either.

Porosity in carbonates differs even more widely. There may be caverns big enough to drive a 10-ton truck through alongside pores which are invisibly fine. Consequently, the behavior of fluids in carbonates is poorly understood. Efforts are usually made to compare reservoir performance in carbonates with that in sandstones. What we really need is an attempt to define quantitatively pore sizes and shapes in a particular rock, and then an attempt to explain the reservoir performance of that particular pool.

It was pointed out by Elkins (1950) that when the dissolved gas comes out of solution it forms continuous threads, first in the channels and then later in the interstitial pores. This means that gas starts to flow at higher oil saturation than in the case of sands. Therefore, gas drive is less efficient in fractured and vuggy carbonates than it is in sandstones. It is supposed that the only way to get the oil out of the interstitial pores is to drop the pressure in the channels so that the dissolved gas in the oil forces it out of the pores into the channels. The secondary sweep of gas or water will only remove oil from the channels, and it will actually block off and isolate the oil in the interstitial pores. For this reason, there have been few successful secondary recovery projects in vuggy or fractured limestones.

The importance of rock compaction in oil recovery has not been sufficiently recognized. It takes place mostly in relatively unconsolidated strata. The degree of consolidation can be determined quantitatively by shale density. It may be suggested here that if the density of shale is less than about 2.2, the possibility of compaction and surface subsidence should be contemplated.

The effect of permeability on recovery is very great. Rocks of low permeability have poor primary recovery. They consequently contain more oil in place and are therefore better prospects for enhanced recovery. On the other hand, the low permeability greatly increases the difficulty and expense of injecting fluids.

5. STRUCTURE

Dip. The efficient recoveries of gravity drainage and gas cap drive can be achieved only when the rocks have comparatively steep dips— probably 15° or more. A steep dip also helps the water drive because the higher density of water tends to keep an evenly advancing front. Very steeply dipping reservoirs with strong water drives are characteristic of the flanks of salt domes.

Faults. Nearly all anticlinal reservoirs are broken by faults. Usually they are vertical and strike at an angle of about 70° to the axis of the anticline. Normally they are seals and break the reservoir into individual units, each with its own pressure and oil-water contact. Obviously, for efficient production each fault block must be considered a separate reservoir.

In many cases, when water was injected into faulted fields, it travelled along the faults. The faults were tightly closed under normal conditions, but when injection pressures exceeded the original reservoir pressure, the fluids opened the faults and travelled along them. If the injection pressure were not cut back quickly, permanent channels would be eroded, and the reservoir would be permanently harmed.

Joints. Joints are naturally occurring vertical fractures. They are usually tightly closed at depth. However, the pressure used to inject water tends to open them. When this happens, serious bypassing takes place.

6. FLUID CONTENT

Oil in place. The most important number to determine in either primary or secondary recovery is the amount of oil in place (Bond, 1979). More enhanced-recovery projects have failed because there was insufficient oil in place than for any other reason. Difficulties in determining oil in place from either cores or electric logs will be treated at length in Chapters 3 and 4.

The original oil in place is sometimes solid tar. Paraffin wax, asphalt, and pyrobitumen can occur in reservoirs. Under core analysis and most wire-line methods of formation evaluation, it is indistinguishable from the original oil. Secondary recovery operations have been started before it was discovered that much or all of the "oil" was solid.

The determination of oil in place is most difficult before primary recovery because of the effect of drilling mud on cores and electric logs. It is critical, and equally difficult to determine, at the end of primary recovery. The success of the secondary recovery project will depend on the difference between the oil at the beginning and end of the project. It is easiest to determine after the end of secondary recovery. Presumably by this time the oil saturation is down to its irreducible minimum, and drilling and coring will not reduce it any more.

Interstitial water saturation. Under primary conditions, the relative amount of oil and water in the pores is determined by the capillary properties of the reservoir rock. If the sand is mostly wet by oil, it is usually supposed that water will be a poor recovery agent. It will go through the bigger pores and will not peel the oil off the surfaces and out of the finer pores. However, the Bradford sand has much lower water saturation than would be predicted by capillary pressure curves, so the sand must be partly oil wet. Waterflooding has been notably successful here. It has not been unequivocally shown that any reservoir is mostly oil wet.

A high water saturation in the pores increases the efficiency of recovery under gas drive (Dickey and Bossler, 1944). This is because the water tends to crowd the oil into the larger pores where it can be pushed by the gas.

Gas content. The more gas there is in solution in the oil, the better the primary recovery. This is because there is a larger amount of gas to push the oil out of the pores by its expansion. Also, large amounts of solution gas are found in oils of low viscosity. However, secondary recovery also works best on low-viscosity oils.

Oils with little gas tend to be heavy and viscous. It is believed that the primary recovery has been as little as 10 percent in some cases. Secondary recovery, except for steam injection, has not been very successful with high-viscosity oils.

The presence of a gas cap can facilitate primary recovery by affording a means for a gas cap drive. However, much oil has been irretrievably lost by depressuring an original gas cap. The oil moves up into it, wetting the sand with oil.

7. PLANNING DEVELOPMENT TO CONFORM TO GEOLOGY

Spacing. In primary recovery in a widespread, uniform sand, one well can drain an entire reservoir. Consequently, the only advantage of additional wells is that the oil is obtained faster.

It is obvious that the closer the spacing the more complete the oil recovery. However, many studies (Craze and Buckley, 1945, Arps et al., 1967) have failed to show any quantitative relation between spacing and ultimate recovery under primary production. However, many fields originally developed on a 40- or 80-acre spacing are now being redrilled more closely, showing that ultimate oil recovery is not, in fact, independent of spacing.

In the case of secondary recovery close spacing is much more necessary. When fluids are injected, either gas or water, the rate of injection must be fast enough to make the oil production rate economic. The spacing will then depend on the permeability. It is also important that both injection and producing wells penetrate the same element of the sand body.

Pattern. In enhanced recovery operations, the pattern must conform to the geology. If a sand body consists of a multiple series of beaches or bars alongside each other, the orientation of these elements should be determined. This can be done with isopach maps and electric log correlation. Often, the electric logs of the sand correlate well in one direction and hardly at all in another. The wells should be spaced closely across the sand bodies and farther apart along their long axes. Pressure readings and interference tests are useful.

Where fractures exist naturally or are produced deliberately or accidentally by high pressures, the pattern must be oriented so the direction from injection to producing wells will be at right angles to the fractures. In the case of enhanced recovery, it is usually better to have the number of injection wells at least equal to the number of producing wells. This fact is not generally recognized. If an injection well is surrounded by a ring of producers, the fluids injected may take off in a particular direction, hardly affecting any of the producing wells. On the other hand, if a producing well is surrounded by a ring of injection wells, the oil will be forced toward it from all directions with no means of escape.

Completion methods. Methods of completing the wells depend on the geology, especially the nature of the rock.

In the case of unconsolidated formations, the sand grains run into the borehole, filling it up and decreasing its productivity. They erode the pumps and tubing connections. In such cases, a fine screen is put in the hole and gravel is packed outside it. Attempts are sometimes made to inject plastic into the sand to consolidate it. None of these methods is entirely successful.

In the case of moderately consolidated formations, casing is cemented in the hole. Bullets or shaped charges of explosive are used to perforate the casing and cement.

If the formations are hard and impermeable, they are hydraulically fractured by pumping in a fluid with sand suspended in it. This opens vertical radial fractures, which are then propped open by the sand. This operation cannot be performed close to the water-oil contact because water will come up the fractures.

REFERENCES

Alpay, D. A., 1971, A practical approach to definition of reservoir heterogeneity: SPE Paper 3608.

Bond, D. C., 1979, Determination of residual oil saturation, Interstate Oil Compact Commission, Oklahoma City, Oklahoma.

Bowen, J. M., 1975, The Brent oil-field, *in* Austin W. Woodland, ed., Petroleum and the continental shelf of northwest Europe: New York, Halsted Press, p. 353–360.

Craig, F. F., Jr., P. O. Willcox, J. R. Ballard, and W. R. Nation, 1976, Optimized recovery through cooperative geology and reservoir engineering: SPE Paper 6108.

Craze, R. C., and S. E. Buckley, 1945, A factual analysis of the effect of well spacing on recovery: API Drill. and Prod. Practice, p. 144.

Dickey, P. A., and R. B. Bossler, 1944, Role of connate water in secondary recovery of oil: AIME Trans., v. 155, p. 175.

——, 1950, Influence of fluid saturation on secondary recovery of oil, *in* Secondary recovery of oil in the U.S., 2nd. Edition: New York, American Petrol. Institute, p. 222–227.

Dowling, Paul L., 1970, Application of carbonate environmental concepts to secondary recovery projects: Soc. Petrol. Eng. SPE Paper 2987, 16 p.

Eckelman, W. R., and R. J. Dewitt, 1975, Prediction of fluvial-deltaic reservoir geometry, Prudhoe Bay, Alaska: Trans. World Petroleum Congress, Tokyo, Japan, v. 2, p. 223–228.

Elkins, L. E., 1950, The importance of injected gas as a driving medium in limestone reservoirs as indicated by recent gas injection experiments and reservoir performance history, *in* Secondary recovery of oil in the U.S., 2nd Edition: American Petrol. Institute, New York, p. 370–382.

Fettke, C. R., 1950, Influence of geological factors on secondary recovery of oil, *in* Secondary recovery of oil in the U.S.: New York, Amercan Petrol. Institute, p. 204–213.

Fettke, C. R., 1950, Water flooding in Pennsylvania, *in* Secondary recovery of oil in the U.S.: New York, American Petrol. Institute, p. 413–443.

Flewitt, W. E., 1975, Refined reservoir description maximizes petroleum recovery: Soc. Prof. Well Log Analysts, Annual Logging Symposium.

Halbouty, Michel T., 1975, Needed, more cooperation between earth scientists and petroleum engineers: SPE Paper 6107.

——, 1977, Synergy is essential to maximum recovery.: Jour. Petrol. Tech., July 1977, p. 750–754.

Harris, D. G., 1975, The role of geology in reservoir simulation studies: Jour. Petrol. Tech., May 1975, p. 625–632.

Hewitt, C. H., 1966, How geology can help engineer your reservoirs: Oil & Gas Jour., Nov. 14, 1966.

Jardine, D., D. P. Andrews, J. W. Wishart, and J. W. Young, 1977, Distribution and continuity of carbonate reservoirs: Jour. Petrol. Tech., July 1977, p. 873–885.

King, R. L., and W. J. Lee, 1976, An engineering study of the Hawkins (Woodbine) field: Jour. Petrol. Tech., February, p. 123–128.

Kingston, P. E., and H. Niko, 1975, Development planning of the Brent field: Jour. Petrol. Tech., October 1975, p. 1190–1198.

Larson, Thomas C., 1974, Geological considerations of the Weber Sand reservoir, Rangely field, Colorado: Soc. Petrol. Eng. SPE Paper 5023. 8 p.

Morgan, J. T., F. S. Cordiner, and A. R. Livingston, 1977, Tensleep reservoir study, Oregon basin field, Wyoming, reservoir characteristics: Jour. Petrol. Tech., July 1977, p. 886–896.

Wayhan, D. A., and McCaleb, 1968, Elk basin heterogeneity—its influence on performance: SPE Paper 2214.

Wilson, Wallace W., 1950, Supplement to Fettke, C. R., 1950, Influence of geological factors on secondary recovery of oil, *in* Secondary recovery of oil in the U.S.: New York, American Petrol., Inst., p. 211.

Zeito, George A., 1965, Interbedding of shale breaks and reservoir heterogeneities: SPE Paper 1128.

2

Examination of Rotary Well Cuttings

1. HISTORICAL

Examination of the fragments is the most basic and important way to determine what kind of rock the drill is penetrating. From the earliest attempts at drilling to the present time, drillers and geologists both have examined the cuttings with interest.

The first scientific analysis of oil-well cuttings was done by John F. Carll of the Pennsylvania Geological Survey in the Pennsylvania oil regions during the early 1870s. He persuaded the drillers of several wells near Titusville to save samples from the surface down. Then, he put them in small vials on a shelf and also described the types of rocks that were penetrated. Next, he plotted the thickness of the different lithologies to scale and correlated them. Finally, he determined the elevation of the several wells above sea level and drew a structure contour map on the top of the third oil sand, which was the best producer.

Carll did not use the word "log," and it is unknown when it was first applied. It derives from the custom of sailors throwing overboard a log, attached to a string, on which were tied knots. From the number of knots that slipped through the hand in a time interval, the speed of the ship could be determined. The observation was recorded in the captain's "log book." The drillers were required to keep a log book in which they recorded the events of drilling the well, such as down time, bit changes, rock changes, etc. The geologist then kept his log book in which the succession of rocks was recorded. When this was plotted on a long, narrow strip of paper, it was called a *strip log*.

In the early days of the industry, drilling was mainly by the cable-tool method. A bit on the end of a heavy bar of iron was suspended in the hole on a cable. It moved up and down, banging the rocks and knocking off chips. Usually about 25 feet of water was kept at the bottom of the hole. After an interval (usually five feet) of penetration, the bit was pulled out of the hole. A bailer, consisting of a pipe with a flap valve at the bottom, was run in. When a tongue on the flap hit the bottom, the water and cuttings rushed into the bailer. The hole was bailed clean and then the bit was run in again. Consequently, cable-tool cuttings represented accurately the five feet of hole that had been drilled.

The cable-tool method had the even greater advantage that the fluid pressure in the hole was always low, like that of a short column of water. Consequently, if a producing sand was penetrated, oil or gas would flow into the hole. The great disadvantage was that if a water sand was penetrated, the hole would fill up with water, impeding the drilling. Cable-tool methods are now seldom used.

Cuttings were the principal source of geological information pre⸱ vious to 1935. Because of their limitations, errors in correlation were made, and many oil-bearing sandstones were cased off.

Electric logs came into widespread use about 1935. Because they were so fast and accurate, they largely displaced sample logs, even though they gave poor indications of some lithologies and unreliable information on oil or gas content. When oil was discovered in the Alberta and Williston basins in the early 1950s, it became evident that the electric logs of that time were unsatisfactory in carbonate sections, and geologists returned to the samples. Sample examination is now included as part of the mud-logging operation and is carried out on most important wildcat wells.

2. THE ROTARY DRILLING METHOD

In the rotary method, the bit is at the end of a pipe which is hung from the derrick and rotated like a twist drill (Figure 2–1). Gas, air, water, or mud is pumped down the inside of the drillpipe and comes up in the annular space outside of it. The cuttings are carried to the surface by this upward flow of fluid (Figure 2–2).

If permeable formations containing water are encountered down the hole, mud is used as a circulating fluid. The fluid pressure at the bottom of the hole is higher than the fluid pressure in the formation, so the water is prevented from entering the hole. Oil and gas are also prevented from entering. It is thus possible to drill right through a commercially productive oil- and gas-bearing formation without even noticing it, and this has happened many times.

The mud and cuttings at the surface are discharged from the surface casing onto a screen which is moved rapidly back and forth, called a

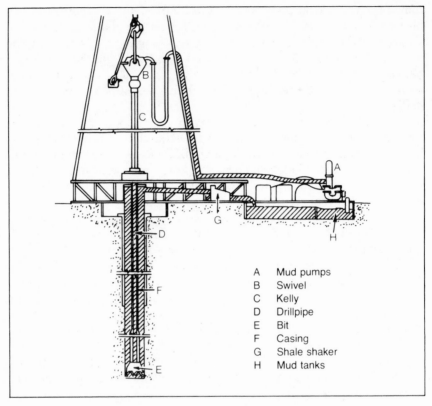

A	Mud pumps
B	Swivel
C	Kelly
D	Drillpipe
E	Bit
F	Casing
G	Shale shaker
H	Mud tanks

Fig. 2—1 Diagram of rotary mud system *(after Moody)*

shale shaker. The cuttings fall off into a pile, while the mud goes on to the tanks.

As the drilling proceeds, a member of the crew periodically takes a sample of the cuttings coming off the shale shaker. The intervals may be 10, 20, or 30 ft (3, 6, or 10 m) of drilled depth. The samples are washed with a hose in a bucket to remove the mud, placed in cloth sacks, and allowed to dry. They are not dried in an oven or near a stove, for this drives off the light hydrocarbons and even carbonizes the organic matter. The heat destroys their usefulness for modern geochemical analysis for source rock potential.

The cuttings take a time to come up the hole, so they are not representative of the rock being cut at the instant they are taken. The *lag time* is determined by dropping some extraneous material, such as grains of oats, into the mud stream and noting the time required for

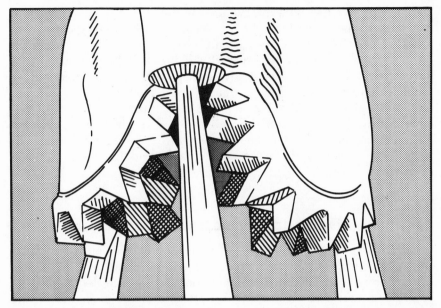

Fig. 2–2 Diagram of a rotary bit *(after Moody)*

them to make a round trip. The lag time varies from a few minutes to several hours. The depth correction depends on both the lag time and the rate of penetration.

Some fragments come up the hole much faster than others. Consequently, the cuttings samples contain a substantial amount of rock that was actually drilled a long time before. Furthermore, pieces of rock from up the hole may fall in and be ground up by the bit. Rotary cuttings, therefore, are very unsatisfactory because they are so contaminated. But they are all the geologist has, so he must make the best of them. He must guess which chips came off the bottom of the hole. The best way to do this is to be familiar with the upper formations and note the first appearance of anything new. Some geologists claim they can distinguish new cuttings from reground cavings by their shape.

When a new formation is penetrated, the new rock type will appear in the samples, but old rock types will also be present. The actual depth to the new rock type can best be determined from the drilling time log, which shows the rate of penetration. The various wire-line logs, such as electric logs, show lithologic changes with an accuracy of a foot or two.

When an important new rock is penetrated, such as an oil sand, it is customary to stop drilling and continue to circulate the mud, taking samples at specified time intervals. This clears the old cuttings out of the mud, so that those from the bottom of the hole can be recognized.

The sacks of samples are picked up by the geologist who often examines them immediately, especially if the well is approaching its objective horizon. Otherwise, they are sent to a separate distributing point. There they are washed again, dried, and put in manila envelopes, with one set of samples going to each person or company authorized by the operator of the well to see them. In many countries, such as Canada, a set may be sent to a government agency that will hold them confidentially for a specified period and then permit anyone to examine them.

3. EXAMINATION AND DESCRIPTION OF SAMPLES

The samples may be examined with a hand lens, but normally a microscope is used. The preferred type is a binocular, with objectives giving from 5 to 50 times magnification, using reflected light. The overall nature of the fragments is best determined at 5 or 10X, the higher powers being reserved for examination of details.

In addition to the microscope, the geologist should have a pair of tweezers and a watch glass. He can then take out a grain and drop acid on it to see if it is calcareous. A dissecting needle is also handy to prod into fragments of shale to see how hard they are and whether they contain much silt (Figure 2–3).

Shows of oil. The most important thing to look for, obviously, is traces of oil. Often green or black oil can be seen between the grains in ordinary light. Oil staining is made much more evident by irradiating the samples with ultraviolet light. Oil fluoresces brightly with various shades of yellow. The fluorescence can be enhanced by a solvent such as chloroform or carbon tetrachloride. This examination is usually made in a box to exclude normal light (Figure 2–4).

Identification of rock types. Fragments of rock should be identified with as much accuracy and detail as possible, given the limitations of time and the knowledge of the geologist. Presumably he has been taught to distinguish different rock types. He should have at hand some textbook to help him name them correctly, such as *Petrography*, by Williams, Turner, and Gilbert, or *Sedimentary Rocks*, by Pettijohn.

When looking at a set of samples for the first time from a new area, the geologist should record everything he sees in a notebook. After he has come to know the section, he will disregard the cavings and other contaminants and watch especially for those rock types which he has come to recognize as significant marker beds.

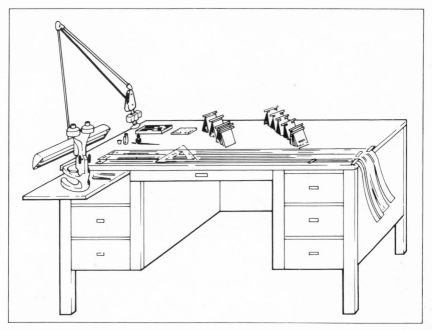

Fig. 2—3 Arrangement of the microscope and other materials for a sample examination. *(after Maher, Oklahoma Geological Survey)*

Sandstones. The first thing to note in a sandstone is its average grain size (Figure 2–5). This must be estimated quantitatively, according to the Wentworth classification (Table 1).

Most useful is a piece of cardboard or slide on which sand grains of different sizes are cemented. A ruler graduated in millimeters can also be used. Whether the grains are mostly the same size or whether there is a range of sizes should be noted.

Next, the mineralogy of the grains should be determined. Most sands are predominantly quartz, but other minerals occur. If feldspar comprises a considerable percentage (more than 10%) of the grains, the rock may be called an *arkose*. If rock fragments comprise a considerable fraction, the rock is called a *graywacke*. Rock fragments most commonly found are chert, schist, and igneous rocks of various types.

There are several different classifications of sandstones; some companies use one and others use another. Unless a clear scheme is given to the geologist, he should simply guess at the percentage of nonquartz grains and say what they are without trying to put a name on the sandstone. It may be misunderstood by people who learned a different classification.

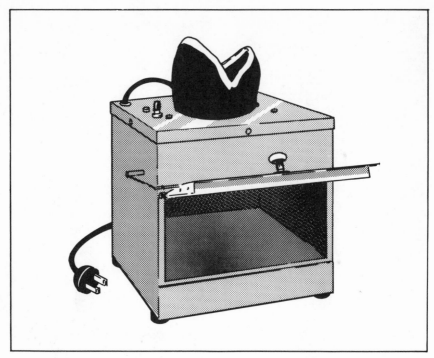

Fig. 2—4 Box with an ultraviolet light to detect oil in cuttings

The shape of the grains should be labeled either *elongated* or *equant*. The surface ranges from rough and *angular* with sharp corners to smooth and *rounded*, with the corners smoothed off. Most consolidated sandstones have overgrowths of secondary silica. These are difficult to recognize.

If the sand grains are not completely disaggregated, an attempt should be made to determine the nature of the cement which sticks the grains together. This is most commonly secondary silica, but it may be clay or calcite.

The pores between the grains can generally be discerned. If they are empty, the sand has a high porosity. They are often filled with *cement,* which is clear silica or calcite, or *matrix,* which is usually clay deposited with the grains or later precipitated in the pores by moving water.

The color of the rocks should be named. Most sandstones are white, but some are gray or red.

Accessory minerals are usually mentioned, although they are sel-

A. Very coarse sand (1-2 mm)

B. Coarse sand (½-1 mm)

C. Medium sand (¼-½ mm)

D. Fine sand (⅛-¼ mm)

E. Very fine sand (1/16-1/8 mm)

F. Coarse silt (1/32-1/16 mm)

G. Medium silt (1/64-1/32 mm)

H. Fine silt (1/256-1/64 mm)

Fig. 2—5 Appearance of different sizes of sand and silt under a microscope with 6.3X magnification. *(after Maher, Oklahoma Geological Survey)*

TABLE I

WENTWORTH CLASSIFICATION OF SAND PARTICLE SIZE

Names of particles		Diameters of particles		
Gravel	boulders........more than 256 mm			
	cobbles.............from 256 mm ....to 64 mm			
	pebbles.............from 64 mm ....to 4 mm			
	granules............from 4 mm ....to 2 mm			
Sand	very coarse.........from 2 mm ....to 1 mm			
	coarse ....from 1 mm..............to 0.5 mm			
	medium.............from 0.5 mm.............to 0.25 mm			
	fine ....from 0.25 mm............to 0.125 mm			
	very fine............from 0.125 mm ....to 0.062 (or .05) mm			
Silt........................from 0.062 (or 0.05) mm .to 0.005 mm				
Clay...less than 0.005 mm				

dom diagnostic. The most common is mica. Carbonaceous fragments, pyrite, various red iron ores and stains, and small concretions also occur. Green glauconite grains may be very abundant; they signify shallow marine depositional environment.

Silt consists of grains smaller than 1/16 mm. They can hardly be seen with the naked eye but are easily distinguished with the microscope. Siltstones may be called sandstones unless some absolute measure of particle size is available.

Most sandstones are partly disaggregated by the rotary bits, and the individual grains go through the shale shaker. Consequently, relatively less sandstone than shale is preserved in the cuttings, and there is actually more sandstone in the rock section than would appear from sample examination.

Shale. Shales consist predominately of particles less than 1/256 mm (5 microns) which cannot be distinguished with the microscope. They consist of fine particles of clay minerals and quartz, but their composition can be determined only with the X-ray spectrograph.

It is desirable to stick a needle into the shale to determine if it is silty or sandy. Pure shale is not gritty when rubbed between the teeth.

The most diagnostic character of shale is its color. Various shades of gray and black are common, but red, green, mottled, purple, and other colors occur. Naming colors is very subjective, and it is useful to use a rock color chart. One was published by the National Research Council in 1948, and some oil companies have made special charts. However, they are not often available.

The appearance and color of shales (and other rocks) is often masked by the dust on the surface of the chips left after the drilling mud is washed off. It is best to determine the color on a wet sample. Some geologists want to look at the whole sample under water, but this involves drying it out again before it can be put back in the envelope. It is easier to pick up a few representative fragments and immerse them in a drop of water on a watch glass.

Accessory minerals in shale are common. They include carbonaceous matter, like carbonized leaves, pyrite, and ferruginous concretions. Small fossils are also common and should be noted. When identified by a paleontologist, they may be useful for correlation.

Carbonates. Limestone and dolomite can be recognized because they are usually white or light gray with a fine or smooth, dense texture. If carbonate is suspected, a fragment should be immersed in a drop of acid on the watch glass. If it is limestone, it will effervesce vigorously. If it is dolomite, it will effervesce weakly, emitting small bubbles at intervals. There are also staining methods for distinguishing limestone from dolomite.

Other rocks. Other types of rock are frequently penetrated in oil wells. Gypsum and anhydrite may be recognized because they are white, soft, and insoluble in hydrochloric acid. Chert is fine-grained silica. It can be black or white, and it is very hard. Salt is often drilled, but it is seldom found in cuttings because of its solubility in water. Coal is sometimes drilled.

Igneous and metamorphic rocks are very important because they usually constitute basement and drilling should be stopped. Often, sediments are found beneath igneous sills or even under overthrust sheets of granite. However, it is always necessary to stop the well and take some time to decide whether the chances of finding oil below warrant the attempt to drill through igneous or metamorphic rock.

Contaminants from outside substances often occur in the cuttings. There may be materials added to the drilling mud to increase its density or to plug fractures in cases of lost circulation. Fragments of steel from the bit are common. Oil is sometimes added to the mud to make it smoother or as a means of releasing stuck drillpipe. Cement is very commonly seen. It effervesces like limestone but consists of needle-shaped crystals and is usually speckled.

The principal things to watch for in examining cuttings are changes in rock character that can be used in correlation and the nature of the rocks which might produce oil or gas.

A skilled geologist familiar with the area can examine 200 samples per day or even more. Beginners or newcomers to an area should expect to go much more slowly.

4. PLOTTING THE LOG

Cuttings descriptions are nearly always plotted on a scale of 100 feet to the inch (1/1200). The mixing and contamination does not warrant a larger scale. Wire-line logs are usually reduced to this scale to facilitate comparison with sample logs.

Blank logs in many styles are available. Figure 2–6 shows a style of heading commonly used. The well identification, location, elevation, and date are recorded.

The different lithologic types are shown by symbols which are pretty well standard throughout the industry, although each company usually has its own scheme. Figure 2–7 is a simple set of symbols, and Figure 2–8 is a sample of a strip log using them. The symbols are often colored: shale, blank; sandstone, yellow; limestone, blue; red shale, red. Maher (1964) suggests a more elaborate symbol and color scheme.

Percentage logs. When examining a well for the first time in a new area, it is better to record the description first in a notebook and then plot it on the strips. The percentage of each rock type should be guessed. For example, an entry might look like the following:

5510–5520   Sandstone, fine, white, angular: quartz, 80%; limestone, white, biomicritic, 10%; shale, micaceous, gray, 10%; traces red shale

5520–30   Same

Appropriate abbreviations should be used. These should not be so short that anyone might have trouble understanding them.

In plotting a percentage log, the symbols should fill in the blank column according to their percent, sandstone from the right and limestone from the left, leaving shale for the middle.

The descriptions of the samples should be neatly lettered in the space provided on the right-hand side of the strip. Formerly a fine *crow quill* pen was used; now tubular type drawing pens may be more practical.

Interpretive logs. Interpretive logs utilize the known stratigraphic sequence, and the first appearances of the new lithologies are plotted as if they constituted the entire sample, even though other rock types from up the hole are present. The exact depth to the top and the thickness of the unit can be determined from the drilling time or electric logs.

In plotting interpretive logs it is not necessary to record the descriptions in a notebook; they are recorded directly on the strip. The strip may be laid out on a desk alongside the electric log, as shown in Figure 2–8.

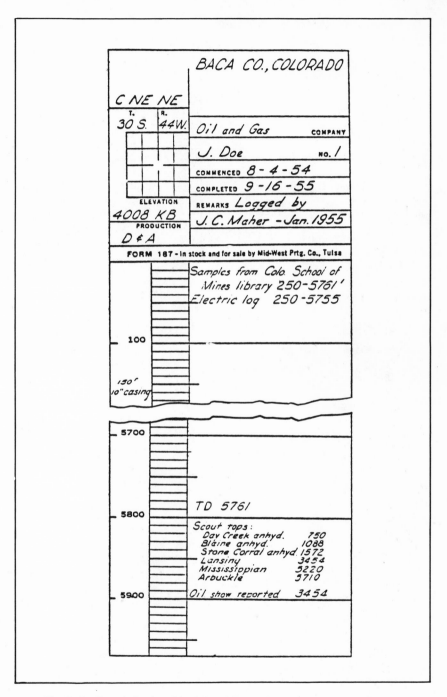

Fig. 2—6 Sample log heading *(after Maher, Oklahoma Geological Survey)*

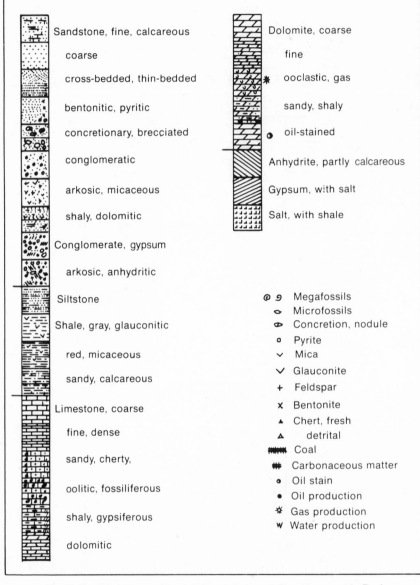

Fig. 2—7 Symbols used in plotting strip logs *(after Haun and LeRoy)*

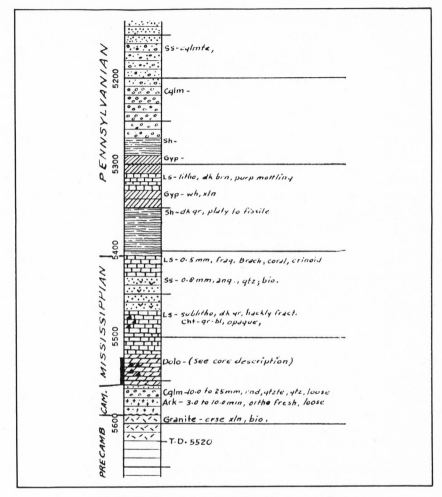

Fig. 2—8 Typical strip log *(after Haun and LeRoy)*

REFERENCES

Carll, John F., 1875, The Oil Regions, Report of progress I, Second geological survey of Pennsylvania, Harrisburg.

Haun, J. D., and L. W. LeRoy, eds., 1958, Subsurface geology in petroleum exploration, Colorado School of Mines, Golden, Colorado.

LeRoy, L. W., 1950, Subsurface geologic methods, 2 ed., Colorado School of Mines, Golden, Colorado.

Low, L. W., 1951, Examination of well cuttings, Colorado School of Mines Quarterly, vol. 46, no. 4, pp. 1–48.

Maher, John C., 1964, Logging drill cuttings, Oklahoma Geological Survey, Guidebook XIV, Norman, Oklahoma.

Moody, Graham B., 1961, Petroleum exploration handbook, McGraw-Hill, New York.

Pettijohn, F. J., 1975, Sedimentary rocks, Harper and Row, New York.

Swanson, R. G., 1981, Sample examination manual, AAPG, Tulsa.

Williams, H., F. J. Turner, and C. M. Gilbert, 1954, Petrography, W. H. Freeman, San Francisco.

3

Analysis of Cores

The only satisfactory way to determine the lithology and reservoir properties of a subsurface rock is to take cores. Unfortunately, in recent years, improvements in wire-line logging techniques have led to a belief, promoted by the logging companies, that cores are less necessary. The most important fact about a rock is whether or not it contains petroleum, and this can be ascertained only with a core. Most other petrographic and reservoir properties can also be determined best by a sample of the rock. Wire-line log responses should be interpreted by comparing them with cores.

1. CHOICE OF NUMBER AND LOCATION OF CORES

It is usual to plan a coring program before starting the well. This is sensible, especially for development wells. Cores are expensive, and they should be taken only when they will provide really important information. Usually, cores are only taken of the productive formation, and the first well may be planned without coring at all. However, it is better to take frequent cores in an important wildcat well, especially in possibly productive formations, for identifying the geological formations penetrated.

It is important to get a continuous core through the entire productive sand. The core will give data on reservoir properties necessary to calculate recoverable oil and gas reserves. These properties can then be estimated in other wells using wire-line logs, which are cheaper. Many features of the sand such as crossbedding, changes in grain size, and minor sedimentary structures can be described. These will give information on the probable size, shape, and extent of the permeable sand body. They also can be extrapolated to other wells by means of the wire-line logs.

The top of the pay sand, or a possible pay sand, is usually picked by a *drilling break*. This is a noticeable increase in rate of penetration. When this happens, it is customary to stop penetrating and circulate the samples from the bottom. If they show indications of an oil or gas reservoir rock, the bit is withdrawn and a core is taken.

2. TYPES OF CORES

Conventional cores. Cores are taken most commonly with a conventional double-tube barrel, run on the end of the drillpipe in place of the bit (Figure 3–1). The cutter head consists of small cones which leave the center of the hole undrilled. This cylinder of rock, usually

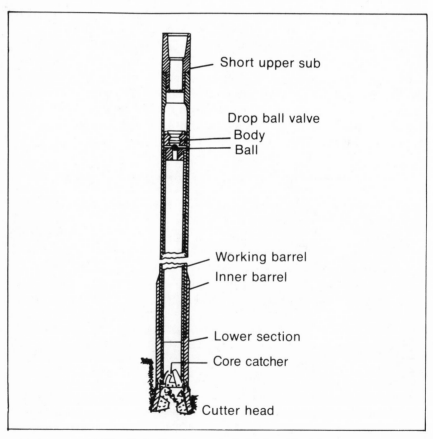

Short upper sub

Drop ball valve
Body
Ball

Working barrel
Inner barrel

Lower section
Core catcher
Cutter head

Fig. 3–1 Rotary core barrel *(after Moody)*

about two inches (5 cm) in diameter, passes up into an inner barrel that does not rotate. It is separated from the outer barrel by ball bearings. The cylinder of rock, now called the core, enters the inner barrel, passing an assembly of springs called a *core-catcher*, that keeps it from falling back out. The inner barrel is usually 20 ft (6 m) long. When 20 feet of hole have been drilled, the pipe is pulled from the hole.

Recoveries with the conventional barrel are good in hard formations like the Pennsylvanian of Oklahoma and West Texas. They are poor in unconsolidated formations.

Diamond cores.   When a diamond-studded bit is used instead of the ordinary core bit, better recoveries can be obtained. The core breaks up less and the barrel is often made 60 ft (20 m) long.

Wire-line cores.   It is possible to run a small-diameter inner barrel down inside the drillpipe where it latches in place (Figure 3–2). The bit

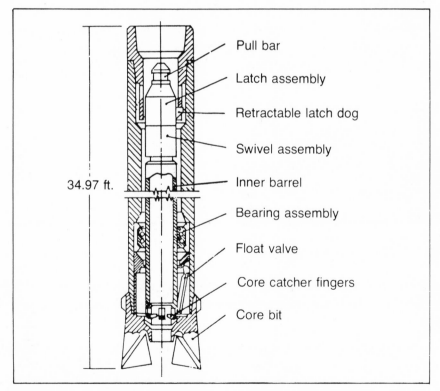

Fig. 3–2 Wire-line core barrel as used by the *Glomar Challenger (after Larson, courtesy JPT)*

cuts the outside of the hole, and the core slips up into the inner barrel, which is usually 15 ft (5 m) long. When it is desired to retrieve the core, the kelly is removed and an overshot socket is run on a wire line that grabs the pull bar and pulls the inner barrel out of the hole. The core is removed and the barrel run back again into the bit. It is not necessary to pull the whole drillstring out of the hole.

Wire-line cores have been used very successfully on the ship *Glomar Challenger*, which has been drilling exploratory holes into the bottom of the deep ocean.

Sidewall cores.   After the well is drilled, it is possible to take small cores out of the sidewall of the borehole. The most common method is to drive a hollow cylinder into the wall with a charge of explosive (Figure 3–3). The cylinders are attached to the tool with a small cable, so they can be retrieved when the tool is pulled out on a wire line. The charges are fired electrically. The cylinders penetrate only in relatively soft formations.

It is very advantageous to be able to choose the location of the core from the electric log. However, the sidewall cores are unsatisfactory because they are so small. The grains of sand are often so disturbed by the cylinders that good values for porosity and permeability cannot be obtained. If they are taken in a permeable sand, it will be thoroughly flushed by the drilling fluid. However, oil staining can be observed. Grain size analysis can be run on the samples. When this is compared with the porosity obtained with the electric log, a rough estimate of the permeability can be made.

Rubber-sleeve cores.   It was formerly impossible to obtain cores of the softer oil sands of the Gulf Coast and Venezuela with conventional tools because the sand simply ran out of the barrel. This problem was solved by inserting a rubber sleeve between the inner and outer barrels, doubled in on itself (Figure 3–4). As the core enters the inner barrel, the rubber tube slips over and encloses it like an inside-out sock being slipped over a foot. Excellent recoveries are often obtained.

3.   DESCRIBING AND PRESERVING THE CORE

The removal of the core from the barrel is remarkably awkward, considering the sophistication of the device. The tool is suspended by the traveling block over the derrick floor. The bit is removed, which releases the core catcher. The core then slides down out of the barrel, bottom portion first. It is laid in a V-shaped trough and the drilling mud is washed off with a hose. A steel tape is laid out along the core. The depth of the hole when the core barrel was run is known, and this

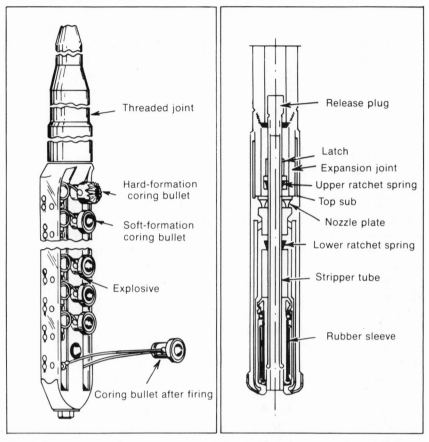

Fig. 3—3 Side-wall coring gun *(courtesy Schlumberger)*

Fig. 3—4 Rubber-sleeve core barrel *(after Christensen Inc.)*

depth is taken as the top of the core. The core is then marked at appropriate intervals with a felt pen. Usually there is less length of core than penetration, and the geologist must estimate as best he can where the loss occurred so that the depth marks are as nearly correct as possible.

Then, the core is immediately examined carefully and minutely. Special attention is given to shows of oil or gas. Sometimes the oil or gas oozes out of the core. This is usually taken as a bad sign, indicating low permeability. Good, permeable sand will have been flushed by drilling mud filtrate and will contain only residual oil. Gas can sometimes be detected by the odor of the core, but only during the first few minutes. Salt water can sometimes be detected by a "dead" smell or a salty taste. The more porous portions of the sand are usually better saturated with oil than the tight streaks, and both types of rock should be distinguished, described, and measured.

Special attention should be given to shale and silt partings. These may be very thin, but they still effectively eliminate vertical permeability and thereby change the behavior of fluids in the reservoir. Fractures also should be noted with much care. Often, cores are fractured by the coring operation. Such induced fractures can be recognized by the fact that they have fresh surfaces, go down the center of the core, and curve out upward. Natural fractures are usually nearly vertical but are not in the center of the core. They are usually tightly closed at depth. If they are open at depth, they are coated with minerals. Such open fractures strongly affect the reservoir behavior, but the closed fractures do not, unless they are opened by excessive fluid pressure as in hydraulic fracturing. In either case, it is extremely important to notice them.

Other geological features like changes in grain size, cross bedding, dip, etc., should be noted, but these can also be described later when the cores are taken to the laboratory.

When a piece of core is picked up for closer examination, it is important to return it correctly without getting the bottom end up. Based on the description, intervals are chosen for core analysis. Representative pieces of core three or four inches (7–10 cm) long are picked out for analysis for porosity, permeability, and oil content. These are usually sealed in tin cans or frozen with dry ice and taken to the laboratory. Sometimes they are immersed in wax in order to preserve the wettability of the pore surfaces.

It is common to wrap the core with an impermeable plastic film and metal foil, and then pack it carefully in wooden boxes for transportation to the laboratory. Here it is often sliced longitudinally with a diamond saw. The flat surfaces so formed are examined again, both rough and after polishing. The textures of limestones are brought out by etching and may be preserved and photographed by means of acetate peels.

Sedimentary structures and textures. Since 1950, much has been learned about the depositional environments which give rise to porous and permeable reservoir rocks. Basically, the properties of porosity and permeability are conferred on a sandstone by the process of winnowing, that is, the removal of the fine material by waves and currents. Two depositional environments (among others) are especially favorable: channels and beaches. The beaches generally are parallel to the ancient shore trends, while channels are usually perpendicular to them. Consequently, it is important to observe those features recognizable in the cores which can be used to determine whether the producing sand was deposited in a channel or on a beach. As many distinctive sedimentary structures as possible should be observed and recorded while the core is laid out on the tray on the rig floor.

Laminations should be noted, including their spacing and angle to the vertical. The direction of crossbedding can possibly be determined later with the dipmeter, and this is extremely important. *Ripple marks* and *sole markings* on bedding planes give clues to depositional environments. *Bioturbation* or disturbance of the laminae by burrowing organisms is very common in oilfield rocks. *Slumps* and *convolute bedding* are less common but significant.

Sedimentary structures in limestones are more easily recognizable and significant than those in sandstones. They may be less obvious on the fresh core, and they are brought out later when the core is sliced down the middle with a diamond saw and etched with acid.

Particular attention should be given to features suggesting a reef environment. These include reef and mound-building organisms such as corals and calcareous algae. Reefs have good porosity, but the pores are mostly secondary vugs; the original porosity has been greatly modified by secondary solution and reprecipitation of calcite. Such obvious features as fossil shells, oolites, pellets, breccias, etc., should be noted, especially their contribution to porosity.

4. ANALYZING THE CORE

Porosity. The most important attribute of a reservoir rock is its porosity, because this determines the amount of fluid it can hold.

Porosity is defined as the "property possessed by a rock of containing interstices, without regard to size, shape, or interconnection of openings. It is expressed as the percentage of total (bulk) volume occupied by the interstices" (API, 1941). Porosity, therefore, is the pore volume divided by the bulk volume. It is expressed either as a decimal fraction or as percent.

It is frequently stated that there are two kinds of porosity, total and effective. The effective porosity is that which is interconnected. It may be doubted whether any ordinary sedimentary rock has pores that are completely sealed off and not interconnected, so this distinction is unimportant. However, a considerable fraction of the connected porosity, 25 to 75 percent or more, is actually ineffective as far as oil and gas are concerned because water is held in the finer spaces of the pores by capillary forces. This water is called *connate water* or *irreducible minimum water saturation*, and it is always present, wetting most of the solid surfaces. Oil cannot be stored nor can it move in these portions of the pore space where this water exists. Roughly speaking, only those pores whose walls are farther apart than 5 to 10 micrometers (0.005 to 0.010 mm) can contain oil.

There is no way to determine accurately the percent of pore volume in an oil or gas sand which is occupied by water, except possibly by taking a core using oil as a drilling fluid. Sometimes it can be estimated from electric logs or from the measurement of capillary properties.

Porosity can be measured in a remarkably large number of ways. The bulk volume can be determined by measurement or by immersion in a liquid that does not enter the pores. The pore volume can be found by cleaning the pores, weighing the sample with the pores empty, and weighing again with the pores filled with a liquid of known density. It can also be determined by drawing out the air from the pores and measuring its volume. Another method is to compress the air in a chamber containing the core and note how much volume there is inside the core, using Boyle's law. The core may be disaggregated and the volume of the grains measured. The bulk volume less the grain volume gives the pore volume. Or the weight of the core may be divided by the density of the grains to give the volume of the grains. It is possible to determine porosity by taking a thin section of the rock and determining the fraction of an area occupied by pores. These methods are accurate to about one porosity percent; that is, there may be a 5-percent error if the rock has a porosity of 20 percent.

The porosity of sandstone ranges from less than 10 percent in a highly cemented rock to more than 30 percent in soft, unconsolidated formations. The presence of clay matrix in the interstices reduces the porosity, but not as much as the secondary silica does because clay itself is quite porous. The pores of clay are so fine that oil is prevented from entering them by capillary forces. Sandstones with a porosity of 0.10 or less are seldom productive.

Some rocks, especially limestones, may have open fractures. These contribute to the porosity if they are wide open, but usually they contribute much more to the permeability. Often, a fractured rock produces abundantly at first, but the production falls off rapidly because the total volume within the fractures is small. The great producing fields in the Asmari limestone of Iraq and Iran produce from open fractures, as do also the Mara and La Paz fields of Venezuela. In both cases, thick limestones are sharply bent over an anticline in what appears to be a tensional environment.

Because of their great effect on reservoir performance, fractures should be minutely described. Their probable width, attitude, spacing, length, presence of asphalt and minerals, etc., should be determined.

Limestones are more or less soluble in water and frequently contain solution channels or *vugs*, which appear to be mostly interconnected. They contribute more to permeability than to porosity, and some productive limestones may have porosities as low as 0.05 or less. The volume of the vugs can seldom be measured on cores, although it is important to estimate it. A good way is to slice the core and trace the outlines of the vugs; then the porosity is calculated from the relative areas of the rock and the vugs.

Interstitial porosity of limestones can be measured by the methods used on sandstones. It is very important to distinguish interstitial porosity from the porosity in vugs and fractures in order to understand and predict the behavior of the reservoir during production.

Permeability. Next to porosity, the most important attribute of a reservoir rock is its permeability, the ability of the rock to conduct fluids. The first person to define and measure permeability was Henry Darcy, who experimented with the flow of fluids through sands. His apparatus is shown in Figure 3–5. The most useful way of expressing the Darcy equation is:

$$Q = \frac{k}{\mu} \frac{A(p_1 - p_2)}{L}$$

Where:

Q	= rate of flow, cc/sec or b/d
k	= constant of proportionality
μ	= viscosity of the fluid, cp
A	= cross-sectional area, cm^2
$p_1 - p_2$	= potential difference (hydraulic head) causing flow, atm
L	= length of flow path, cm

If Q is expressed in cc per second, μ in centipoises, A in square centimeters, $p_1 - p_2$ in atmospheres, the constant of proportionality will be the permeability, expressed in Darcys.

In contrast to the many ways of measuring porosity, there is only one way to measure permeability, and that is to pass a fluid through the sample. The most common type of apparatus is shown in Figure 3–6. A small cylinder of rock is taken out of the core, usually parallel to the bedding. It is inserted into a rubber stopper that is squeezed around it in a core holder. Air is put through the core at an input pressure that is measured on a gauge. The outlet is usually at atmospheric pressure. The rate of flow is measured with a flowmeter.

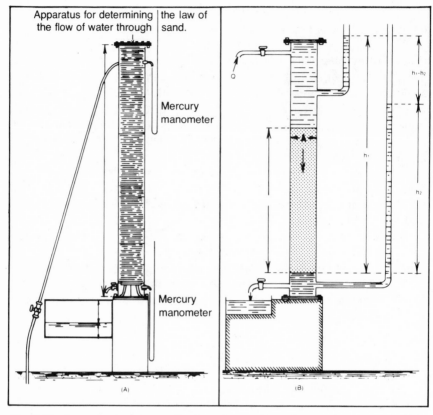

Fig. 3–5 Apparatus to measure the flow of water through sands *(courtesy the Univ. of Chicago)*

Air has a low viscosity and reacts hardly at all with the solid, so it is most commonly used. Gas is compressible so that the rate Q must be the flow in the center of the core at the algebraic mean pressure. Water reacts with the rock because it makes the clays move and plug the pore throats. Also, the water close to the solid surfaces takes on a structured character and does not obey Darcy's law. Permeability to water, therefore, is usually less than to gas, although the values theoretically should be the same.

Relation between porosity and permeability. There is no direct relation between porosity and permeability. A medium consisting of billiard balls will have the same porosity as that consisting of small, round shot, but the billiard balls will have the greater permeability. Dimensional analysis shows that permeability has the dimensions of

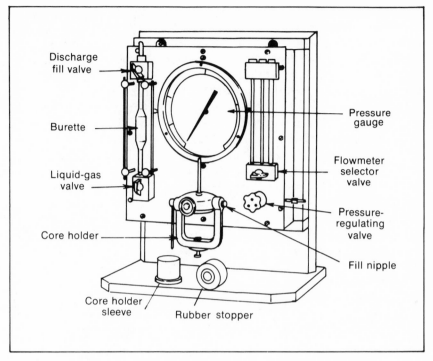

Fig. 3—6 Ruska-type permeameter *(after Frick, courtesy SPE)*

length squared, and it varies as the square of the pore diameter. The smaller the pore diameter, the greater the surface area. The greater the surface area, the lower the permeability. Wyllie showed that the permeability is related to the surface area as follows:

$$k = \frac{\phi^3}{k_o S_p^2} \left(\frac{L}{L_a} \right)^2$$

where:

k = permeability of the porous medium
k_o = a shape factor
S_p = surface area per unit pore volume
ϕ = porosity
L = length of the sample
L_a = actual length of the flow path

S_p depends on the grain size, so there is a relation between permeability and porosity that depends on the characteristics of the sample.

Early attempts to relate permeability to porosity were largely given up. Recently they have been revived, however, because porosity can be determined from some wire-line logs. If there is a reliable relation between porosity and permeability, then permeability can be at least estimated. Figure 3–7 shows the relation between porosity and permeability in cores from the Bradford field, Pennsylvania, and Figure 3–8 shows the relation in the Loudon field of Illinois. Note that in both cases the relation is exponential and the permeability drops to 0.1 millidarcy at about 8-percent porosity. Attempts have also been made to estimate permeability from grain size analyses of sidewall cores.

Permeability shows large variations from one inch of rock to the next. The error of measurement is small, but the average permeability of a unit of rock can only be estimated to about a factor of two. This being the case, estimating permeability from log-derived porosity using an empirical porosity-permeability relation, like those in Figures

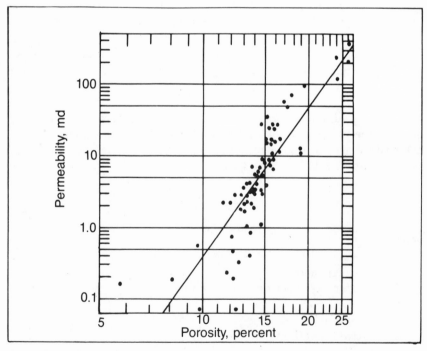

Fig. 3–7 Permeability plotted against porosity in Bradford field *(after Fettke, Pennsylvania Geological Survey)*

3–7 and 3–8, may give values as reliable as those obtained by direct measurement on cores.

Knowledge of permeability is necessary in estimating thickness of pay sand. Usually, any sand with a porosity below an arbitrary cutoff value is considered to be noncontributing. This cutoff value ought really to be taken as a fraction of the maximum permeability. Thus, if a rock has a maximum permeability of 100 md, any portion of the pay sand with a permeability of less than 10 md may be considered non-commercial. If it has a maximum permeability of 10 md, then sand with less than 1 md is noncommercial.

This low-permeability cutoff is more important in secondary than primary production. In the case of primary production, some oil will bleed out of the tight streaks. In the case of secondary recovery, the flushing medium will bypass the less permeable beds and flow through the more permeable for a long time before it drives the oil out of the tight streaks.

Fluid saturation. The filtrate from the drilling mud usually enters the oil or gas-bearing formation ahead of the core bit. It flushes out the oil ahead of it, so the oil in the core is not the oil originally present in

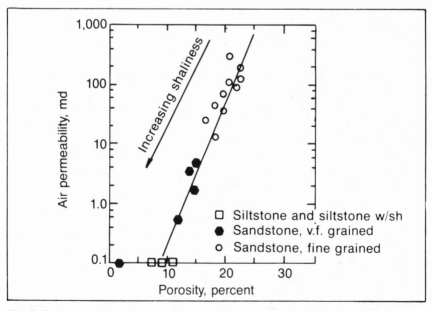

Fig. 3–8 Permeability plotted against porosity in Loudon field, Ill. *(after Harris, courtesy SPE)*

the rock, but rather the irreducible minimum oil saturation left after water-flooding. Even in cable-tool drilling when the pressure in the hole is low, reliable oil saturation values cannot be determined. However, oil saturation is regularly determined on cores because it gives at least an indication of the productivity.

As the core is withdrawn from the hole, the fluid pressure on it decreases. Gas comes out of solution in the oil and water and tends to drive these liquids out of the core. As a result, there is no way of determining the amount of oil or gas in the rock originally in place. Some of this effect can be mitigated by the use of a core barrel that seals the pressure in the inner barrel while the core is withdrawn.

Attempts have been made to get at the true fluid content in place by spiking the drilling fluid with tracers. Also some drilling fluids have powdered substances like rubber or gilsonite added which plug up the pores and minimize the flushing action. Bubbling nitrogen through water used as drilling fluid reduces the weight of the column and reduces the flushing effect.

There are two principal methods of determining the fluid content of cores. The most popular is to heat the rock in a retort which vaporizes the oil and water. They are condensed and caught in a graduated cylinder below the condenser. Another method is to extract the core with a solvent. The water is measured and the oil content is determined by the difference in weight before and after extraction. The volumes of oil and water are expressed as percent of the pore volume and are called oil saturation (S_o) and water saturation (S_w).

The retort method is subject to error because the heavier fractions of the oil do not distill over but crack, forming coke and light fractions. Correction factors for oils of different types are applied. Both methods can be misleading if there is solid hydrocarbon, wax, asphalt, or pyrobitumen naturally in the pores. These substances are quite commonly present, but they are seldom detected. If they are computed as if they were liquid and recoverable oil, as they are, serious errors in the estimation of recoverable oil will be made.

REFERENCES

American Petroleum Institute, 1941, Glossary of terms relating to reservoir behavior: Proceedings, v. 22, sect. IV, Prod. Bull. 228, p. 86–96.

Anderson, Gene, 1975, Coring and core analysis handbook. PennWell Pub. Co., Tulsa, 200 p.

Beckman, Heinz, 1976, Geological prospecting of petroleum: New York, John Wiley, 183 p.

Campbell, John M., 1973, Petroleum reservoir property evaluation: John M. Campbell, Norman, Ok., Chap. 12, Core analysis, p. 300–331.

Fettke, C. R., 1938, The Bradford oilfield: Penn. Geol. Surv. Bull. M-21, 225 p.

Frick, Thomas C., ed., 1962, Petroleum production handbook, v. II, Reservoir engineering: Soc. Petrol. Engineers, Dallas.

Harris, D. G., 1975, The role of geology in reservoir simulation studies: Jour. Petrol. Tech., May 1975, p. 625–632.

Hubbert, M. K., 1940, The theory of ground water motion: Jour. of Geol., v. 48, p. 785–944.

Kelton, F. C., 1953, Effect of quick-freezing on saturation of oil well cores: Trans. AIME, v. 198, p. 21–22.

Mattax, C. C., R. M. McKinley, and A. T. Clothier, 1975, Core analysis of unconsolidated and friable sands: Jour. Petrol. Tech., December 1975, p. 1423–1432.

Muskat, M., 1937, The flow of homogenous fluids through porous media: New York, McGraw-Hill, 763 p.

Rose, W. D., and W. A. Bruce, 1949, Evaluation of capillary character in petroleum reservoir rock: AIME Jour. Petrol. Tech., May 1949.

Sangree, J. B., 1969, What you should know to analyze core fractures: World Oil, April 1969, p. 69–72.

Scheidegger, A. E., 1957. The physics of flow through porous media: New York, MacMillan.

Wyllie, M. J. R., and G. H. F. Gardner, 1958, The generalized Kozeny-Carman equation: World Oil, March 1958, p. 121–128, and April 1958, p. 210–228.

4

Mud Logging

1. HYDROCARBON SHOWS IN MUD

Beginning in the 1950s, efforts were made to monitor the returning mud stream continuously in order to detect traces of oil or gas from possible producing formations as they were penetrated. It was usual to house the equipment in a trailer that could be kept at the well site, and it was operated by semiskilled people. Recently, mud loggers have become quite sophisticated. Besides checking the mud stream for hydrocarbons, the operators examine and describe the cuttings. More or less automatic instruments monitor and record many of the drilling variables. They are now staffed by well-trained geologists and engineers (Figure 4–1).

Fig. 4–1 Modern mud logger *(courtesy Magcobar)*

Theory. A producing formation contains only a small amount of hydrocarbon compared to the volume of mud that is circulated past the bit while it is being drilled. For example, 1 ft (30 cm) of oil sand drilled by a 9-in. (22-cm) hole contains about 0.75 gals (3 liters) of oil and possibly 4 cu ft (100 liters) of dissolved gas. If the drilling rate is 2 minutes per ft (6 minutes per meter) and the circulation rate is 400 gpm (1,500 liters per minute), then the 0.75 gals (3 liters) of oil is picked up by 800 gals (3,000 liters) of drilling mud. However, most of the oil will be flushed out ahead of the bit by the mud filtrate, or only about 50% of the oil and gas will be drilled up. The concentration of oil in the mud will thus be 0.75/800 or 750 ppm. Four standard cu ft (110 liters) of methane weigh about 80 grams that, entrained in 800 gals (3,000 liters) of drilling mud, would amount to about 26 ppm. It is said that as little as 10 ppm oil can be detected. However, actual conditions vary, and the indications of oil and gas are very nonquantitative. They are usually expressed as arbitrary "units."

Trap. Located near the shale shaker is the trap, which removes the gas from the mud. It usually consists of an enclosed box which floats on the mud or through which a portion of the mud stream passes. A propeller or paddle wheel agitates the mud, and the gas that comes off is sucked out and transmitted through a plastic tube to the laboratory in the trailer.

It is extremely difficult to say how much of the gas in the mud is removed by the trap. If the trap is poorly installed, it may not remove any. Clearly, the rest of the apparatus depends on the proper functioning of the trap, but it is often a comparatively crude and ineffective device.

Much more quantitative is the steam still, used in some installations. A jet of steam is passed through a sample of the mud, and this removes nearly all of the gas and much of the lighter hydrocarbons in the oil. These are passed on to the detection equipment.

Detection of gas and oil in the mud. The stream of gas drawn off the mud is passed over a hot-wire gas detector (Figure 4–2). This instrument consists of a Wheatstone bridge in which two of the four arms consist of platinum filaments that are heated by the flow of electric current through them. If a mixture of air and methane is passed over one of these filaments, the methane burns and the additional heat raises the temperature of the filament. This causes its resistance to increase and the Wheatstone bridge to become unbalanced. The instrument is quite sensitive.

If the temperature of the filament is held slightly below that necessary to ignite methane, it will still ignite the heavier gaseous hydro-

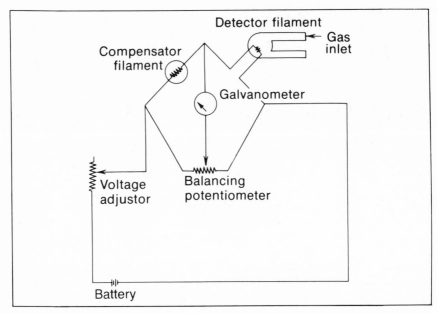

Fig. 4–2 Schematic diagram of hot-wire gas detector (after Pearson, courtesy McGraw-Hill)

carbons. It is common to have two hot-wire gas detectors, one of which indicates methane and the other heavier-than-methane hydrocarbons.

More definitive analyses can be made with a gas chromatograph (Figure 4–3). It consists of a tube which is packed with powdered material capable of adsorbing hydrocarbon gases. A stream of helium is passed through the tube, which is called a column. A small quantity of hydrocarbon is injected into the helium stream and, as it passes over the powder, it is all adsorbed. After a while, however, the hydrocarbons begin to come off the column—not all at once, but one at a time, beginning with the lightest. The helium is passed over a hot-wire detector or a flame ionization detector. This consists of a gas flame burning in an electrostatic field. As the hydrocarbon gases come by, they ionize and become conductive. This actuates the recorder, which gives a record of detector response versus run time (Figure 4–4). The peaks are identified and made quantitative by injecting a known mixture of hydrocarbons and noting the reaction of the instrument.

Oil in the mud is detected by treating the mud to reduce surface tension and gel strength so the oil separates and then examining it for fluorescence under UV light.

Detection of hydrogen sulfide. A piece of paper coated with a lead

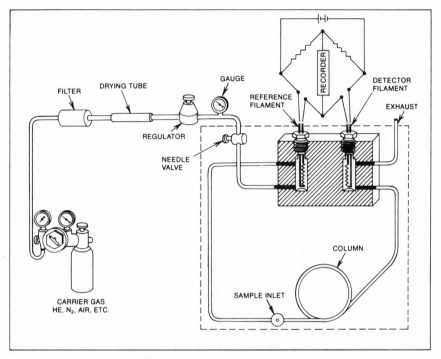

Fig. 4–3 Gas chromatograph

salt is installed in the gas-air line from the trap. If hydrogen sulfide appears in the stream, it blackens the paper. This color change is detected by a photocell and recorded. Sometimes the hydrogen sulfide sets off an alarm.

Factors affecting detection of oil and gas in mud. It is, unfortunately, common practice to put diesel fuel or asphalt in the mud to improve its lubricating properties. The oil and gas from the formation tend to dissolve in the diesel fuel, so that they are not picked up at the trap. Some companies claim that they can still get significant gas shows when there is diesel fuel in the mud because it contains no light hydrocarbons. If crude oil containing dissolved gas and gasoline is put in the mud, it makes detection of oil shows impossible. Some foreign oil companies make a practice of never putting any kind of oil in the mud of an exploratory well, and this seems to be a most desirable practice.

The weight of a column of water is 0.434 psi per foot (9.8 kPa per meter) if the water is pure or more if it is salty. Water weighs 8.33 lb per gal (1 kg per liter). Therefore the weight of a mud column in psi per

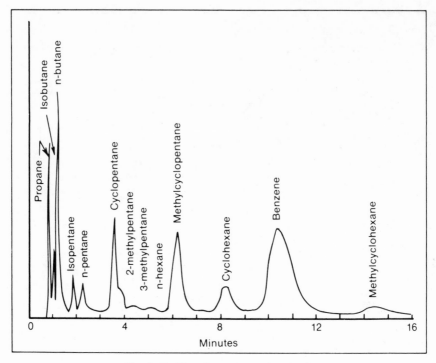

Fig. 4—4 Gas chromatograph record

ft is 0.434/8.33 or 0.052 × weight in pounds per gal (in kPa per meter it is 9.8 × mud density). Ordinary drilling mud has a weight slightly greater. Mud weighing 9.0 lb per gal exerts a pressure of 0.468 psi per ft (10.6 kPa per meter). In many areas, the pressure in the pore fluids is between 0.45 and 0.50 psi per foot of depth (10.18 and 11.3 kPa per meter) and this is called normal. If the pressure in the pore fluids is higher than the mud weight, gas will come out of solution and become dissolved or entrained in the mud. If it is not removed in the tanks, it will be recirculated, so a background gas response will be built up. If too much gas is entrained in the mud, its density decreases and the well may become underbalanced.

When the pressure in the mud is greater than the pressure of the fluids in the formation, the situation is called *overbalanced*. When a permeable sand is drilled with the mud overbalanced, the filtrate from the mud enters the sand ahead of the bit, flushing the oil or gas ahead. The extra pressure from the jets in the bit increases this effect. Only very small amounts of oil or gas may be left in the sand to be drilled up and enter the mud stream. An overbalanced situation, therefore, may completely eliminate a gas show in the mud. The better and more permeable the sand, the more the flushing effect. Many, if not most, gas sands of the Rocky Mountains, Mid-Continent, and eastern U.S. have abnormally low reservoir pressures, ranging down to 0.3 psi per foot (7 kPa/m). In these areas, even the lightest drilling mud will always result in overbalance.

The drop in pressure, or rather in hydraulic head, in the mud as it rises through the annulus is considerable. When the pumps stop, therefore, the pressure in the mud in the lower part of the hole decreases correspondingly. If the drillpipe is raised, it causes a swabbing (suction) effect on the mud of about the same magnitude—more if it is raised too fast.

When the top of the kelly has moved down to the rotary table, it is necessary to add another joint of pipe to the string in order to continue drilling. The pumps are stopped, and the kelly is pulled up until the uppermost collar on the drillpipe appears. The collar is then held by the slips while another joint is picked up off the pipe rack and stabbed into it. The kelly is then screwed on, the drillstring is lowered to bottom and drilling resumes. This operation is called *making a connection*. It always causes a substantial drop in the pressure of the mud at the bottom of the hole. This drop in pressure causes gas to come out of the cuttings and the formations in the wall of the hole. The gas was most probably in solution in the pore water, and the drop in pressure causes it to come out of solution. Therfore, any time the pumps shut down and circulation is stopped and the drillpipe is raised, the gas content of the mud at the bottom of the hole increases. After circulation has resumed and this mud reaches the trap some time later, there will be a short-lived show of gas. This is called *connection gas* or *trip gas*. The number of pump strokes between the time the pumps were started again and the time the gas appears on the mud logger record gives the volume of the annulus and the lag time.

Shale containing large amounts of organic matter (one percent or more of organic carbon by weight) is likely to contain both methane and heavier hydrocarbons. These are released to the drilling mud and cause gas indications, even when no reservoir rock is being drilled.

Interpretation of mud logs. Although it would be foolish to drill an expensive wildcat well without carefully monitoring it for shows of hydrocarbons, the mud logger cannot be depended on completely. If the well is overbalanced, it can drill right through a producing sand without recording any show. On the other hand, organic-rich shales may make strong hydrocarbon shows where there is no producing sand.

Any oil in the mud naturally masks shows of oil or gas from the formations.

Much depends on the quality of the personnel. The trap may be operating poorly, mud may get in the tubes and block them, oil from lubricants may give false shows, or the instruments may get out of calibration.

There are two principal methods of interpreting mud log shows. In the first method, the well is drilled to total depth, logged electrically, and cased. The mud-log shows are then evaluated along with the electric logs and sidewall cores. The possible producing horizons are gun-perforated and tested. The disadvantage of this method is that the drilling mud on the face of the sand tends to plug the formation and a good sand may give a very poor test.

The second way is to stop drilling at any good show, take a core, and run a drillstem tester in the open hole. This has the disadvantage that the electric logs are not available to help evaluate the show.

2. OTHER DRILLING PARAMETERS RECORDED BY MUD LOGGERS

The trailer with the mud logging equipment which is manned continuously by trained people provides a very advantageous opportunity to record a number of other significant parameters associated with drilling. These are mainly of interest to the drilling engineer.

Mud properties.   The properties of the mud must be controlled and recorded. The most critical is its density, usually expressed in the U.S. as pounds per gallon. In metric countries it is given as density, which is grams per cubic centimeter. The mud weight can be expressed conveniently as pounds per square inch per foot of depth [or kPa per meter (1 psi/ft = 22.62 kPa/m)] in terms of the pressure it exerts at the bottom or at any depth in the hole. This pressure is then compared with the pressure of the pore fluids on the one hand and the *fracture pressure gradient* on the other. The fracture pressure gradient is the estimated pressure that will cause the walls of holes to fracture. If the mud pressure exceeds this, the mud will run out into the fracture. There will be no returns of mud coming up the annulus, so this situation is called *lost circulation*.

Other properties of the mud that are periodically recorded are gel strength, viscosity, percent solids, percent sand, percent chloride, alkalinity, and pH. Knowing these properties, it is possible to calculate not only the pressure at the bottom of the hole but also the pressure changes due to laminar and turbulent flow in the annulus, and the swabbing and surge pressures caused by raising or lowering the pipe.

Rate of penetration. The rate of penetration in minutes per foot is recorded at two-foot intervals and plotted on a log at 5 in. equals 100 ft. Changes in rate of penetration are very important to the geologist because they indicate a change in lithology. An increased penetration rate is called a *drilling break* and may indicate a sand. It may also be the first indication of an abnormally high-pressure zone that might cause a blowout.

Weight on bit and rotary table speed. The driller always has before him a dial showing the weight on the derrick (weight on the hook). Subtracting the weight of the drillstring, this gives the weight on the bit which is periodically recorded. The rotary table speed is recorded. These two parameters affect the drilling rate. The torque on the drillstring is also sometimes recorded.

Mud system. The pump strokes are recorded cumulatively and also as strokes per minute. The mud pressure into the kelly is recorded. The level of mud in the tanks is monitored closely and recorded. If it gets too high it indicates an intrusion of gas, oil, or water into the hole. If it gets too low it indicates lost circulation. Either involves trouble, so an alarm rings at both high and low levels.

Figure 4–5 is an example of a log prepared by a mud logger. The curves on the left side are the electric logs, to be described in the next chapter. A kick to the right on the resistivity curves indicates a limestone or sandstone.

The rate of penetration was quite steady at about 6 minutes per ft until 9,280, when it increased to two minutes per foot. This corresponds, although not exactly, to a sand indication on the resistivity log. This was the Tonkawa Sand. The middle column is a graphic log of the lithology as determined from the cuttings, and the next column gives their description. The two lines next to the right are the gas shows in the mud; methane is the solid line and hydrocarbons heavier than methane the dashed line. Gas shows occurred at 9,232 and 9,278. The shows are corrected for lag. At 9,421 they pulled the drillpipe, ran electric logs, and drillstem tested with poor results. These round trips produced a big kick on the gas log.

Drilling was resumed and normal (9.0 to 9.4 ppg) mud was used to 15,170 ft into the top of the high-pressure Morrow shale. Logs were run, drilling was resumed to 15,185 ft, where 10¾-in. casing was set. The mud was then weighted up to 14.6 ppg. A sharp drilling break occurred at 15,496 ft and the mud weight was increased to 17.1 (density 2.05). Drilling continued at one minute per foot, and at 15,508 ft the well kicked, the blow-out preventers were closed, and the pressure built up to 600 psi (42 kg per sq cm) on the drillpipe and 800 psi (56 kg

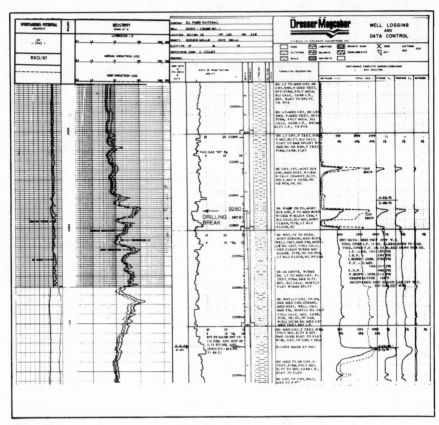

Fig. 4–5 Log of a gas well *(prepared by Dresser Magcobar, courtesy El Paso Natural Gas Co.)*

per sq cm) on the casing. The mud weight was raised to 17.7 (density 2.12) to kill the well. Drilling was continued with 16.8 to 17.0 mud but at a very slow rate—20 minutes per ft (65 minutes per meter). The well was finally completed in the Upper Morrow sand, producing 12 million cu ft (0.34 million cu m) of gas per day from 15,496 to 15,522 ft (4,724 to 4,732 meters).

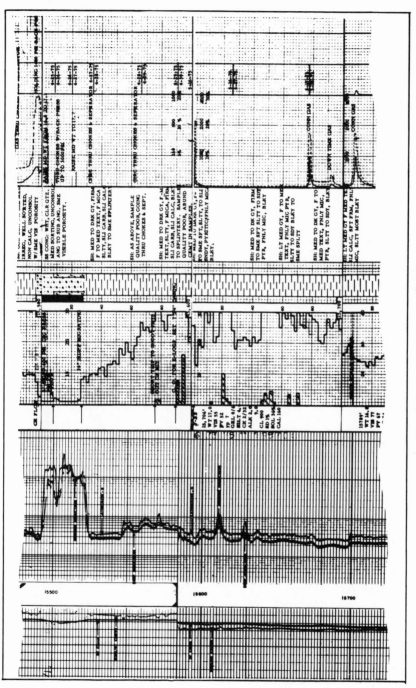

Fig. 4–6 Continuation of a mud log

REFERENCES

Choate, Lee R., 1963, Baroid well logging service—Mud analysis logging: Baroid Division, National Lead Co., Houston, Texas, 76 p.

Imco Services, 1974, Practical subsurface evaluation: The Halliburton Company, Dallas, Texas, 135 p.

Magcobar, various manuals. Division of Dresser Industries, Inc.

Moore, Preston L., 1974, Drilling practices manual: Tulsa, Oklahoma, PennWell Publishing Company, 446 p.

Pearson, A. J., 1961, Miscellaneous well logs, *in* Graham B. Moody, ed., Petroleum exploration handbook: Chapter 21, p. 11.

Wilson, Tom, 1976, The quantification of mud log gas shows: Soc. Prof. Well Log Analysts 17th Annual Logging Sym. Trans., Pap. FF, p. 10.

5

Electrical and Other Wire-Line Logs

1. ELECTRICAL LOGGING

In the 1920s attempts were made to utilize the electrical properties of rocks in searching for ore bodies. These methods were not very successful, mainly because the effects of near-surface changes in rock properties could not be removed and they tended to mask the changes in rock character at depth. It occurred to two French brothers, Conrad and Marcel Schlumberger, to lower the electric devices they had been using on the surface down a borehole. The results were very successful because the near-surface effects were almost the same up and down the walls of the hole, so the variation in electrical properties which were observed resulted only from changes in the rocks.

Efforts were at once made to relate the electrical properties which were measured to the geological and reservoir properties of the rocks. These attempts were not entirely successful, and even today it is usually impossible to determine the rock type, porosity, permeability, and fluid content from the electrical and other wire-line logs alone. It must always be remembered that the logs measure electrical and other *physical* properties which often only are indirectly related to the reservoir properties. If a set of cores is obtained and compared with the electrical logs of the same well, then the interpretation of logs of other wells in the same area is much more ensured.

In spite of their limitations, we depend on electric logs more than on anything else to tell us about the geology and reservoir properties of oil pools. Many special courses in well log interpretation are offered by the logging companies, oil companies, consultants, and universities. A production geologist must take one or more of these courses. Not enough space is available in the present manual to treat the topic adequately, considering its extreme importance.

Self-potential. The self-potential log is also called spontaneous-potential or SP log. The device consists simply of a nonpolarizing electrode which is lowered down the hole on a cable. Another electrode is placed in the ground (on the surface), and the difference in potential is measured on a recording galvanometer (Figure 5–1). The actual absolute potential difference depends on the nature of the electrodes, the soil water, and natural earth currents, and it is meaningless. What is recorded are the variations in potential difference as the electrode is lowered down the hole. The potential when the electrode is opposite a bed of shale is called the *shale base line* and it is considered zero. The variations from the shale base line are usually negative, and they are plotted to the left (Figure 5–2). They may amount to 100 mv or more.

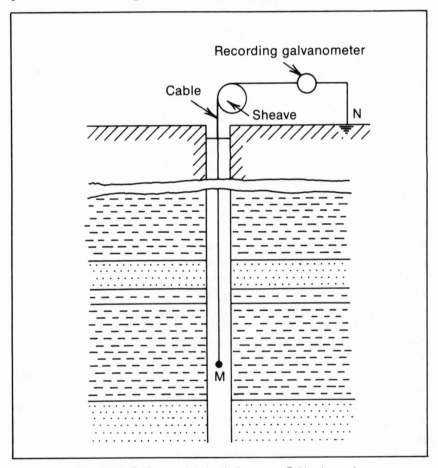

Fig. 5–1 Self-potential circuit *(courtesy Schlumberger)*

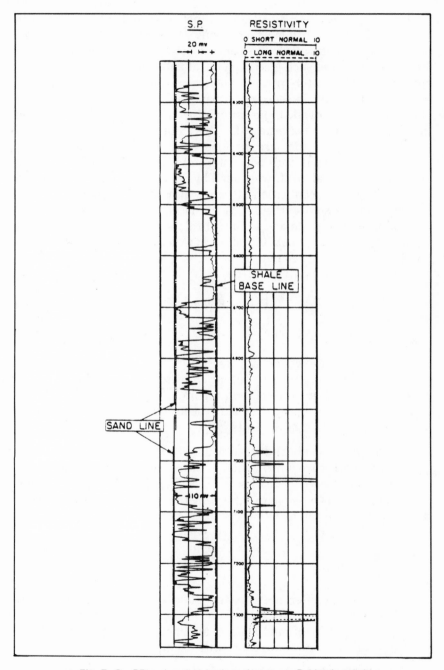

Fig. 5—2 SP and resistivity logs *(courtesy Schlumberger)*

The causes of the SP voltage are not well understood. They are electrochemical potentials which result when solutions of electrolytes of different concentration and composition are in contact with each other. Two or three such electrochemical potentials are believed to be present. (1) The water in the mud is generally less concentrated in salts than the water in the sand. A potential difference will thus exist at the edge of the invaded zone, at point A in Figure 5–3. This is called the liquid junction potential. (2) The shales act as semipermeable membranes because they are more permeable to Na^+ cations than to Cl^- anions. There will thus be a potential between B and C, called the membrane potential. (3) It was found by Schmidt (1974) and Baharlou (1975) that the concentration of pore water in shales is usually less than that in the adjacent sands. A liquid junction potential will then exist between the shale and sand regardless of whether there is drilling mud in the hole or not. The effect of this potential on the SP curve has not yet been worked out.

The sum of the first two effects is

$$E_c = -K \log \frac{a_w}{a_{mf}} \qquad (5\text{--}1)$$

where:

E_c = total electrochemical potential
a_w = chemical activity of formation water
a_{mf} = chemical activity of mud filtrate
K = a coefficient proportional to the absolute temperature and the composition of the salts.

For sodium chloride and fresh water it is 71 at 25° C (77° F). The chemical activities are proportional to the concentration, and the resistivity of an electrolyte is inversely proportional to the concentration. Consequently, it is possible to determine the resistivity of the water in the sand from the SP, as follows:

$$SP = -K \log \frac{R_{mf}}{R_w} \qquad (5\text{--}2)$$

where:
SP = voltage difference between shale and sandstone, millivolts
K = a constant, near 71
R_{mf} = resistivity of mud filtrate, ohmmeters
R_w = resistivity of formation water, ohmmeters

Water flowing through a porous medium gives rise to an electromotive force (EMF) or voltage called the *streaming potential*, so the filtration of the mud filtrate into the sand might affect the SP. This effect is very small.

The drilling mud in the hole is quite conductive, and its effect is to provide a path through which the current which is generated by these EMFs flows (Figure 5–3b). If an insulating plug were placed in the hole, as in Figure 5–3a, the SP curve would be sharp, as shown on the graph on the left side of the figure. However, the curve actually observed is rounded, as shown in Figure 5–3b. The maximum SP will be less than the static SP (SSP), depending on the thickness of the bed. The top and bottom of the bed are marked by the points of inflection on the curve.

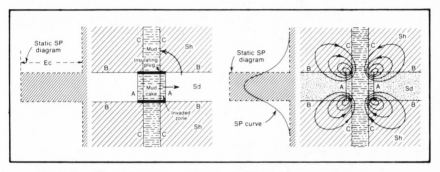

Fig. 5–3 Effect of drilling mud in hole on SP log *(courtesy Schlumberger)*

The presence of interstitial clay in the sand decreases the SP. If the sand is very shaly, the SP may be reduced practically to zero. If the sand contains thin shale beds, these will reduce the SP. It is generally difficult or impossible to tell whether a sand has a low SP because it contains interstitial clay or shale interbeds. Either condition gives a wrong value for the water resistivity (equation 5–2). The water appears to be less concentrated (more resistive) than it really is. If the sand is highly resistive the SP will be reduced.

The SP curve has little meaning in limestones. If the limestone is low porosity, it will have a high resistivity, which weakens the SP. If the limestone contains very little clay, the SP will tend to be high negative, as in sand. But if it contains considerable clay, the SP will be low (Figure 5–4).

If the mud filtrate is equal in concentration to the water in the sands, which is often the case if there is salt in the mud, the SP curve is wiped out and there is no deflection opposite sands. If the concentration

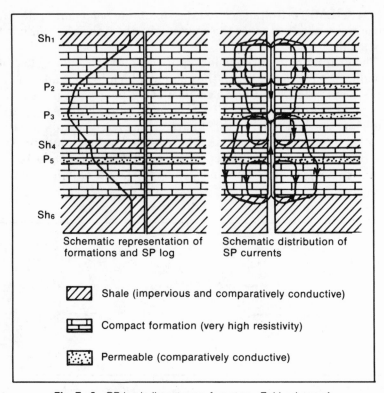

Fig. 5—4 SP log in limestones *(courtesy Schlumberger)*

of the mud filtrate is more than formation water, the SP curve will be reversed. This is often seen in the case of fresh-water sands in the upper part of the hole.

Resistivity. The conventional resistivity device consists of a generator which feeds current into the earth through electrodes A and B as shown in Figure 5–5. The current flows out at A, radiates in all directions in the conductive earth, and eventually finds its way back to point B.

The potential field surrounding the electrode will be described by equipotential surfaces which are spheres. Suppose we insert two other electrodes, M and N, into the field for the purpose of measuring it (Figure 5–5). The voltmeter is connected to M and N. The resistance of the shell of a sphere between the surfaces located by the electrodes M and N is:

$$R = \frac{E}{I} \text{(Ohm's law)} \qquad (5–3)$$

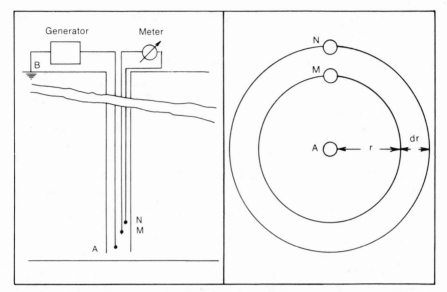

Fig. 5—5a Circuit to measure resistivity (*courtesy Schlumberger*)

Fig. 5—5b Configuration of electrodes to measure resistivity of shell or radius (r) and thickness (dr)

where:

> R = the resistance of the circuit
> E = the potential difference between M and N
> I = the current flowing through A

The resistance will depend on the resistivity ρ of the rock material as follows:

$$R = \frac{\rho}{4\pi} \int_{r=a}^{r=b} \frac{dr}{r^2} \tag{5--4}$$

where:

> ρ = resistivity of the rock
> a = radius of the inner sphere
> b = radius of the outer sphere
> r = distance from the center to the equipotential surface.

This reduces to

$$R = \rho \frac{b-a}{4\pi ab} \tag{5--5}$$

The values of b and a depend on the configuration and spacing of the electrodes, which for any particular tool are fixed. Then, if the current I is held constant, E, the voltage between M and N will depend only on ρ, the resistivity of the rock. The circuit is arranged so that resistivity is plotted against depth on the right-hand side of the strip.

Resistivity is the specific resistance of any material. In the case of rock, it is expressed in ohmmeters. A cube of rock one meter square which transmits one ampere of current under a potential difference of one volt is said to have a resistivity of one ohmmeter.

The mineral constituents of a rock (except for the metallic ores) have very high resistivities (electrical insulators are made of baked clay). The electricity is conducted through rock almost entirely by ions. In the case of a clean sandstone, the ions are those dissolved in the pore water. In the case of a shale or a shaly sand, the ions are those dissolved in the pore water and, in addition, the exchangeable ions on the surfaces of the clay minerals.

The resistivity of a clean sandstone depends on its porosity and on the resistivity of the water in the pores. The latter depends on the concentration of salts in the water. Subsurface water ranges from quite fresh to very salty. Porosities of sandstones range from a few percent to more than 30. Consequently, the resistivities of sandstones range widely.

The resistivity of shales also ranges widely depending on the porosity, water salinity, and nature of exchangeable cations. It tends to increase with increasing compaction and consolidation. At any one locality, the resistivity of shales does not change much—not nearly as much as the resistivity of the interbedded sandstone.

The resistivity of a clean sand fully saturated with water depends on the resistivity of the water as follows:

$$R_o = F\,R_w \qquad (5\text{--}6)$$

where:

R_o = resistivity of the rock fully saturated with water
R_w = resistivity of the water
F = a factor depending on the porosity and pore shape and size

The factor F is called the *formation factor*. It was shown by Archie (1942) that

$$F = \frac{a}{\phi^m} \qquad (5\text{--}7)$$

where:

  a = a constant depending on the rock
  m = another constant depending on the rock
  ϕ = porosity

For clean, consolidated sandstones, the constant a is usually near one and the constant m is near two. So the formation factor F is usually

$$F = \frac{1}{\phi^2} \qquad (5\text{--}8)$$

Oil and gas are nonconductors of electricity. If they are present in the pores of a rock, there will be less water, so the resistivity of an oil or gas-bearing rock will be higher than that of one containing only water. It was also determined experimentally by Archie that:

$$R_t = \frac{F\,R_w}{S_w^{\;n}} = \frac{R_o}{S_w^{\;n}} \qquad (5\text{--}9)$$

where:

  R_t = resistivity of the rock partly saturated with water
  S_w = percent of porosity containing water
  n = a constant depending on the rock

The constant n is near 2. Consequently, the water saturation S_w can be calculated as follows:

$$S_w^{\;2} = \frac{F R_w}{R_t} \qquad (5\text{--}10)$$

From equation 5–6

$$F\,R_w = R_o$$

which is the resistivity of the rock when fully saturated with water of resistivity R_w. Therefore,

$$S_w = \sqrt{\frac{R_o}{R}} \qquad (5\text{--}11)$$

The oil or gas saturation is $(1 - S_w)$.

The value of R_o can be determined from the resistivity of water-saturated sand in the same or a nearby formation. It can also be estimated from the SP log using equations 5–2, 5–6, and 5–8. Under favorable circumstances (by no means always), the oil or gas saturation of a formation can be calculated from the SP and resistivity measurements.

The presence of clay in the sand seriously affects this calculation. If the SP is lower than it would be in a thick, clean sand, it makes R_w too high. The S_w comes out too high, and the calculated oil saturation is lower than true. If the conduction by the ions on the clay surfaces makes the resistivity of the rock lower than it would be in a clean sand, it makes the S_w seem too high and the oil saturation lower than it really is. This is called the *shaly sand problem*. The effect can result in passing up oil- or gas-producing sands. If they contain much clay, they will appear to the resistivity log as if they were saturated mostly with water, even though they may produce clean hydrocarbons.

When drilling overbalanced, the filtrate from the mud invades the permeable oil or gas sands and drives the oil or gas away from the borehole. This invasion may extend farther than the effective radius of exploration of the resistivity device.

Figure 5–6 is a horizontal section through a permeable oil-bearing bed, showing the borehole, a filter cake (mud cake) on the wall of the hole, an invaded zone, a zone of mixing, and the uncontaminated zone.

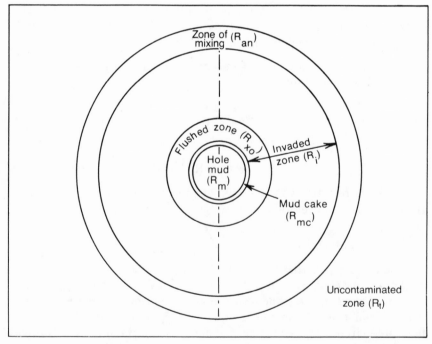

Fig. 5–6 Horizontal section through a permeable oil-bearing bed, showing zones of mud filtrate contamination *(from Schlumberger)*

The resistivities of these zones are, respectively, R_m, R_{mc}, R_{xo}, R_{an}, and R_t. Figure 5–7 shows the radial distribution of fluids. The mud filtrate, which is nearly fresh, also displaces the formation water, which is usually salty. Figure 5–8 is a radial plot of the resistivity changes away from the hole.

The invaded zone will have the resistivity of a rock partly saturated with oil, and mostly with water of resistivity R_{mf}. Its effective resistivity is called R_{xo}. The advancing drilling mud filtrate, which is fresh, pushes the salty connate water ahead of it by the process of miscible displacement. The zone of mixing will have the resistivity of a rock saturated partly with oil and mostly of water with a resistivity of R_w. The uncontaminated zone will have the resistivity of a rock saturated mostly with oil and partly with water of resistivity R_w. Its resistivity is called R_t (t means "true"), and this is what is needed to evaluate the formation with equation 5–10.

In order to determine the oil saturation of a possible producing formation, it is necessary to get a reliable value for R_t, its true resistivity. The diameter of the invaded zone is seldom known; it may range between 0 and 100 in. (3 m). At short electrode spacings (say 16 inches or 40 cm for the short normal resistivity curve), the effect of the invaded zone may be so great that the tool reads resistivity values close

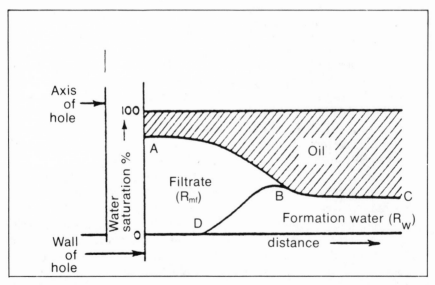

Fig. 5–7 Radial distribution of fluids in the vicinity of the borehole, qualitative (after Schlumberger)

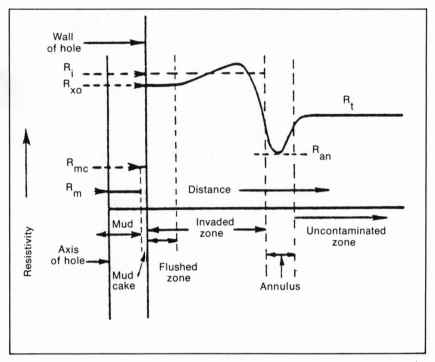

Fig. 5–8 Radial distribution of fluids and resistivities in an oil sand resulting from the invasion of mud filtrate *(courtesy Schlumberger)*

to R_{xo}. In order to get greater penetration, the electrodes may be spaced farther apart. A common spacing is 18 feet, 8 inches or 5.7 m (the lateral). Unless the bed is at least as thick as the spacing, the resistivity of the beds above and below has a large effect and the readings will not give true values for R_t. Also with electrodes spaced this far apart, the top and bottom of the beds are not well defined. Most older logs show the SP, short normal, and lateral curves.

To avoid these difficulties, two other devices are now used. The *induction* log consists of two (or more) coils as shown in Figure 5–9. A high-frequency current flows through the transmitter coil. This gives rise to induced ground-loop (eddy) currents in the surrounding rock. The more conductive it is, the larger will be the currents. The effects of the induced currents can be picked up by a receiver coil. The directly coupled effects of the transmitter are balanced out.

The induction log works in holes with no mud or with oil-base mud. It also works well in ordinary, fresh-water muds. The response is less

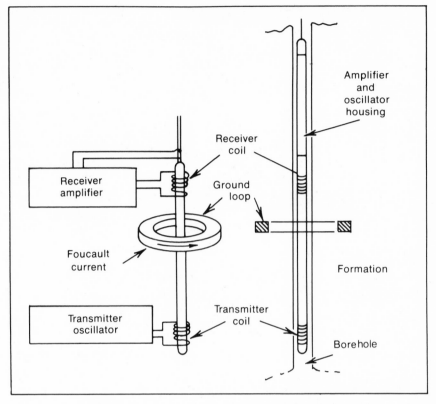

Fig. 5—9 Schematic of induction log *(after Schlumberger)*

affected by the high-resistivity flushed zone; therefore, it gives good values for R_t.

Another device that was developed to penetrate more deeply is the laterolog, shown in Figure 5–10. Two guard electrodes are maintained at the same potential as the current electrode A_o. The current is held constant, so the potential varies with the resistivity of the formation. The guard electrodes focus the current so it penetrates more deeply. The laterolog gives the best values for R_t in salty muds.

The laterolog can also be made smaller so that it is affected mostly by the invaded zone. It is then called the laterolog 8.

Figure 5–11 is an example of a curve showing two induction logs with different spacings, IL_d and IL_m, and a short penetration laterolog, LL8. The IL_d and IL_m gives good values for R_t (about 0.8 ohms). The LL8 gives a value of 5.0 ohms for the invaded zone (R_{xo}).

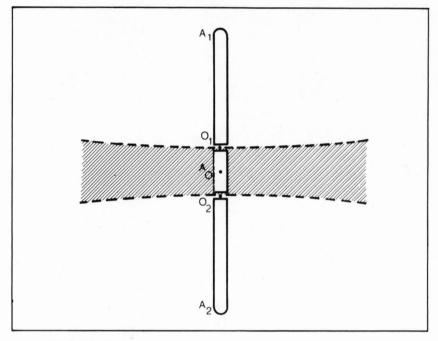

Fig. 5–10 Schematic of guard electrode device *(after Schlumberger)*

Knowing the spacing of the electrodes, the diameter of the invaded zone d_i can be estimated. When d_i and R_{xo} are known, they can be used to correct the deep resistivity reading to get a better value for R_t. The basic formula says:

$$S^n_w = \frac{F R_w}{R_t} \qquad (5\text{–}12)$$

Similarly,

$$S^n_{xo} = \frac{FR_{mf}}{R_{xo}} \qquad (5\text{–}13)$$

where S_{xo} is the water saturation of the flushed zone. Dividing equation 5–12 by equation 5–13 and making n = 2, we get

$$\left(\frac{S_w}{S_{xo}}\right)^2 = \frac{R_{xo}/R_t}{R_{mf}/R_w} \qquad (5\text{–}14)$$

This equation gives the ratio of S_w to S_{xo}.

If the ratio is 1, the water saturation of the flushed zone is the same as that of the uninvaded zone. Therefore no residual oil is present, and

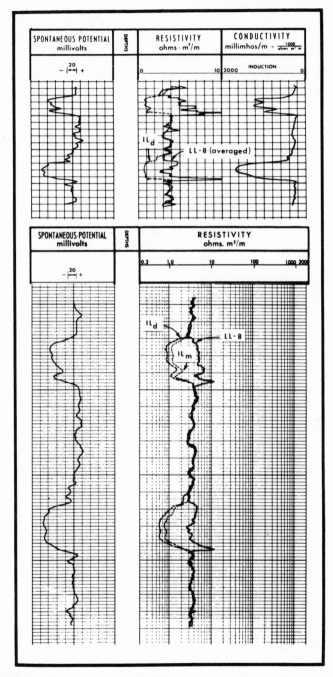

Fig. 5—11 Induction-electric log with SP, two induction logs, and the laterlog 8 *(courtesy Schlumberger)*

the sand is 100 percent saturated with water. If the ratio S_w/S_{xo} is 0.7 or less, movable hydrocarbons are present.

If we know S_{xo} from a knowledge of the porosity ϕ or the formation factor F, then we can calculate S_w from equation 5–14. It has been found that empirically $S_{xo} = S_w^{1/5}$, approximately. Inserting this in equation 5–14,

$$S_w = \left(\frac{R_{xo}/R_t}{R_{mf}/R_w} \right)^{5/8} \tag{5–15}$$

The SP curve gives an approximation of R_{mf}/R_w.

2. RADIOACTIVE LOGS

The gamma-ray log. Certain elements are radioactive; that is, they disintegrate spontaneously with the emission of alpha, beta, and gamma rays. The radioactive elements most common in natural earth materials are potassium, with an atomic weight of 40, uranium, and thorium. Gamma rays are high-energy electromagnetic waves, like X-rays, that can go through steel pipe.

Potassium is most abundant in clay minerals, so shales emit more gamma rays than clean sandstones or limestones. Uranium occurs in measurable amounts in black, organic shales, so they are especially radioactive. Potassium salt deposits or uranium ores may be still more radioactive.

The sonde consists of a scintillation counter that notes the natural gamma rays which hit it. These come at irregular intervals, so the instrument has to move slowly to permit smoothing circuits to average the gamma-ray counts. A speed of 1,800 feet per hour is slow enough. The gamma-ray sonde is calibrated in units of radioactivity prescribed by the American Petroleum Institute.

The gamma-ray curve somewhat resembles the SP curve because it indicates mostly the shaliness of the formation. It is often substituted for the SP when it cannot be obtained, as in cased holes when the hole is empty or when the mud is salty. It is excellent for correlation because kicks in shale beds can sometimes be traced long distances. It is less satisfactory than the SP for picking the top and bottom of sands or limes. It is run in cased holes together with a collar locator in order to locate accurately the perforating gun.

Neutron log. Neutrons are electrically neutral particles with a mass about the same as that of a hydrogen atom. They are emitted by specially prepared substances like radium and beryllium. They leave the source with a high velocity, but they are slowed down by collisions with other atoms. The lighter the atom, the more the energy of the neutrons is reduced and the more they are slowed. In a few microseconds, they are going so slowly that they can be captured by a hydrogen atom. When this happens, the hydrogen atom emits a gamma ray that is detected by a scintillation counter.

The sonde consists of a neutron-emitting substance spaced about 15 in. from a scintillation counter. If there is much hydrogen in the rock, the neutrons are slowed down when they are only a short distance from the source. The instrument therefore reads higher when there is less hydrogen. Hydrogen composes a considerable fraction of both water and hydrocarbons, which are present in the pores of a rock. The neutron log thus indicates mainly porosity.

In one variety of neutron log, the tool is pressed against the wall of the hole in order to eliminate the effect of variations in hole diameter. This is called the sidewall neutron porosity device, or SNP log, and it records porosity directly on the log. It is calibrated by inserting it in a block of water-filled limestone of 19-percent porosity.

Although it cannot be used quantitatively, the neutron log indicates porosity, although it cannot distinguish between oil and water. Gas, of course, has a lower hydrogen concentration than oil or water, and sometimes the neutron log indicates a gas-oil or gas-water contact. Figure 5–12 is an example of gamma ray, neutron, and SNP logs.

Density log. A radioactive source emits gamma rays into the formation. These collide with electrons in the elements in the rock, which causes a back-scattering of gamma rays. The more the electrons, the more the back-scattered rays, which are picked up by a detector. The number of electrons per cubic centimeter in the rock is nearly proportional to its density, depending on the atomic number and molecular weight of the elements which make it up.

The effect of the mud cake is considerable, so the sonde is pressed tightly against the wall of the hole and the recorder automatically applies corrections.

Porosity is related to formation density by the following formula:

$$\phi = \frac{\rho_{ma} - \rho_b}{\rho_{ma} - \rho_f} \qquad (5\text{--}16)$$

where:

ϕ = porosity
ρ_{ma} = density of the matrix
ρ_b = bulk density
ρ_f = density of the fluid filling the pores.

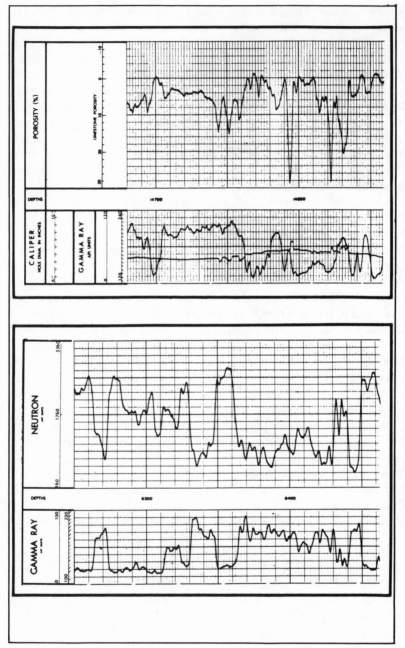

Fig. 5-12 Gamma-ray, neutron, and SNP logs *(courtesy Schlumberger)*

Common values of matrix density are:

quartz sands	2.65
limy sands	2.68
shales	2.71
limestones	2.71
dolomite	2.87

With the shallow depth of penetration of this tool, the fluid filling the pores will be mostly mud filtrate, with a density near 1.0, depending on the temperature and salinity. The density of the residual hydrocarbons is less than that of water, depending on the molecular weight of the oil. Residual gas naturally has a much lower density.

Reliable density and porosity measurements are, of course, extremely important in determining the lithology and fluid content of oil reservoirs. The density log is, therefore, a very great advance in well logging.

Thermal decay time (TDT) log. The time required for the decay of thermal neutrons in a formation depends largely on the amount of chlorine. Chlorine, of course, occurs in the salt water commonly associated with oil. A burst of neutrons is shot into the formation. Then for a few microseconds the secondary gamma rays are measured during two time intervals and the exponential rate of decay of thermal neutrons is determined. Knowing the nature of the rock and the chlorine content of the interstitial water, it is possible to calculate the water saturation.

The TDT log can be run through casing. It responds only to salt water, so it gives no useful information in uncased holes where the drilling mud filtrate has invaded the formation. Neither is it useful unless the interstitial water is salty.

The TDT log is especially useful in picking oil-water contacts. Run successively at intervals of months or years, it is possible to follow the rise of the oil-water contact as oil is produced.

3. OTHER LOGS

Sonic logs. The sonic velocity log was first developed to measure the velocity of acoustic waves through the earth so that reflection seismic travel times could be converted to true depth. It proved to be useful in indicating lithology, porosity, and the presence of fractures.

A transmitter makes a sound which travels through the rock. Two detectors pick it up, and the delay time from one to the other is recorded as milliseconds per foot, Δt. More recent devices use two transmitters and two sets of detectors (Figure 5–13).

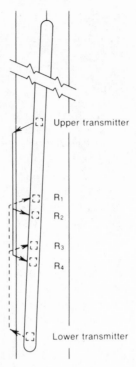

Upper transmitter

R₁

R₂

R₃

R₄

Lower transmitter

Fig. 5–13 Schematic of seismic velocity tool *(after Schlumberger)*

The velocity of acoustic waves in rocks ranges widely, from about 6,000 ft per second (2 km per second) for shales to over 20,000 fps (6.09 km per second) for limestones. Correspondingly, the Δt ranges from 166 to 50 microseconds per ft (50 to 15 microseconds per meter). The sonic log gives good indications of lithology, distinguishing between shale, sandstone, and limestone. Increasing porosity decreases the velocity, so the results can be used to estimate porosity. At any one locality, the travel time of shale decreases with depth because the shales become more compacted. An abnormally long travel time of shale indicates undercompaction, which suggests abnormally high pressure in the pore fluids.

A variation of the sonic log records the acoustic waves as they arrive on variable-density film (Figure 5–14). The first waves to arrive are the compressional body waves, followed by the shear wave and finally by the surface wave (boundary wave). From the arrival times, the values for elastic moduli can be determined. From these and the

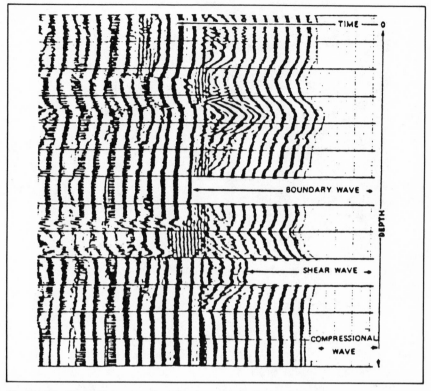

Fig. 5–14 Variable-density film showing arrival times of different waves *(courtesy Seismograph Service Co.)*

appearance of the records, zones of porosity and fractures can be detected.

Another application of acoustic waves is the borehole viewer. Waves are emitted and reflected back from the wall of the hole, and they are recorded on variable density film. This gives a picture of the inside of the borehole (Figure 5–15). A magnetometer in the tool senses the earth's magnetic field and marks north, south, east, and west on the film.

Dipmeter logs. The dipmeter sonde consists of 3 (or 4) arms 120° apart which contact the walls of the hole. Contact is made with rubber pads, in each of which is a microresistivity device. Each trace is recorded separately as the sonde is pulled up the hole, and each gives a very detailed log of the minor lithologic variations. If there is any dip to the bedding, the kicks on one or two of the traces will lag behind the

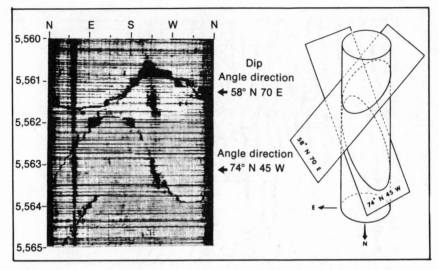

Fig. 5–15 View of interior of borehole processed from acoustic waves showing two fractures with different attitudes *(courtesy Seismograph Service Co.)*

third. The diameter of the hole (caliper) is also recorded, as well as the compass direction of one of the arms. Another device in the sonde records the deviation of the hole from vertical.

The three traces may be matched visually and, knowing the diameter of the hole, the dip may be calculated from the lag of each with respect to the others. Knowing the compass direction of one of the arms, the dip direction may be calculated. This value must be corrected for the amount and direction of the deviation of the hole from the vertical.

Calculations may be made mathematically using spherical trigonometry, by nomographs, or by stereographic nets. If the responses are recorded digitally, then the matching of the curves and the calculations may be made on a computer.

Figure 5–16 is an example of an optical recording. On the right is the caliper curve showing that the hole is 8 inches in diameter. The next three curves to the left are the micro-resistivity records. Electrode No. 1 shows a kick 4.3 in. above a datum, No. 2 shows the same kick 3.2 in. and No. 3, 2.4 in. The logs on the left side give the azimuth of the No. 1 electrode, the hole inclination, and the azimuth of the inclination.

The dips are plotted on the side of a log as shown in Figure 5–17.

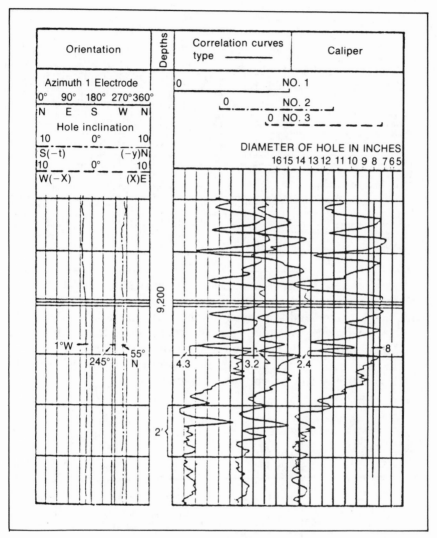

Fig. 5–16 Example of a dipmeter log *(after Pearson, courtesy McGraw-Hill)*

The direction of dip is shown by the arrow, and the amount of dip by the distance of the point to the right of a zero base line.

The value of the dipmeter to geological interpretations is immediately obvious. Structural dips of less than a degree or so hardly show in the borehole and can best be determined from well-to-well correlations or seismic sections. Greater dips can easily be determined

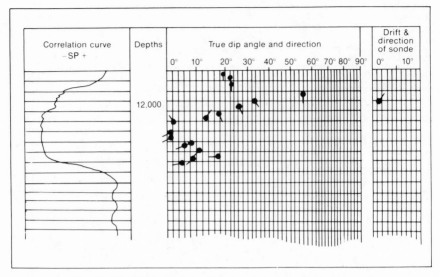

Fig. 5–17 Method of plotting dips *(courtesy Schlumberger)*

with the dipmeter and are added to other data to work out the structure. In a discovery well, the dip of the producing formation indicates the direction in which the second well should be drilled.

The dipmeter locates angular unconformities quite well (Figure 5–18). Sometimes the interpretation is ambiguous.

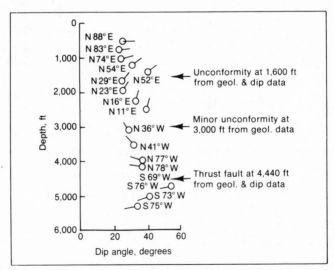

Fig. 5–18 Dipmeter records across unconformities *(after Pearson, courtesy McGraw-Hill)*

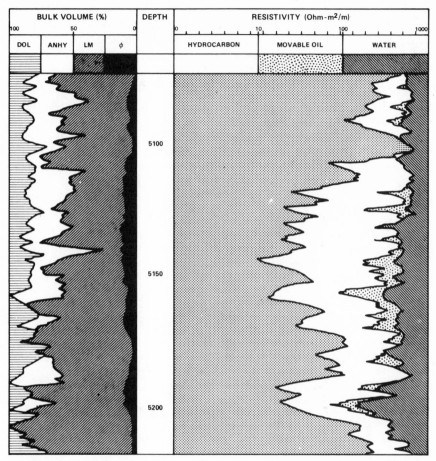

Fig. 5—19 Typical computer-processed log *(courtesy Seismograph Service Co.)*

Offshore wells in deep water, especially in the North Sea, are drilled from platforms in clusters. Some of the holes may be deviated as much as 40 degrees from the vertical. The data on hole deviation and the dip of the strata becomes both difficult and critical during field development.

Corrections and computations. Quantitative calculations of reservoir properties, such as porosity, water saturation, oil saturation, movable oil, etc., require corrections for such effects as borehole size, mud resistivity, formation temperature, etc. These are most conveniently made by means of charts. Books of such charts are supplied by the logging companies.

In many cases, cross-plotting two logs gives more reliable data than one log alone. The density, sonic, and neutron logs are all affected by porosity and lithology, and it is sometimes illuminating to plot the readings against each another at the same depth.

Calculations of oil and water saturations and movable oil are made from the several different resistivity and SP curves. These can be done using charts, nomographs, or slide-rule devices. However, if the basic data is recorded digitally, these calculations can all be made on a computer. An example of a computed log is shown in Figure 5–19. These are interesting and attractive. However, there are so many sources of error and uncertainty in well-log interpretation that computed logs should be viewed with some distrust. The interpretations that come out well in one area may turn out to be quite wrong in another. A good production geologist will look at each log individually and ask himself what it means qualitatively, with reference to the nature of the rock as he has seen it from cores and cuttings. Then he can apply quantitative calculations, tempered by his knowledge of the geology.

REFERENCES

Archie, G. E., 1942, The electrical resistivity log as an aid in determing some reservoir characteristics: Petrol. Techn., v. 5, no. 1.

Baharlou, Alan, and P. A. Dickey, 1973, Chemical composition of pore water in some Paleozoic shales of Oklahoma: Soc. Prof. Well Log Analysts 14th Ann. Logging Symposium, p. 1–7, Paper M.

Dickey, P. A., 1945, Natural potentials in sedimentary rocks: Trans. AIME, v. 164, p. 256–266.

Doll, H. G., 1948, The S.P. Log, theoretical analyses and principles of interpretation: Trans. AIME, v. 179.

Guyod, Hubert, and Lemay E. Shane, 1969, Geophysical well logging: Houston, Hubert Guyod, 256 p.

Holt, Olin R., 1975, Log quality control: Soc. Prof. Well. Log Analysts, 16th Ann. Logging Symposium, Pap. BB, 19 p.

Lynch, E. J., 1968, Formation evaluation, 2nd. Ed.: New York, Harper and Row, 422 p.

Pearson, A. J., 1961, Miscellaneous well logs, *in* Graham B. Moody, ed., Petroleum exploration handbook: p. 21–30.

Pirson, Sylvain J., 1963, Handbook of well log analysis: New York, Prentice-Hall, 326 p.

Schlumberger C., and M. Schlumberger, 1934, A new contribution for subsurface studies by means of electrical measurements in drill holes: Trans. AIME, v. 110.

Schlumberger, Ltd., 1970, Fundamentals of dipmeter interpretation: New York, 277 Park Ave., Schlumberger Ltd., 145 p.

Schlumberger, Ltd., 1974, Log interpretation, vol. I, Principles: New York, 277 Park Ave., Schlumberger Ltd., 109 p.

Schmidt, G. W., 1973, Interstitial water composition and geochemistry of deep Gulf Coast shales and sandstones: AAPG Bull., v. 57, no. 2, p. 321–337.

Waxman, M. H., and L. J. M. Smits, 1968, Electrical conductivities in oil-bearing shaly sands: Soc. Petrol. Eng. Jour.

Wyllie, M. R. J., 1963, The fundamentals of well log interpretation, 3rd edition: Academic Press, 238 p.

6

Environments where Reservoir Sandstones are Deposited

For many years, beginning with the discovery of the Cushing field in 1912, petroleum geologists were preoccupied with structure as the principal geological feature which controlled oil occurrence. However, it came to be recognized that most oil fields were in sandstones that had been deposited in a near-shore marine environment. In order to understand depositional environments better, the American Petroleum Institute (about 1950) began a large research project (Shepard, 1959) to study the recent sediments of the northern Gulf of Mexico. At the same time, several oil companies, notably Shell, Gulf, and Exxon, also began to examine the features of the modern Mississippi delta.

Since that time, an enormous amount of work has been done on sedimentary depositional environments, both recent and ancient, especially those associated with deltas. We have come to recognize the subelements of the deltaic environments and the diagnostic characteristics of the sediments deposited in them. With this knowledge, we can also identify the different components of sandstone oil reservoirs deposited in ancient deltas. Knowing their characteristic size and shape, we can predict where and how they will extend themselves.

Most of this study has been applied to the discovery of new oil and gas fields. However, the techniques can be applied equally, and perhaps more successfully, to (1) the prediction of the eventual size and shape of a newly discovered reservoir and (2) to the subdivision of the reservoir into elements with similar porosity, permeability, and other parameters.

The attributes of a reservoir rock are porosity and permeability. Both are conferred on a sand by the process of "winnowing," which is the washing of the grains by currents. The fine particles are removed and deposited elsewhere. There are two depositional situations that are especially conducive to winnowing by water waves and currents:

river channels and beaches. Large river deltas are places where sediments are being deposited in the sea. Both channels and beaches are characteristic features of deltas. The rapid sedimentation and shifting course of the rivers cover the sands with fine-grained sediments, so that they will be preserved and can become oil reservoirs. Submarine fan and turbidity current deposits form in deep water off the front of deltas; they sometimes contain oil and gas.

We have recently come to realize that a majority of the world's great oil fields in sandstones were formed in the deltas of ancient rivers. Consequently, it is desirable to look at modern deltas and note what kinds of sediment are being deposited in the different sub-environments. The most thoroughly studied modern delta is that of the Mississippi River.

1. THE MODERN DELTA OF THE MISSISSIPPI RIVER

During the Pleistocene ice age, about 20,000 years ago, sea level was 200 meters lower than at present. A deep valley was cut by the ancient Mississippi River. This valley extended from the Gulf of Mexico to the mouth of the Ohio River in Illinois. About 10,000 years ago, the glaciers melted and the sea filled up to its present level, flooding the valley. The Mississippi River started to fill the inundated valley with sediments. The present delta system of the Mississippi, south of Baton Rouge, has all been constructed within the last 3,500 years.

The Salé-Cypremort delta was built about 3,500 years ago in the western part of the delta, in the vicinity of Atchafalaya Bay. After it was built up to a few feet above sea level, the river shifted its course a long distance upstream. The new course followed the west side of the new alluvial plain, crossing it and building a new delta farther east. This old river course is now the Bayou Teche. Several other deltas were built out farther east. The modern delta is building southward. It is called the Balize delta (7 in Figure 6–1). In the meantime, the Salé-Cypremort delta has subsided below sea level, partly by compaction of the underlying mud and partly by crustal bending.

The present course of the river from Baton Rouge southeast is much longer than if the river flowed straight south toward Atchafalaya Bay. If left alone, it would, by this time, have abandoned the modern delta and be building a new delta far to the west of the Balize delta on top of the old Salé-Cypremort deltas. It is prevented from doing so by the U.S. Corps of Engineers because, if it did, the city of New Orleans would be left to one side.

As soon as the river abandoned each delta, two things happened. First, the waves attacked the old coastline. They washed the clay away

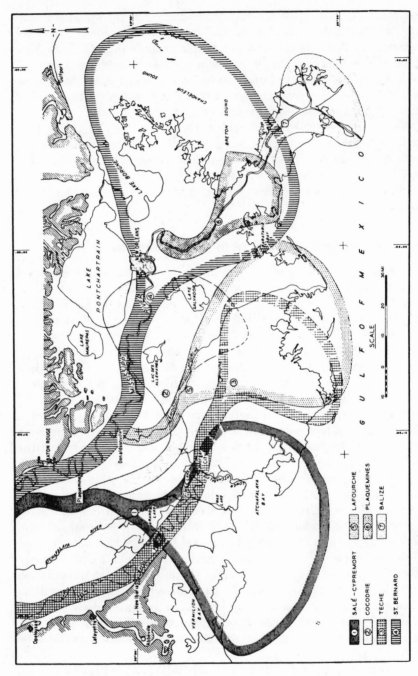

Fig. 6–1 Successive deltas of the Mississippi River *(courtesy Houston Geological Society)*

from the sand, carrying the clay out to sea and piling the sand up into a beach.

Second, the recently deposited muds* and clays started to compact. Initially they had a porosity of 80 percent; the pores filled with water. The weight of the overlying muds caused the deeper muds to compact, reducing their porosity and expelling the interstitial water. This caused the surface to settle.

Both processes tend to cause the marshy surface to settle below the sea, which thus *transgresses* over the land. In a delta complex, then, the sea regresses where the new delta is building seaward and at the same time transgresses where the old deltas are sinking below sea level, and the shoreline moves landward. The overall construction of the delta, of course, involves regression.

The whole pattern of the delta may be subdivided into several sedimentary subenvironments. Some are built up in salt or brackish water and are called *marine*, whereas others are built in fresh river water and are called *fluvial*.

2.   OTHER MODERN RIVER DELTAS

Other modern river deltas resemble the Mississippi in some respects and differ in others. They differ greatly in size. The depth of water and pattern of waves and currents are different for various deltas, so there are several basis for classification. Sneider suggests a classification based on the sediment carried by the river.

(1) Mud-type deltas bring down relatively small amounts of sand compared to the clay-size sediment. The Mississippi and Orinoco deltas are examples. There are many distributary channels which bifurcate, and they are usually quite straight. The shorelines are in some places sandy and in others muddy.

(2) Sand-type deltas deposit large amounts of sand in the distributary channels and along the shore. There are few active distributaries which frequently meander. Examples are the Nile and Niger deltas.

(3) Coalescing deltas at the foot of a mountain range are built by short parallel streams which deposit sediment into a nearby basin. Few of these have been studied. Examples are the Catatumbo delta system of Venezuela (Hyne and Dickey, 1977) and the deltas on the northeast coast of Sumatra.

Another widely used classification is wave-dominated or river-dominated deltas. The tidal influence may be large or small. Figure 6–2

Mud is used to denote clay and sand which have not been separated by wave action.

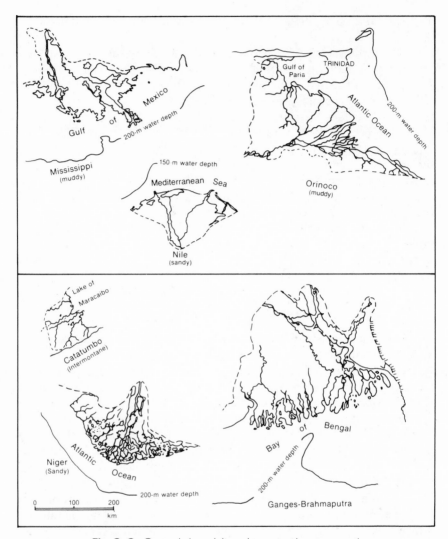

Fig. 6–2 Several river deltas drawn to the same scale

shows, at the same scale, diagrams of examples of the different types of deltas.

3. ENVIRONMENTS OF CLASTIC SEDIMENTATION

Different authors describe sedimentary depositional environments in different ways and use different terms. Figure 6–3 is a diagram

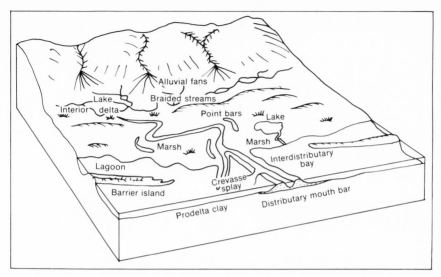

Fig. 6–3 Diagram of sedimentary environments

showing several environments with the terms in most common use in the oil industry.

Alluvial fans and braided streams. At the foot of any high mountain range, there are deposits of coarse material, poorly sorted, consisting of conglomerates, sands, and clays. The streams are steep where they issue from the mountains, and their gradients flatten as they go out on the plain. Consequently, such streams deposit sediment all the way. As each stream builds up its bed, it eventually falls out of it to a new course, sweeping back and forth sideways and building a conical deposit. Consequently, individual beds of conglomerates, sands, silts, and clays are of small thickness and lateral extent. All deposition takes place in an oxidizing environment. Figure 6–4 is a diagram showing the characteristic features of braided-stream deposits.

Alluvial fans are seldom preserved in the subsurface because high mountains tend to be eventually eroded down. The sediments are reworked and transported by longer rivers to the sea. Occasionally, oil and gas do occur in this type of deposit. The Quiriquire field of Venezuela and the Elk City field of Oklahoma (Sneider et al., 1977) are in alluvial-fan and braided-stream deposits. The Prudhoe Bay Field of Alaska also contains braided-stream deposits (Eckelman et al., 1975).

Deltaic plain. As the river descends to the plain, it reaches a level below which it cannot excavate. Sometimes the level is determined by

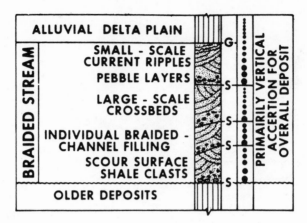

Fig. 6—4 Vertical facies characteristics of braided-stream deposits *(after Swanson)*

a sort of natural dam that causes the gradient to become nearly flat. When this happens, the river starts to deposit its load of sediment. There are large areas in South America where big rivers are depositing sediment in the interior of the continent, usually as a result of recent tectonic uplift which interrupted their course to the sea. Alternating wet and dry seasons result in intense oxidation of the recently deposited sediment. This continental deposition has not been carefully investigated. These deposits seldom contain oil, probably because they normally lack permeable contact with the organic marine shale that is the source rock of petroleum.

Eventually, the river reaches the seacoast. When it does, the current stops and drops its load of sediment, forming a delta. When the pile of sediment reaches sea level, the river has to cross it in order to reach the sea, where it dumps most of its sediment over the front edge of the delta. Essentially, the process is similar to that of a bulldozer building an embankment, which goes back, picks up a load of dirt, and pushes it over the flat surface until it reaches the edge where it rolls down the slope. Similarly, the rivers deposit most of their material over the delta front.

Usually, the delta plain subsides slowly as a result of compaction of the soft, recently deposited muds. Sometimes there is more rapid subsidence of the delta area, caused by tectonic forces deep in the earth's crust. Consequently, the river continues to deposit some sediment on the surface of the delta.

Fluvial environments. When a river is no longer digging the bottom of its bed, it tends to flow in great sweeping curves called *meanders*. In times of flood, the river overflows its banks, depositing a mixture of mud, silt, and sand over its *flood plain*. Within its banks, the river is flowing very fast; but as soon as it spills over, it slows down. The most sediment is deposited adjacent to the banks, which therefore are built up higher than the rest of the flood plain. Downtown New Orleans, for example, is 20 feet below the level of the water in the river. These superelevated banks are called *natural levees*.

The river is continually undercutting and digging its banks away on the outside of the meander bends. As a result, the meanders tend to move slowly downstrean. The flood-plain deposits are mostly clay with some silt and sand. The river picks up the flood-plain deposits, dropping the sand on the inside of the next bend, but carrying the clay and silt on downstream (Figure 6–5). The inside bend, where sand is deposited, is called a *point bar*. The point bars tend to enlarge as the meanders migrate. They may be buried by later floods and preserved as bodies of clean, porous sand. The reservoir rocks of many oil fields are sandstones deposited as point bar deposits by an ancient river.

The current is fastest on the outside of the bend; here coarse sand, pebbles, and chunks of the natural levee deposits are rolled down the stream bed. As the meander moves sideways, the sands deposited in shallow water cover the bottom. They may exhibit large-scale festoon type cross beds. As the main stem moves farther away there are small scale cross beds, and finally the natural levee or marsh deposits cover the sands. As the meanders move from side to side and downstream, they are able to clean up sizable areas of sand. Point bars may be

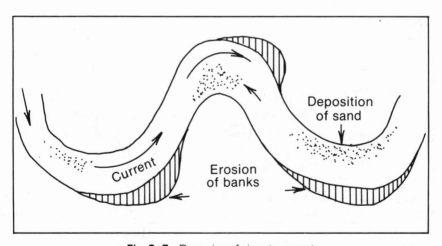

Fig. 6–5 Dynamics of river transport

lunate (crescent-shaped) or they may form irregular belts of permeable sand several miles wide and many miles long. This lateral motion is diagrammed in Figure 6–6.

As the meanders migrate downstream, they form multiple point bars, as shown in Figure 6–7.

The low parts of the flood plain are called *back swamps*. The are occupied by aquatic vegetation of various sorts, including cypress trees. Occasionally, a meander bend is cut off and becomes an *oxbow lake*.

Marine environments.   Farther downstream where the tides are able to enter twice a day and cover part of the delta, the swamps are salty and are called *marshes*. These are spotted by lakes and penetrated by tidal channels.

As the river approaches the coast, it usually divides into several distributaries. During times of flood, the distributaries excavate their beds into the previously deposited sediments to a depth of 3 to 30 m or more. In low water, the currents slow down and fill the trench by depositing first coarse and then finer sand, washed clean of mud by the force of the current. If the river abandons the channel, as by the cutoff of a meander or upstream diversion, the current finally stops completely. Eventually, the abandoned sand-filled channel is buried by flood-plain deposits.

Channel deposits can be recognized by their erosional base, which truncates older stratified deposits and causes an abrupt change in lithology to a coarse sand (Figure 6–8). In the lower part are frequent chunks of clay, apparently pieces of the stream bank which fell in.

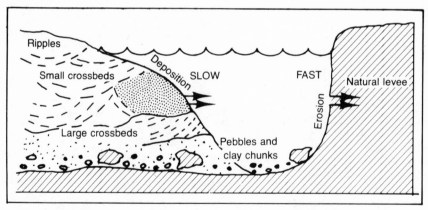

Fig. 6–6   Diagrammatic cross section showing lateral migration of meanders

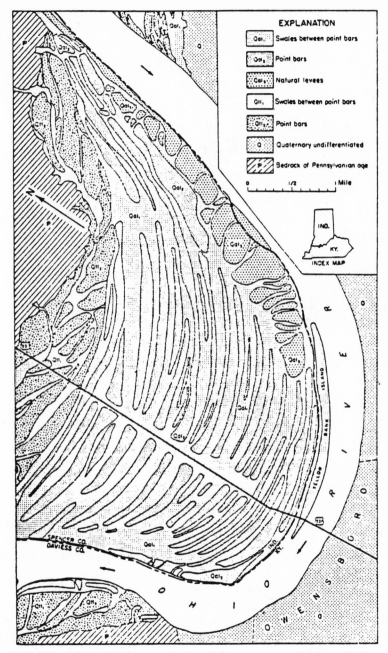

Fig. 6–7 Migration of meander builds multiple-point bars *(courtesy Illinois Geological Survey)*

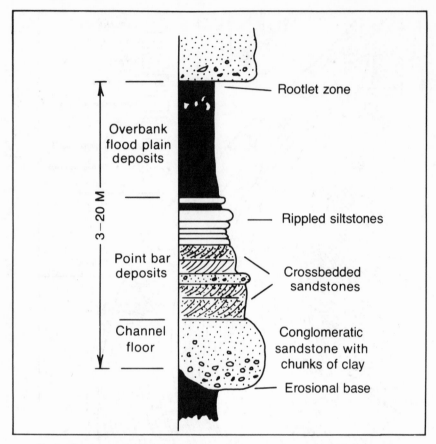

Fig. 6—8 Channel sand deposit *(after Pettijohn, courtesy Harper and Row)*

Overlying the coarse sand at the base is a series of crossbedded sands. The crossbeds come in sets concave upward, called troughs (Figure 6–9). These are frequently overlain by crossbedded and ripple-marked sands, the crossbeds being on a much smaller scale. Animal burrows are scarce or absent, but pieces of carbonized wood and leaves are common.

Channel sands can often be recognized on the electric log using the gamma-ray, SP, and short-spacing resistivity curves. The base is abrupt. Usually, both SP and resistivity are a maximum in the lower layer because the sand contains less clay and has a higher porosity. The overlying beds contain more clay, both interstitial and in laminae, so both SP and resistivity decrease in amplitude and become serrated

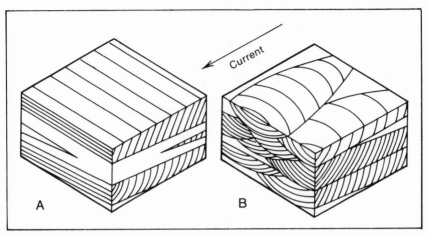

Fig. 6—9 Types of crossbedding *(after Pettijohn, courtesy Harper and Row)*

near the top. Finally, the curves approach the shale base line. Figures 6–10a and 6–10b are idealized logs of channel sands. If chunks and pebbles of clay are abundant at the base of the sand, they will pull the gamma-ray and SP curves in and the bell shape will be less evident.

Channel and point bar sands range in thickness up to 30 m, occasionally more. The width of the meander belt and therefore of the point bar complex is 15 to 50 times the channel width. The width of the permeable sands in the point bar complex ranges from a few hundred meters to a kilometer, sometimes more. The width of the sand body is usually 200 times its thickness.

Delta fringe deposits. Right at its mouth, each distributary deposits its load of sand, forming a distributary-mouth bar (Figure 6–11). Although the river channel may have been 20 or more meters deep upstream, the water is shallow over the bar—from 2 to 5 meters, depending on the strength of the currents. This bar is usually half-moon shaped. An active distributary channel builds out rapidly, adding to the seaward side of the bar. The central part of the older bars is excavated by the current. Sand is deposited right at the mouth, but farther out the sand is mixed with silt and clay. Consequently, a river-mouth bar has a lower contact that grades into the underlying clay and an abrupt upper contact. The electric log shows weak SP and resistivity at the base with shale interbeds, increasing and becoming more constant upwards, sharply dropping at the top of the sand (Figure 6–10a).

When a distributary is abandoned, it fills up with sand whose lithology is the usual channel type described above, coarser at the bottom.

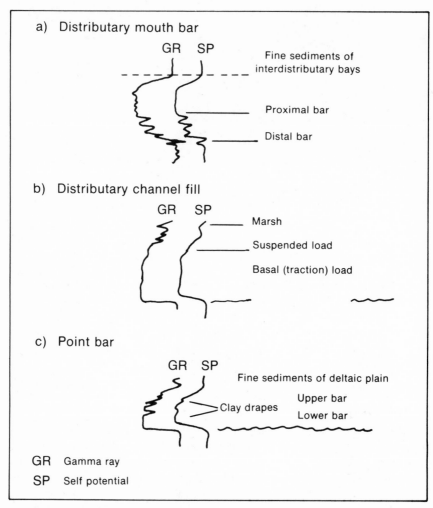

Fig. 6—10 SP and gamma-ray logs of different sand types *(after Zamora)*

Between the active distributaries are bays of shallow, quiet water where mostly mud is deposited. Occasionally when the river is in flood, it breaks through its natural levee and spreads sandy material out over the interdistributary bays. These sands are called *crevasse splays.*

The waves attack the sediments deposited at the shoreline. They winnow out the mud and wash the sand clean, usually leaving a sandy beach. The waves are at right angles to the wind, which seldom blows directly on to the shore. As a result, the swash of the waves has a long

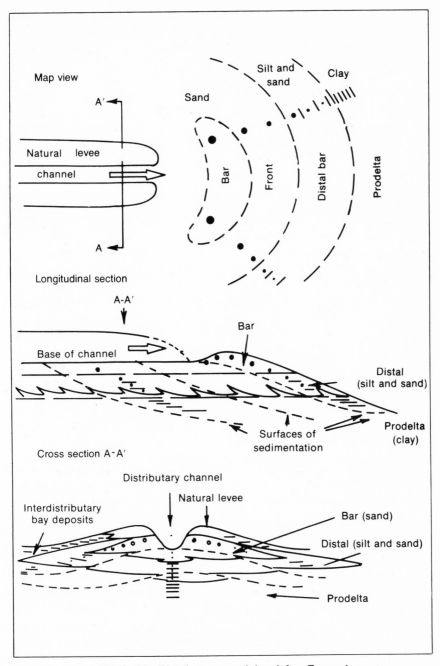

Fig. 6—11 Distributary mouth bar *(after Zamora)*

shore component. The sand grains are spread out parallel to the shore, and the beach has a smooth front. Sometimes the water deepens steadily off the beach front, but more often there is a sand bar parallel to the beach but 100 meters or so offshore.

Shoreward of the beach, there is usually a shallow lagoon. Consequently, beach deposits are often called barrier island deposits. The lagoon has very irregular shorelines. Mostly mud is deposited there, but occasionally thin beds of sand are deposited when winds and storm waves wash sand over the barrier island and spread it out in the lagoon. Figure 6–12 is a cross section of a barrier island along the coast of the Gulf of Mexico. Barrier islands have a width-thickness ratio of about 200.

Beach or barrier island sands differ markedly from channel sands (Figure 6–13). The beach builds out seaward over the offshore sand bar deposits, which in turn lie on prodelta clays. Consequently, the base of a beach sand is gradational, with thin interbeds of silty sand and shale. Burrowing and reworking (*bioturbation*) by animals is often intense in this zone. The sand becomes cleaner and coarser upwards. Burrows are also common near the top. Ripple marks are common, but crossbeds are not.

Vertical stacking of delta deposits. Each of these sedimentary environments builds out seaward on top of its next seaward neighbor.

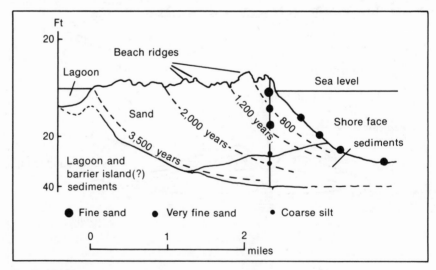

Fig. 6–12 Cross section, Galveston Island: a prograding barrier island (after Bernard, courtesy Geological Society of America)

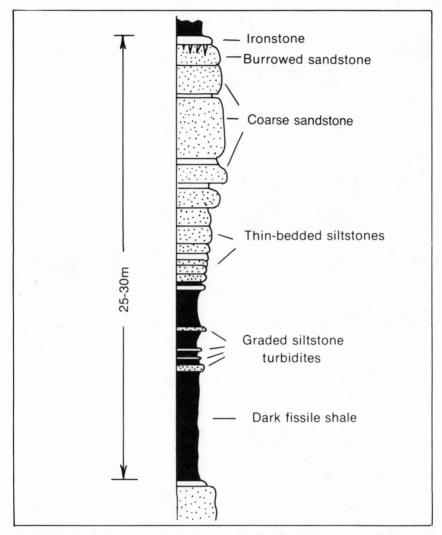

Fig. 6—13 Typical beach or barrier bar sand *(after Pettijohn, courtesy Harper and Row)*

Under the marsh deposits lie the previously deposited interdistributary deposits with river-mouth bars, cut by channel sands. These in turn lie on older beach deposits, which lie on still older prodelta clays (Figure 6–14). In general, then, sediments deposited in shallow water tend to lie on top of sediments which were previously deposited in deeper water.

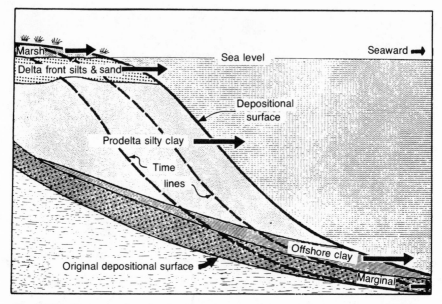

Fig. 6–14 Cross section showing how the environments migrate seaward over each other in a delta. The younger landward deposits lie over the older seaward deposits (after Scruton)

As a distributary builds out to sea, the adjacent older deltas subside because of the compaction of the underlying muds. There is a tendency for the sea to transgress over the older deltas. Sometimes the surf waves wash the uppermost deposits, laying down a rather thin bed of clean sand. After the older deltas have subsided a few meters, the river usually flips out of its channel and starts building out a new delta on top of the older one.

The transgressive shales and sands are extremely important because they offer the only means of correlating through a delta complex. The sands do not correlate. The marine shales contain marine fossils. The sands are usually thin (10–40 ft; 3–10 m) and contain much calcareous material, which gives them a high resistivity.

Consequently, delta deposits often consist of a series of repeating cycles stacked vertically. The lowermost lithologic unit is the marine clay or transgressive sand. On top of this will be laid down the delta fringe deposits; that is, the beaches, offshore bars, and river-mouth bars. On either side of the river-mouth bars are interdistributary muds and occasional crevasse splays. Above these are natural levee deposits. Channels filled with sand cut deeply into this whole succession. This complex will be overlain by natural levee deposits. These are overlain by marsh deposits, represented in the subsurface by coal seams, which end the cycle. The next overlying bed will be the marine shale or sand, representing transgression of the sea due to subsidence. The cycle may repeat itself several times.

4. DEEP-WATER DEPOSITS

Most of the load of a river is carried across the delta and deposited on the steeply sloping (0.5–5.0°) delta front. The material contains fine sands and muds, poorly sorted because it is deposited below wave base. In this zone, underwater slumps and slides are common.

Sometimes a mass of recently deposited sediment breaks loose, becomes incoherent liquid mud, and runs down the front of the slope as an underwater stream called a *turbidity current*. When it reaches the foot of the slope in deep water, a few hundred to a thousand meters or more, it spreads out, its speed slackens, and the suspended sediment drops to the bottom. It forms an underwater fan, something like an alluvial fan, but of much finer sediments and flatter slope. When it has stopped flowing, the suspended sediment load settles out, first the sand and then the mud. The resulting deposit therefore consists of sand at the base, grading up and into silt and shale. After awhile another turbidity current comes rushing down the same course and spreads out on top of the sediment from the first. Turbidity-current deposits therefore consist of successive layers, a few decimeters to several meters in thickness. Each is sandy at the base and grades upward into shale.

Turbidity-current deposits generally are thin and fine and are separated by shale beds, so they do not normally make good reservoir rocks. On the other hand, the currents normally spread out over a large area, so they are laterally more extensive than ordinary shallow-water coastal and fluvial reservoirs. Figure 6–15 is a sequence of turbidity deposits characterized by progradation of the fan. The lower beds were distant from the shore, but as time went on the delta moved seaward and the upper beds contain material that came down the main channel. Each individual bed is finer grained at the top, but the sequence as a whole gets coarser upward.

The parts of the fan farthest from the slope are called *distal*. The sandy beds are thin, fine-grained, and low in porosity and permeability. Those near the slope are much thicker and coarser and often contain pebbly layers. It was pointed out by Bouma (1962) that normally the lowest layer (A) (Figure 6–16) is massive, with grain size diminishing upward, deposited rapidly. Above that (B) are sands with

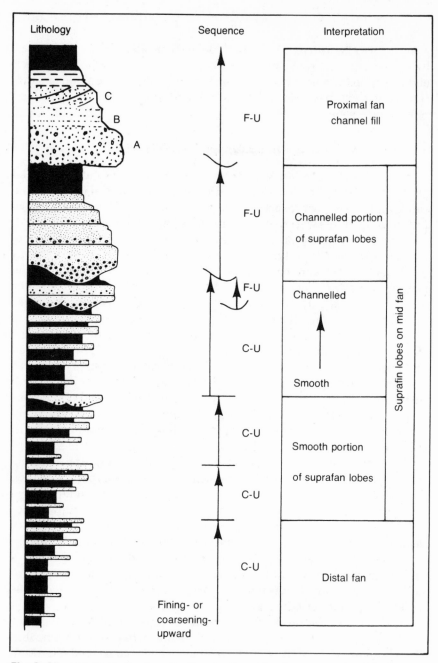

Fig. 6–15 Hypothetical submarine fan stratigraphic sequence produced by fan progradation *(after Walker)*

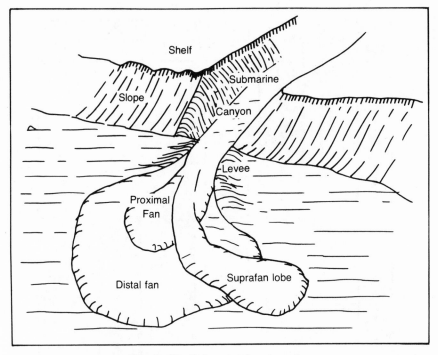

Fig. 6—16 Submarine fan deposits

parallel laminae, and above these are ripple-marked and small-scale crossbeds (C). The consistent decrease in grain size upward is called *graded bedding*, and it is characteristic of turbidity-current deposits.

Oil and gas are occasionally found in turbidity deposits. Examples are the sands of the Ventura basin, California, and the Permian Spraberry sand of West Texas.

5. AEOLIAN DEPOSITS

In desert areas of the world like the Sahara and the Mohave, wind blows sand into great dunes. These piles of sand are clean and well sorted and occur in thick beds characterized by large-scale crossbeds with dips often as much as 30°. They are sometimes associated with other desert deposits, such as salt and anhydrite, and often are red in color.

Such thick sands seldom contain oil or gas. Partly this is because they are continental and may not be connected with marine, organic

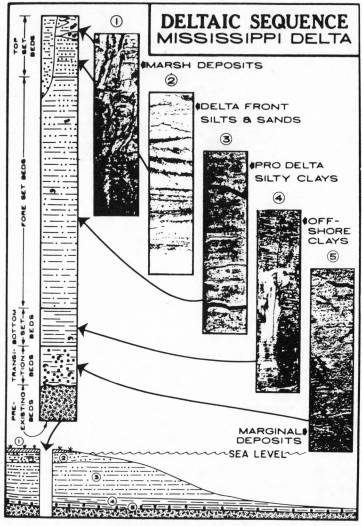

Fig. 6–17 Characteristic lithologies of modern Mississippi delta *(after Scruton)*

source rocks. More likely, these thick, permeable sands do not contain enough shale beds to filter out and retain oil and gas during their primary migration.

The most notable petroleum-bearing aeolian sands are the Permian Rotliegendes sands of the British and Dutch North Sea gas fields.

6. SEDIMENTARY STRUCTURES AND TEXTURES

Each depositional environment is characterized by certain minor sedimentary structures and textures. Among these, crossbeds, ripple marks, and worm burrows have already been named. They were first described by Scruton whose examples are in Figure 6–17.

Along with other evidence, sedimentary structures and textures are helpful in determining the depositional environment of the rock. However, their description and classification is a rather specialized subject. They are available only when cores are taken, as indeed they should be in the pool development, either during a new discovery, or with the application of secondary recovery.

REFERENCES

Bouma, A.H., 1962, Sedimentology of some flysch deposits, Elsevier, Amsterdam, 168 p.

Coleman, J.M., and L.D. Wright, 1975, Modern river deltas, *in* M. L. Broussard, editor, Deltas, models for exploration, Houston Geological Society, pp. 99–149.

Friedman, Gerald M., and John E. Sanders, 1978, Principles of sedimentology, John Wiley, New York, 792 pp.

Hyne, N.J., William A. Cooper, and P.A. Dickey, 1979, Stratigraphy of intermontane lacustrine delta, Catatumbo River, Venezuela, AAPG Bull. v. 63, no. 11, pp. 2042–2057.

Kolb, C.R., and Jack R. Van Lopik, 1966, Depositional environments of the Mississippi deltaic plain, *in* M. L. Shirley, editor, Deltas in their geologic framework, Houston Geol. Soc., 251 pp.

Le Blanc, R.J., Sr., 1977, Distribution and continuity of sandstone reservoirs: Jour. Petrol. Techn., July, pp. 776–804.

MacPherson, B.A., 1978, Sedimentation and trapping mechanism in Upper Miocene Stevens and older turbidite fans of southeastern Joaquin Valley, California AAPG Bull v. 62, pp. 2243–2274.

Pettijohn, F.J., 1975, Sedimentary rocks, 3rd ed., Harper & Row, New York, 628 pp.

Potter, Paul E., 1962, Shape and distribution patterns of Pennsylvanian sand bodies in Illinois, Illinois Geol. Survey Circ. 339, 36 pp.

Scruton, P.C., 1960, Delta building and the deltaic sequence, *in* F.P. Shepard, editor, Recent sediments of the northern Gulf of Mexico, AAPG Memoir, 394 pp.

Walker, Roger G., 1978, Deep-water sandstone facies and ancient submarine fans: models for exploration for stratigraphic traps, AAPG Bull v. 62, pp. 932–966.

Zamora, Lucas G., 1977, Uso de perfiles en la identificacion de ambientes sedimentarios del Eoceno del Lago do Maracaibo: V Congreso Geologico Venezolano, Memorias, Tomo IV, Caracas, Venezuela, pp. 1359–1376.

7

Oil Fields in Different Types of Sand Bodies

1. FLUVIAL SAND BODIES

Oil fields occurring in ancient buried river channels are very common. Usually, these are multiple channels alongside and on top of each other. When the channels are single and surrounded on all sides with shale, they form long, narrow oil fields called "shoestrings." Sometimes they are crooked like a river, and sometimes they are quite straight.

A good example is the Bush City field of Kansas (Figure 7–1). It is about 300 m wide and more than 25 km long. It was first developed in the early 1900s. In the 1930s, secondary recovery was successfully applied. Three closely spaced water injection wells were drilled across the channel. Then about 300 m farther along, 3 oil-producing wells were drilled. Another 300 m farther, 3 more water injection wells were drilled. The water front was confined to the channel, so that the oil was squeezed out of the producing wells and it was not necessary to install pumps.

The Waltersburg Formation in Saline County, Illinois, is a channel sand about a mile (1,500 m) wide and 50 feet (15 m) thick at its thickest place (Figures 7–2 and 7–3). The width of the sand body suggests that it consists not of a single channel, but of a series of adjacent point bars. An example of how they pile against one another, forming swells and swales, is shown in Figure 6–5.

The base of the Vienna limestone is the reference datum for this cross section, and it actually was deposited nearly at sea level. The shales on either side of the channel (blank in cross section) are probably natural levee or interdistributary deposits. These shales were cut out by the river to make the channel. Later, the sea advanced over the delta and deposited the Menard limestone. When it was first deposited, it was flat, parallel with the Vienna limestone. As sedimentation continued, however, the shales compacted more than the Waltersburg

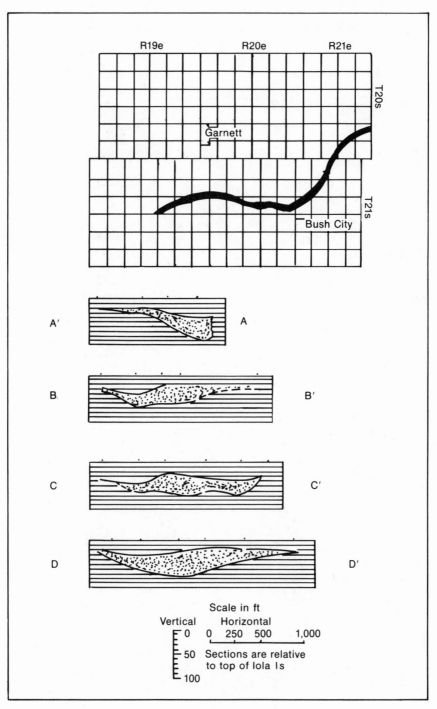

Fig. 7–1 Map of Bush City pool, Kansas *(after Charles)*

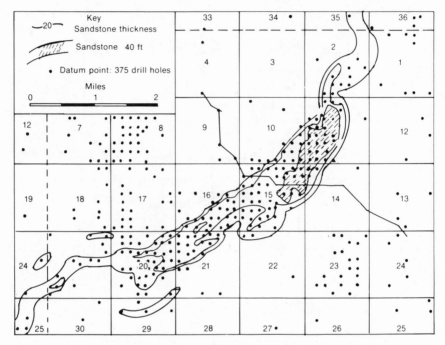

Fig. 7–2 Waltersburg pool, Saline County, Ill. *(after Potter)*

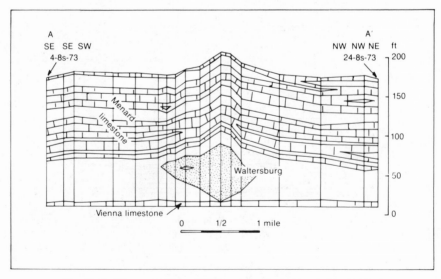

Fig. 7–3 Cross section of Waltersburg formation *(after Potter)*

sandstone. This caused a *draping* of the overlying Menard limestone over the channel fill.

The many different reservoirs of the Seeligson field of Texas were formed by the distributary system of a large river which built a delta into the Gulf of Mexico during the Oligocene (Figures 7–4 and 7–5). The sand is coarser at the bottom of the channel than at the top. This is

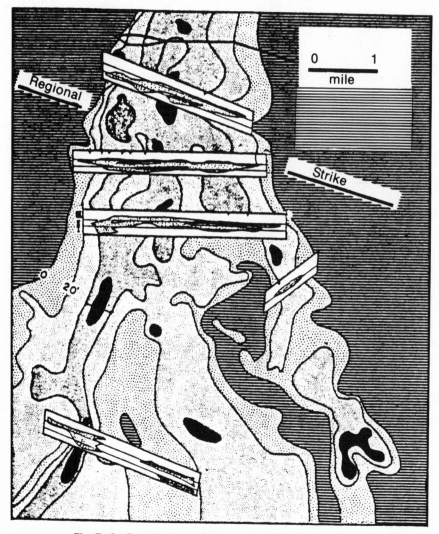

Fig. 7–4   Sand bodies in Seeligson field, Texas *(after Nanz)*

because the sand at the bottom was deposited while the river was still flowing. As the current went slower and slower, it was able to transport only finer and finer sand particles, which were deposited near the top of the channel.

Figure 7–6 shows diagrammatically the characteristics by which a channel sand body may be recognized. The wiggly vertical pair of lines in the center represents the SP and resistivity electrical logs. Where the lines are close together the rock is shale–the farther apart the cleaner the sand. They are farthest apart on the bottom, indicating that the sand is cleaner at the bottom of the channel. The dotted symbol on either side represents the sand body; note that it truncates beds in the adjacent shale. Chunks of shale several inches to several feet in diameter are often found near the sides of the sand body. These are pieces of the stream banks that caved in when the current undercut them. The bedding often shows festoon crossbeds that dip in the downstream direction of the current. These are the internal characteristics of a channel.

From the point of view of external geometry, channels range from several hundred feet (or meters) in width and 10 to 100 ft (3 to 30 m) in thickness. They are often sinuous or meandering. The general trend is perpendicular to the ancient coastline. A generalized but useful rule is that the meandering channels of point bars range from 50 to 500 times

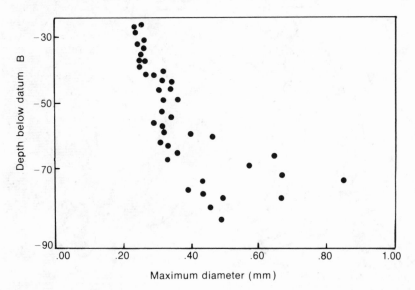

Fig. 7–5 Plot of maximum diameter of sand grains vs. depth below a datum horizon (after Nanz)

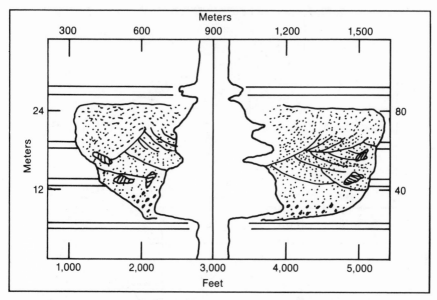

Fig. 7–6 Diagrammatic cross section of a typical channel sand body

as wide as they are thick. An average ratio is 250. Straight distributary channels range from 15 to 100 times as wide as they are thick, averaging 50 times.

2. BEACH-TYPE SAND BODIES

The waves of the ocean continually strike the shore. At the front of a delta, the waves winnow the clay from the sand, building up a beach of clean sand. Such beaches are often buried by layers of mud as the delta builds outward. They thus become completely enclosed in shale and form stratigraphic traps for oil. The rush of the waves and alongshore currents smooth out the coastline so that beaches and barrier islands are often nearly straight. Behind the barrier island is a quiet lagoon where both sand and mud are deposited. The grain size and cleanness of the sand thus decrease away from the ocean toward the land.

Bell Creek, Montana, is an example of an ancient barrier island that was preserved and collected oil. Figure 7–7 shows the producing wells and dry holes in the field.

Figure 7–8 shows the reconstructed depositional environments. Channel sands are present in the Rocky Point and Black Bank fields,

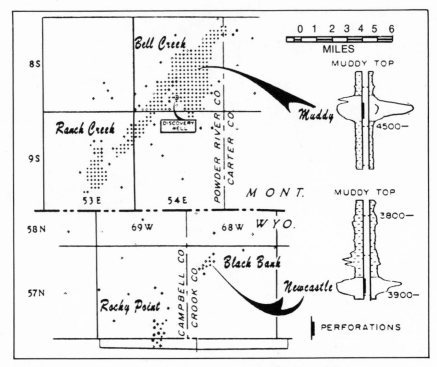

Fig. 7–7 Showing Bell Creek and nearby fields *(after McGregor and Biggs)*

while the Ranch Creek and Bell Creek fields are composed of a series of barrier bars.

In the Venango District of Pennsylvania there is a series of oil sands at several different horizons (Figure 7–9). The oldest is the Third Venango Sand, which appears to be a series of barrier islands formed at the edge of the Devonian Catskill delta. The sea lay to the northwest and the land to the southeast. The Knox Third Sand was laid down at a slightly later time when the sea transgressed. The Third Stray Sand may be contemporaneous but was deposited under water (Figure 7–10). Then the delta pushed the sea back, and the broad beach so formed became the Second Sand.

The Bisti field of Cretaceous age in the San Juan basin of New Mexico is another example of a barrier island sand body (Figure 7–11). The field is about 3 miles (5 km) wide and is a maximum of 50 ft (15 m) thick. The pool is very straight. The sea lay to the northeast and the delta to the southwest.

Beaches and barrier-island sand bodies have the cleanest and

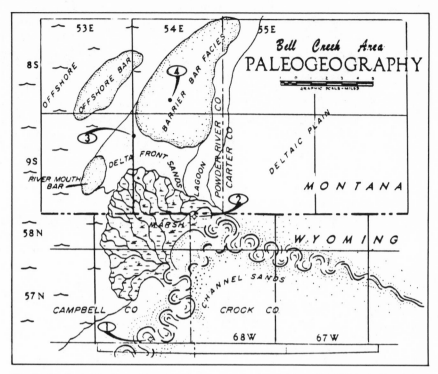

Fig. 7–8   Probable depositional environments in the vicinity of Bell Creek field *(after McGregor and Biggs)*

coarsest sand in the upper part, where it is most washed by the waves. The sand is often stirred up by burrowing organisms such as clams. The seaward pinch-out is abrupt, smooth, and straight. The landward side toward the lagoon tends to be transitional, the sands becoming dirty and interfingering with the shale (muds) of the lagoon. The lagoon side is irregular. Figure 7–12 is a simplified diagram illustrating these features.

3.   TURBIDITY-CURRENT DEPOSITS

The first oil sands to be identified clearly as turbidity-current deposits were those in the Ventura Valley of California (Hsü, 1977). They are more than 10,000 ft (3,000 m) thick and were deposited in a narrow, deep-sea trough. They were subsequently folded into a sharp anticline with a complicated pattern of faults. The thickest and best sands were deposited in the bottom of the trough. There are many sand

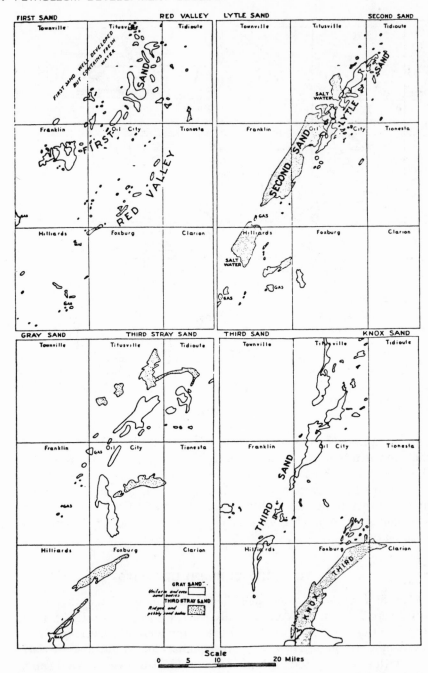

Fig. 7–9 Four maps of Oil City area, Pennsylvania, showing the oil fields producing from several different sandstones

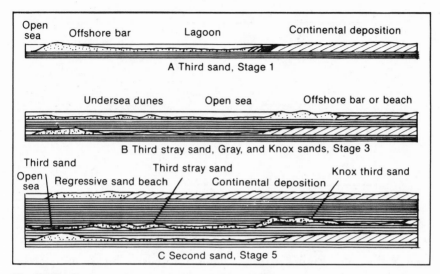

Fig. 7–10 Cross section showing the regressions and transgressions of the sea at the edge of the Devonian Catskill delta

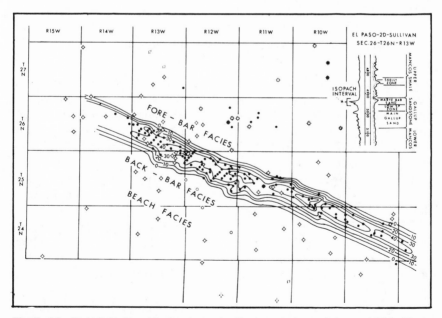

Fig. 7–11 Bisti field, New Mexico, isopach of bar sands, isopach interval 10 ft (3 m). Squares are 6 mi (9.6 km) *(after Sabins)*

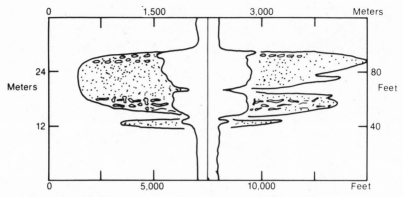

Fig. 7–12 Typical beach-type sand: straight, parallel to coast; fine on bottom, coarse on top; bottom gradational, burrows common

bodies; the AO_1 (Figure 7–13) is one of the best. Thickness patterns show that the currents came from the east and ran parallel to the axis of the basin.

In the San Joaquin Valley, the Stevens sands are also turbidity-current deposits (MacPherson, 1978). In the North Sea in the Viking graben, the Forties field is in deep-sea fan deposits.

4. OIL FIELDS CONSISTING OF SEVERAL DIFFERENT SAND-BODY TYPES

Niger delta. An application of the identification of depositional environments to production has been made by Shell-BP Petroleum Development Co. of Nigeria (Weber, 1971). The Niger delta is large and complex with many depositional environments and a series of growth faults. Figure 7–13 shows the overall structure. The growth faults rotated the sand bodies backward toward the land, which gave them structural closure and interrupted the flow of fluids landward. Consequently, most of the fields are immediately downdip from growth faults.

The location of the individual oil fields, of which there are more than 150, is determined by the growth faults. These can be located by seismic methods. However, both the shape and the internal character of the fields is determined by the stratigraphic characteristics of the sand bodies.

The sand bodies were deposited in a coastal environment similar to the one that still exists in the Niger delta. The important subenvironments are shown in Figure 7–14. Offshore there is a *marine clay*.

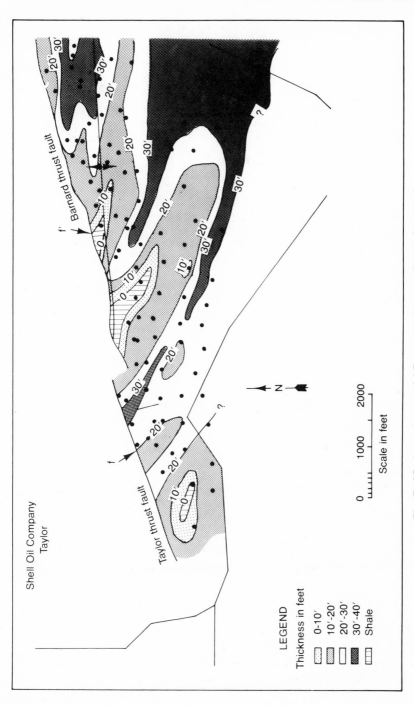

Fig. 7–13 Isopach map of AO₁ sand in Ventura field *(after Hsu)*

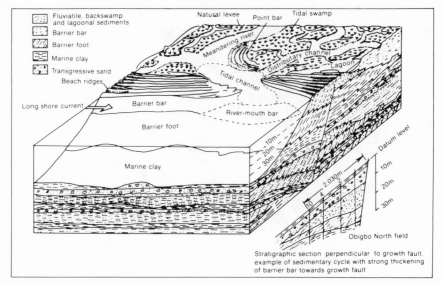

Fig. 7–14 Sedimentary environments in the Niger delta *(after Weber and Daukoru)*

Approaching the beach there is a *barrier foot* zone, consisting of laminated clay, silt, and fine sand in a water depth of 10 to 30 m. The beaches are formed on *barrier bars* which are built up by waves using sand brought in by long shore currents. As the bar builds seaward, successive *beach ridges* are formed. Back of the barrier bar is a tidal coastal plain consisting of tidal flats and marshes with dense mangrove vegetation. The sediments are clays, silts, and peats. These are cut by meandering *tidal channels*, which contain medium to coarse sand. Farther back upstream, the alluvial valley of the rivers consists of clay backswamp deposits and point bar sands near the channels.

The same subenvironments can be recognized in the oil fields. In addition to the electric logs, sidewall cores are taken. These provide information on grain size distribution, sedimentary structures, and authigenic minerals such as glauconite. The depositional environments of the clays are determined by clay mineralogy, faunal content, and pollen analysis.

An example of a Nigerian oil field is shown in Figure 7–15. The principal oil sands may be barrier sands, channel fills, or point bars. Four successive cycles are represented in this field, during each of which the sediment built seaward. These are separated by thin shaly zones, deposited during a period of transgression. The section is paral-

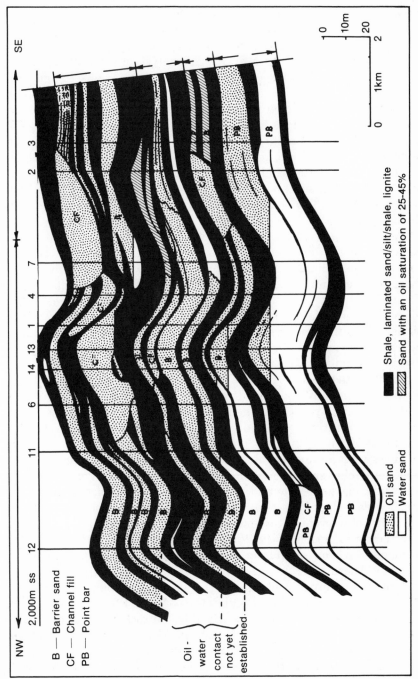

Fig. 7–15 Cross section of a typical Nigerian oil field *(after Weber)*

lel with the boundary fault. A more detailed section of cycle III is shown in Figure 7–16. There are two separate barrier bar sands, separated by the shales of the barrier foot, correlated by the gamma-ray logs. Figure 7–16C shows the grain size analysis of sidewall cores. The sand coarsens upward (except for the lower sand of wells 13 and 14).

Figure 7–17 shows cycle IV. Here there is a barrier bar which has been cut into by a channel which later filled with sand. Note the more uniform gamma-ray logs in 7–17A. Figure 7–17B shows the NE-SW direction of the channel at right angles to the coastline and boundary fault. Figure 7–17C shows that the sand gets finer upward, which helped to identify the sand body as a channel. Permeability was estimated from grain size and porosity, and the most permeable intervals were perforated. The channel sands are much more permeable than the barrier bars.

As soon as a new field is discovered, the component sand bodies are identified in the discovery well. Experience has shown that barrier bars average about 2 km wide and 8 km long, parallel with the coast. Tidal channels run at right angles to the coast. They can be up to 30 m thick and usually are about 50 times as wide as their thickness. This information is added to the seismic to give a preliminary picture of the oil field from the very first well. As additional wells are drilled, more information becomes available, but each well is located to develop the field to best advantage.

Elk City, Oklahoma. The Elk City field was discovered in 1947 by Shell Oil Company as a result of a seismic survey. It consists of a uniform, symmetrical anticline about 10 miles long and 3 miles wide (Figure 7–18). It was originally developed by 310 wells on a 40-acre spacing and will ultimately produce about 100 million bbl of oil. It produces from an interval about 500 ft (150 m) thick containing sandstone, siltstone, shale, conglomerate and limestone of Pennsylvanian age. Only the sandstone and conglomerates are productive, and they range widely in grain size and character. They were not deposited in the delta of a great river like the Mississippi, but rather by short rivers, close to the foot of the Wichita granite mountains (Figure 7–19.)

Cores had been taken from 26 wells representing 1,700 ft (500 m) of interval. These were analyzed for porosity, permeability, grain size and sorting, pore size, and cement. In general, the finer the grain size, the better the sorting and the higher the porosity. Efforts were made to use the electric logs to recognize the different rock types. Those available were the SP, short normal, and microlog. The SP indicated the sandstones and conglomerates. The resistivity curve did not indicate hydrocarbons because the pores had been invaded by drilling mud

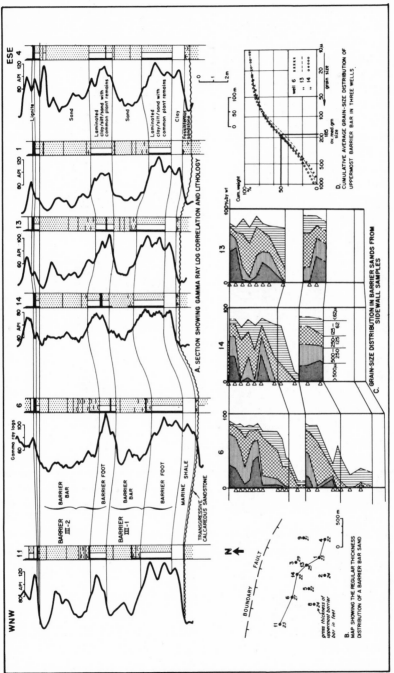

Fig. 7–16 Detailed cross section of cycle III of Fig. 7–15 *(after Weber)*

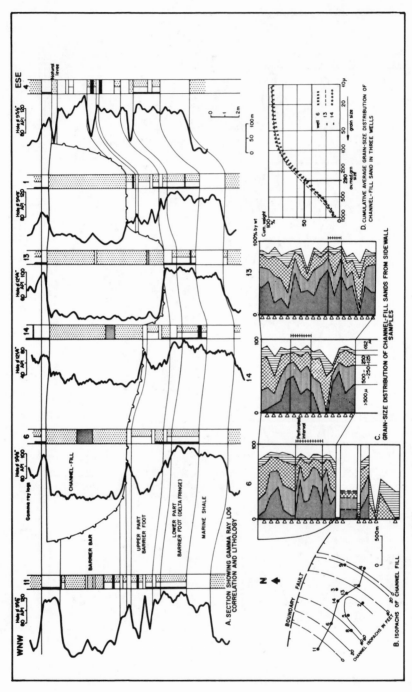

Fig. 7–17 Detailed cross section of cycle IV of Fig. 7–15 *(after Weber)*

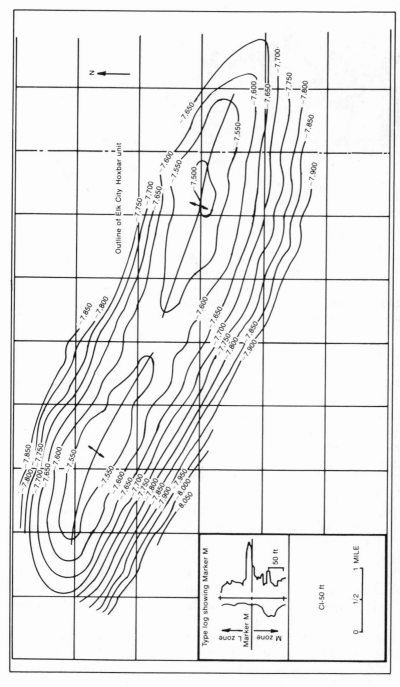

Fig. 7–18 Structure map of Elk City field, Oklahoma *(after Sneider et al., courtesy JPT)*

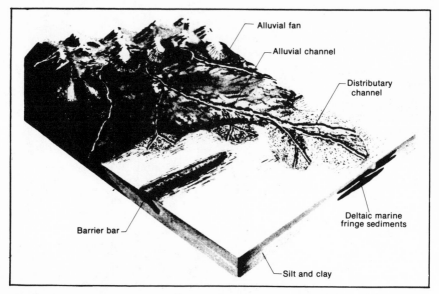

Fig. 7–19 Schematic illustration of the depositional environments in which the different types of sand bodies in the Elk City field were formed (after Sneider et al., courtesy JPT)

filtrate. The conglomerates had lower porosity and permeability and therefore higher resistivity. The shales and silts also had higher resistivity, but a low SP and a "hashy" microlog.

The rock types thus identified were then plotted on detailed cross sections and their depositional environment was reconstructed. Besides the lithology, account was taken of the vertical sequences, fossils, and orientation with respect to depositional strike.

The *barrier bar* deposits characteristically increase in grain size upward (Figure 7–20). They trend parallel with the old shoreline and lack shale breaks. Their thickness is fairly uniform at about 40 ft (12 m). Fluids will flow easily parallel with the long axis of the bar deposits.

Alluvial channel deposits consist of sandstones and conglomerates deposited by swift braided streams. There are many lateral and vertical changes in grain size, formed as the channels shifted back and forth. They are relatively straight, trend at right angles to the bar deposits, and cut through underlying previously deposited sediments.

Deltaic marine-fringe deposits consist of shale, fine, and coarse conglomeratic sandstones, interbedded, with a general upward increase in

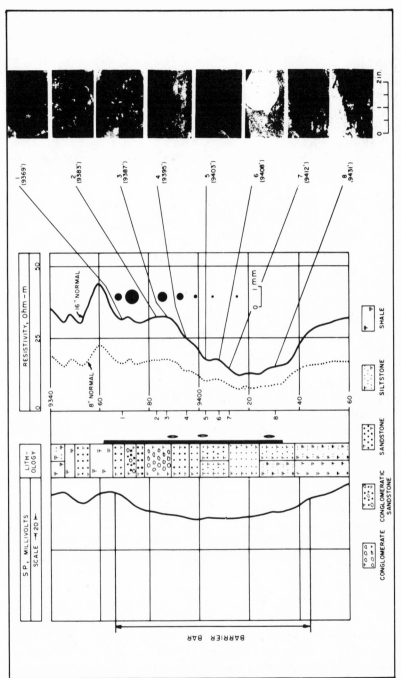

Fig. 7–20 Vertical sequence of rock type and log of a response barrier bar, Elk City field *(after Sneider et al., courtesy JPT)*

grain size. There may be two or three sequences stacked on top of each other, separated by impermeable shales. The grain size and permeability decrease basinward.

Correlation sections were made and the producing interval was subdivided into contemporaneous time-rock units (Figure 7–21). There are a total of nine time-rock units, each consisting of one or several different genetic types of sand body. They are separated by impermeable strata except where the channel deposits cut down into an underlying reservoir. Separate maps were drawn for each of the nine reservoir units. For example, the genetic types in unit L_3 are shown in Figure 7–22. A barrier bar (black) strikes EW in the north-central part of the field. South of it is a broad area of deltaic fringe deposits. These have been cut through by two channel deposits which strike north-south.

Using the depositional environment map as a guide, a sand thickness map was prepared for each unit (Figure 7–23). These two maps show the amount of pore space and the preferred directions of permeability.

It has recently become customary to attempt to identify the types of sand bodies in the discovery well of a new pool. Such identifications were made in the case of the multibillion-barrel Brent field in the North Sea, as outlined in Chapter 1. The Exxon geologists similarly identified the component sand bodies of the even larger Prudhoe Bay field shortly after its discovery (Eckelman et al., 1975). They decided that the Sadlerochit formation consists mainly of a braided-stream complex made up of coarse sands that prograde over more typical delta and delta plain environments (Figure 7–24).

5. ELECTRIC LOG PATTERNS

The first step in understanding a reservoir is to break it down into individual sand bodies. In most cases this has been accomplished by tracing shale beds through the reservoir. Those that go all the way through a deltaic complex were probably deposited by a marine transgression after the river abandoned a delta, which then subsided below sea level.

The next step is to identify the depositional environments. This requires a good deal of specialized knowledge and experience. Many criteria are used, including grain-size variations, sedimentary textures, fossils (especially pollen and spores), trace fossils, and other features that can be observed only in cores. What is one to do, then, if cores are not available—only electric logs? The obvious thing to do,

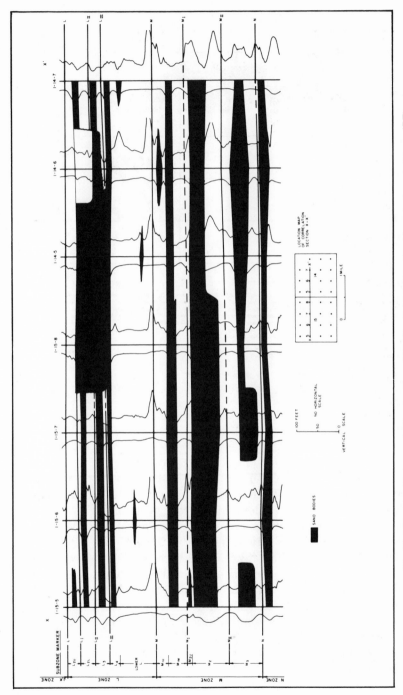

Fig. 7-21 Detailed correlation section, Elk City field *(after Sneider et al., courtesy JPT)*

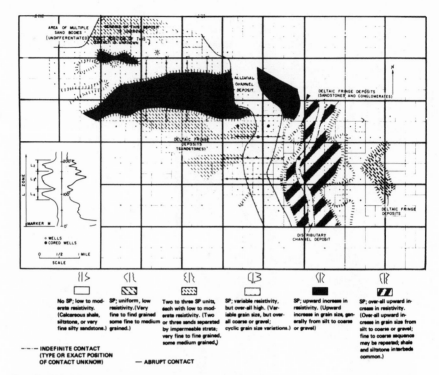

Fig. 7–22 Map of the distribution of sequence of rock types, subzone L₃, Elk City, showing the distribution of the genetic types of sand bodies *(after Sneider et al., courtesy JPT)*

either in a new development or an old field redevelopment, is to core the next well completely. Then the electric log responses can be correlated with the cores, as was done in the case of the Elk City study.

Many environmental studies have now been made, and a wide variety of electric log patterns have been correlated with the depositional environment as determined by external geometry and core data. Swanson (1980) has presented a useful compilation of electric log patterns (Figure 7–25). The alluvial-fan and braided-stream deposits show as stacks of sand with thin shale beds. The point bars nearly always show the abrupt base and narrow top (bell shape), while the stream-mouth and barrier bars show the broad, abrupt top and gradational base (funnel shape). The turbidites show stacked sand bodies separated by shale beds.

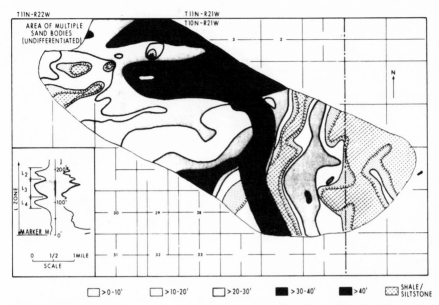

Fig. 7–23 Thickness map of net sand, subzone L₃, Elk City *(after Sneider et al., courtesy JPT)*

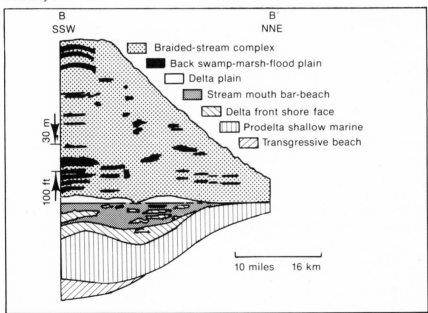

Fig. 7–24 Sedimentary sequences at Prodhoe Bay, Alaska *(after Eckelman et al., courtesy Exxon)*

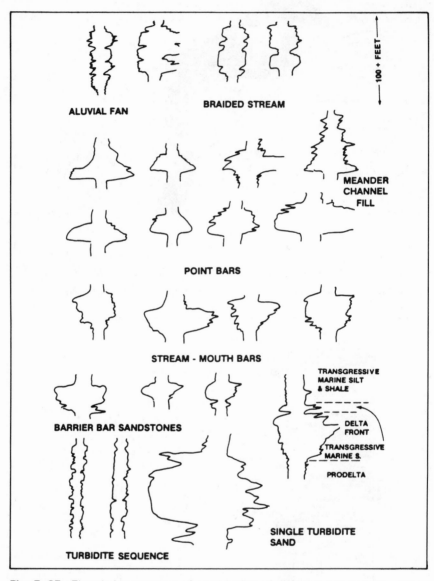

Fig. 7—25 Electric-log patterns of sand bodies of different environments *(after Swanson)*

REFERENCES

Bass, N.W., 1934, Origin of Bartlesville shoestring sands, Greenwood and Butler Counties, Kansas, AAPG Bull. v. 18, p. 1313.

Busch, D.A., 1974, Stratigraphic traps in sandstones—exploration techniques: AAPG Memoir 21, 174 pp.

Charles, H.H., 1941, Bush City oil field, Anderson County, Kansas, in A. I. Levorsen, editor, Stratigraphic type oil fields, AAPG, pp. 43–56.

Conybeare, C.E.B., 1976, Geomorphology of oil fields in sandstones. Elsevier, New York, 341 pp.

Dickey, Parke A., et al., 1943, Oil and gas geology of the Oil City Quadrangle: Pennsylvania Topo. and Geol. Survey, Bull. M25, Harrisburg.

Eckelman, W.R., and R.J. Dewitt, 1975, Prediction of fluvial-deltaic reservoir geometry, Prudhoe Bay field, Alaska, World Petrol. Congress, Tokyo, Proc. v. 2, pp. 223–228.

Hsü, Kenneth J., 1977, Studies of Ventura Field, California, I: Facies geometry and genesis of Lower Pliocene turbidites, AAPG Bull. v. 61, pp. 137–168.

Kingston, P.E., and H. Niko, 1975, Development planning of the Brent field, Jour. Petrol. Techn., October, pp. 1190–1198.

Le Blanc, R.J., Sr., 1977, Distribution and countinuity of sandstone reservoirs: Jour. Petrol. Techn., July, pp. 776–804.

McGregor, A.A., and C.A. Biggs, 1968, Bell Creek field, a rich stratigraphic trap, AAPG Bull. v. 52, p. 1869.

Nanz, R.H., Jr., 1954, Genesis of Oligocene sandstone reservoir, Seeligson Field, Jim Wells and Kleberg Counties, Texas, AAPG Bull v. 38, p. 96.

Potter, Paul E., 1962, Shape and distribution patterns of Pennsylvanian sand bodies in Illinois: Illinois Geol. Survey Circ. 339, 36 pp.

——, 1963, Late Paleozoic sandstones of the Illinois Basin, Illinois Geol. Survey Report of Investigations 217, 32 pp.

——, and F. J. Pettijohn, 1963, Paleocurrents and basin analysis, Academic Press, New York, 296 pp.

Sabins, F.F., Jr., 1963, Anatomy of stratigraphic trap, Bisti field, New Mexico, AAPG Bull v. 47, p. 193.

Scruton, P. C., 1960, Delta building and the deltaic sequence, in F. P. Shepard, editor, Recent sediments of the northern Gulf of Mexico, AAPG, pp. 82–102.

Selley, R. C., 1978, Ancient sedimentary environments, 2nd edition, Cornell Univ. Press, 287 pp.

Sneider, R.M., et al., 1977, Predicting reservoir rock geometry and continuity in Pennsylvanian reservoirs, Elk City Field, Oklahoma, Jour. Petrol. Techn., July, pp. 851–866.

Swanson, D.C., 1980, Handbook of deltaic facies, Lafayette Geol. Soc., Subsurface clastic facies workshop, April 2, 1980.

Weber, K.J., 1971, Sedimentological aspects of the Niger delta, Geologie en Mijnbouw, v. 50, pp. 559–576.

——, and E. Daukoru, 1975, Petroleum geology of the Niger delta, World Petrol. Congress, Tokyo, Proc. v. 2, pp. 209–222.

8

Reservoir Properties of Sandstones

Any rock which has the attributes of porosity and permeability can store oil and gas. If it does, it is a *reservoir rock*. Porosity can be formed by the interstices between grains in sandstones, by solution of grains of unstable minerals in sandstones, by solution channels or *vugs* in limestones, and by cracks or *fractures* in rocks that may or may not also have *matrix* porosity.

The most common reservoir rocks are sandstones which have interstitial porosity between the grains. By no means are all sandstones permeable; the interstices may be filled with clay or crystalline calcite or silica cement that blocks the flow of fluids.

The first oil men in Pennsylvania in the 1860s noticed that oil was found in sandstones that disaggregated under the drill bit and came up in the bailer like sand. Since that time, it has been customary to call oil-producing sandstones *sands*.

Grains. To characterize a sandstone, the grain size must be specified. Grain size is measured in millimeters, and the different grain sizes are named according to the Wentworth scale. They range from clay, which is less than $\frac{1}{256}$ mm (4 micrometers), to coarse sand, which is 1 mm (see Chapter 2).

In any reservoir sandstone, there are grains of different sizes. The relative amount of each size is determined by disaggregating the grains and passing them through a set of sieves of different meshes. Each fraction is weighed and expressed as percent of the total. Particles finer than 0.06 mm cannot be separated by sieves. These fine fractions are suspended in water and separated by their settling velocities, which are governed by Stokes' law (Folk, 1974). This data is plotted in the form of a histogram, Figure 8–1, and is treated by the

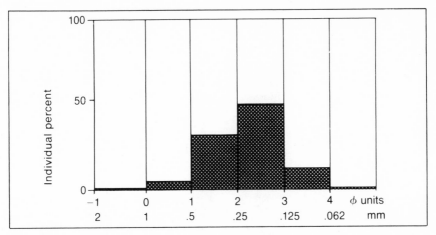

Fig. 8—1 Histogram of typical sandstone. Note that the size scale on the X axis is logarithmic and the sizes decrease to the right *(after Folk)*

methods of mathematical statistics. The same data may be plotted as cumulative frequency curves (Figure 8–2).

Permeability depends on pore diameter, which in turn depends on grain size and grain size distribution. Krumbein and Monk (1942), using artificial mixtures of sand grains, showed that permeability varied inversely as the square of the average grain diameter. Their experimental data gave the relation:

$$K = 760 \ (Dm)^2 \times e^{-1.31SD}$$

where:

K   = permeability, darcys
Dm = mean grain diameter, millimeters
SD = standard deviation, phi units (a measure of sorting)
        a phi unit is diameter in millimeters raised to the -2 power (1 mm, phi = 0; ½ mm, phi = 1; ¼ mm, phi = 2; etc.) That is, phi = $-\log_2 Dm$

Permeability decreases as the sorting becomes poorer because the finer grains fill the interstices between the big grains. Efforts have been made to determine permeability from grain size analysis using Krumbein and Monk's formula, which gives reasonable approximations for clean sands. If there is much clay in the pores, the permeability is reduced so much that these formulas no longer apply. The constant 760 is lower in consolidated natural sands.

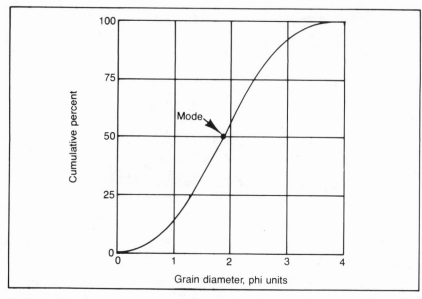

Fig. 8—2 Cumulative frequency curves of the sediment shown in Fig. 8—1 *(after Folk)*

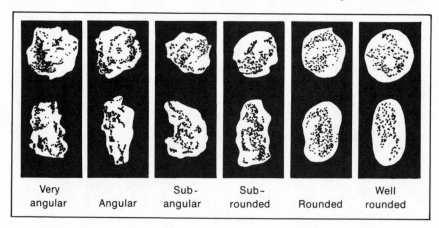

| Very angular | Angular | Sub-angular | Sub-rounded | Rounded | Well rounded |

Fig. 8—3 Sphericity and roundness in sand grains *(after Powers)*

Sand grains may be nearly equidimensional or elongated. They may have smoothly rounded corners, or they may have angular corners and sharp edges. The meaning of the different terms is shown in Figure 8–3. Sand grains derived from igneous or metamorphic rocks are often very angular. When sand is derived from previously deposited sandstones, the sand grains become quite rounded.

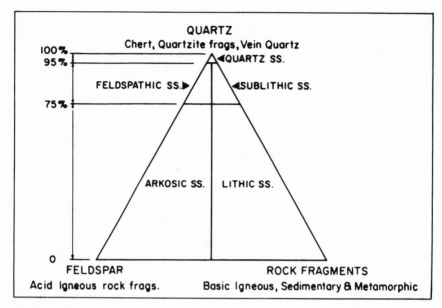

Fig. 8—4 Classification of sandstones by mineralogy of grains *(after R. G. Swanson)*

Matrix. Matrix is the fine-grained particulate material between the grains. It usually consists of clay-sized particles. The fine material may exceed the coarse grains in volume, in which case the grains are supported by the matrix and do not touch each other. The grains resemble the raisins in raisin bread, while the matrix is the dough.

Classification of sandstones. Sandstones are classified according to the minerals that make up the grains. Most classification schemes are based on the relative amounts of the three most common grain types: quartz, feldspar, and rock fragments (Figure 8–4). However, both the words used and the limits of the different classes vary widely from one text to another. Friedman and Sanders (1978) make the sensible suggestion that sandstones should be named after their principal component, that is, quartz sandstone, feldspar sandstone, and rock-fragment sandstone. Other words in common use are *arkose* for sandstone containing much feldspar and *graywacke* for sandstone containing many rock fragments.

Sandstones are usually studied in thin sections; that is, a small sample of rock is attached to a piece of glass and then ground down to a

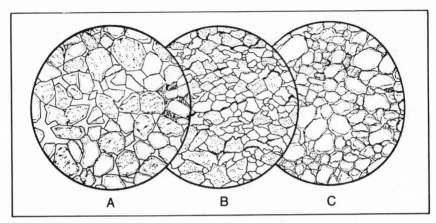

Fig. 8—5 Quartz sandstones *(after Williams, Turner, and Gilbert)*. **A:** Woodbine sandstone (Cretaceous) 6,000 ft (1,800 m) below surface. Houston county, Texas. Diam—2 mm. Well-sorted, porous. Grains mostly quartz with secondary silica overgrowths. Pore surfaces are clean and smooth. **B:** Same as **A** but from a different stratum. Almost no porosity, grains in complete contact with each other. Probably the result of precipitation of secondary silica in the pores. **C:** Coconino sand-stone (Permian), Grand Canyon, Arizona. Diam—2 mm. Well-sorted, rounded quartz grains. Low porosity as a result of secondary silica overgrowths.

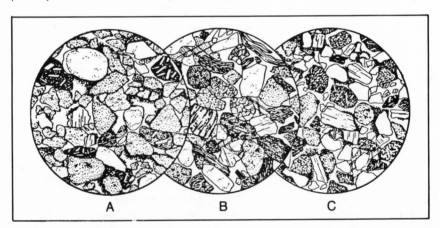

Fig. 8—6 Graywackes *(after Williams, Turner, and Gilbert)*. **A:** Belly River sandstone (Cretaceous) near Calgary, Alberta. Diam 1.8 mm. Fragments of chert and rock fragments (heavily stippled), quartz (clear or lightly stippled), and a little feldspar (with cleavage). Considerable porosity. **B:** Molasse, Bach, Switzerland. Diam 1.2 mm. Well-sorted sand containing rock fragments and mineral grains in about equal amounts. **C:** Eocene sandstone (Wilcox), 3,000 ft (1,000 m) below surface, Central Louisiana. Diam 1 mm. Fine-grained sandstone, loosely packed and porous. The grains are rock particles (chert, slate, phyllite, and volcanic rocks), together with abundant quartz (clear) and fresh feldspar (with cleavage), largely plagioclase.

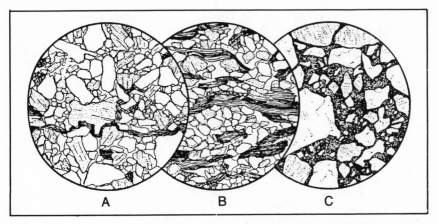

Fig. 8–7 Arkoses *(after Williams, Turner, and Gilbert).* **A:** Miocene arkosic sandstone, 10,000 ft (3,000 m) below surface near Simmler, California. Diam 2 mm. Low-porosity angular and subangular grains: not well sorted but free from clay. Consolidated by compaction without cement. Plagioclase, orthoclase, and microcline (all lightly stippled) and quartz (blank) are about equally abundant. Note pinched and contorted mica. **B:** Micaceous arkosic sandstone (Triassic), Portland, Connecticut. Diam 2 mm. Grains of feldspar (lightly stippled) and quartz (blank); abundant parallel oriented flakes of muscovite and chloritized biotite, larger than other grains, lie parallel to the bedding. **C:** Arkose (Triassic), Mt. Tom, Massachusetts. Diam 3 mm. Unsorted angular-to-subangular grains of quartz and turbid feldspar in a very abundant matrix of ferruginous clay.

thickness of 30 micrometers (0.03 mm). Figures 8–5, 8–6, and 8–7 are examples of orthoquartzites, graywackes, and arkoses, respectively.

Orthoquartzites make excellent reservoir rocks, except when their pores are filled with secondary silica or calcite. The Oklahoma City pool produces from the Simpson sand, which consists of clean, round quartz grains. Permeability is excellent.

Graywackes are often very dirty and because the pores are filled with clay and fine sand particles, they lack porosity and permeability. However, many important producing sands are graywackes. They include the Bartlesville sand of Oklahoma and the Bradford sand of Pennsylvania.

Arkoses contain a large proportion of feldspar grains. Feldspars are easily weathered and will not withstand transport very far. Arkoses are therefore found within a hundred kilometers or so of an outcrop area of igneous rock, usually granite. Arkoses produce oil in southwestern Oklahoma where they are called *granite wash.*

Fig. 8–8 Outcrop of sandstone showing bedding *(after Pettijohn, Potter, and Siever)*

Sedimentary structures. Sandstones are far from homogeneous. They were mostly laid down by running water whose currents varied continually in velocity and direction. All sandstones are, therefore, laminated with alternating layers of coarse and fine grains. Often, thin beds of shale are intercalated in the sandstones. These laminations effectively block movement of fluids across the bedding. Figure 8–8 shows an outcrop of sandstone with the layered character common to most sedimentary rocks.

The laminae are often too thin to show on electric logs, and they are sometimes not noticed in cores because they may be paper-thin layers of clay or organic matter. However, even thin and discontinuous shale laminae are effective in stopping the movement of fluids across bedding planes. This lack of vertical permeability has profound effects on

Fig. 8—9 Large-scale crossbedding *(after Pettijohn, Potter, and Siever)*

reservoir behavior. In waterflooding, most sand bodies act as if they were a pile of separate units, each behaving independently of the others.

Another effect of the laminae is that underlying water very seldom moves vertically upward to the well bore. If a well starts to produce water, it is usually not coming from below (*coning*). Rather, it is coming laterally from the edge water through an especially permeable bed (*fingering*).

Water flowing downstream in a river picks up sand grains and sweeps them over riffles where they come to rest on the downstream slope. The laminations then have an angle of 20° or 30° to the bedding. This is called *crossbedding*. A common type is caused by the shifting channels of a stream (Figure 6–7). The laminations are concave upward and are called *trough* crossbedding. Figure 8–9 shows large-scale crossbedding typical of wind-blown sand.

Ripple marks are commonly seen on the surface of bedding planes in sandstones. They are of many sizes and shapes and resemble the current ripples seen in shallow water along streams and beaches (Figure 8–10).

Diagenesis of sandstones. When originally deposited, clean beach sand has a porosity of about 40 percent and a permeability of several darcies. After it has been buried to a depth of several thousand feet, both porosity and permeability are drastically reduced. The conversion of soft beach sand to rock is called *lithification* or *diagenesis*. The mechanisms of diagenesis are (1) compaction, (2) cementation, (3) recrystallization, and (4) solution. All result from the flow of water through the sand bed.

As additional overburden is deposited on a sediment, the clays compact, expelling large volumes of water. Unlike clays, sand grains are equidimensional and rigid, so they compact very little. It is frequently said that with increasing overburden pressure, sand grains become fractured and distorted, but this may be questioned.

Porosity in sandstones is mostly destroyed by the deposition of secondary silica on the grains. It has long been thought that the extra silica comes from the solution of the quartz grains themselves at points of stress concentration where they touch. This process is called *pressure solution*. There is, however, some evidence that the majority of the secondary silica is brought in from outside or is produced by the solution of silicate minerals other than quartz. Sands are millions of times more permeable than shales, so the water squeezed out of the shales must pass slowly through the sands in a generally shoreward direction toward the edge of the basin where it can escape. As the depth of burial

Fig. 8–10 Ripple marks in sandstone *(after Pettijohn, Potter, and Siever)*

becomes greater, the temperature of the water rises, and it is able to dissolve substantial amounts of silica and alumina from the shales. As the waters move laterally to the shallower parts of the basin, they reach zones of less pressure and temperature and the dissolved material precipitates.

In some sands, calcite is the first secondary mineral to be precipitated in the pores. The seashells in the shales (and especially in the sands) dissolve in the pore water and reprecipitate when the water reaches shallower environments. Later, as the water gets hotter, silica becomes the principal secondary cement to fill the pores. Sometimes the previously precipitated calcite dissolves.

The secondary silica coats the original sand grains, which grow outward into the pores (Figure 8–11). The newly deposited quartz is crystalline, so the interior surfaces are like the faces of a quartz crystal. The area of contact where the grains originally touched each other becomes larger as the pores fill. The new contacts are sinuous, resembling the sutures in the bones of a skull. The process of pore filling can,

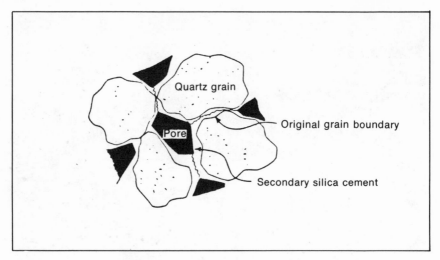

Fig. 8–11 Secondary cement in sandstone

and often does, continue until the porosity is almost completely destroyed. Sometimes a sandstone that has good porosity at one locality is totally cemented only a few kilometers away.

There is also some evidence to suggest that the presence of oil or gas in the sandstone inhibits silica cementation. Below the water-oil contact, the sandstones sometimes lack permeability. This suggests that the cementation took place after the migration and accumulation of oil.

Sometimes the continued flow of water through reservoir rocks actually dissolves away certain components and increases the porosity and permeability. Calcite cement and feldspar grains are especially likely to be dissolved. In the case of certain reservoirs, it is believed that some of the porosity is secondary.

Clays in sandstones. Clay particles within the pores of sandstones have tremendous effects on their reservoir properties. These have only recently been recognized and are still not entirely understood.

The clay minerals in the sand have a porosity of their own. Capillary forces will not permit oil to enter pores finer than about 10 microns wide. The pores between the clay platelets are much finer than this. Consequently, the pores between the clay flakes remain saturated with water and are not available for oil or gas. In some cases, the porosity filled with water may be 75 percent or even more of the total porosity. This water is immobilized by its attachment to the clays, so it will not move. The sand will then produce clean oil. Quantitative calculations of S_w from the resistivity log may make it look as if the sand contained too little oil or gas to be productive.

In many cases the clays were deposited along with the sand grains. Each sand grain lies in a bed of clay. The total porosity of the rock is that of the clay because the pores of the sand are all filled with it. The permeability is virtually zero. There is no effective porosity for oil or gas because the pores are filled with water, held there by capillary forces. Figure 8–13 is a thin section and scanning electron microscope picture of a sandstone whose pores are filled with clay.

The migrating waters also deposit clay minerals in the pores. Usually they are in the form of delicate flakes, sometimes separated and sometimes aggregated into books (Figures 8–12, 14). Occasionally, they form honeycomb-like or fibrous structures. These clays are called *authigenic* because they were not swept in from elsewhere like the sand grains but precipitated where they are now found.

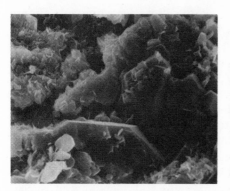

Fig. 8–12 Scanning electron microscope pictures of secondary quartz and clay in pores of sandstone.

Fig. 8–13 Poorly sorted sand with abundant clay matrix *(after Wai)*

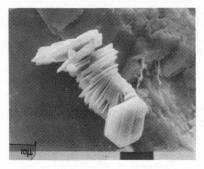

Fig. 8–14 Clay minerals that crystallized in place inside the pores of a sandstone (courtesy Dowell)

The word "clay" is sometimes used to denote a particle less than 4 microns in diameter and sometimes to designate a group of minerals, called the clay minerals. These are all characteristically very fine with a platy structure resembling mica. Because they are so fine, they are difficult to identify—in fact little was known about them until X-ray diffraction was developed.

The clay minerals are hydrated alumino silicates. They basically consist of two types of layers of atoms, like a cheese sandwich (Figure 8–15). One is a layer consisting of silicon and oxygen in the form of tetrahedra that link together in a hexagonal network. The other is a layer of alumina and oxygen (or hydroxyl) in an octahedral form.

There are four principal types of clay minerals. The kaolinites consist of one silica tetrahedral layer and one alumina octahedral layer linked together. They adsorb water only around the edges, not between the layers; so they do not swell. The chemical composition of kaolinite is $(OH)_8Al_4Si_4O_{10}$.

The montmorillonite group, now called smectite, consists basically of three layers: one octahedral alumina layer with a tetrahedral silica layer on each side of it. Water can get in between the tetrahedral layers. With one layer of water, the c-dimension of the lattice is 9.6 Angstrom units, and with more water the c-dimension increases to 21.4 Angstrom units. Smectite therefore swells when it is placed in water. It also tends to disperse into tiny platelets. Smectite in drilling mud adds to the viscosity, and it makes a fine impermeable filter cake, decreasing the water loss of the mud. The chemical composition of smectite is $(OH)_4Al_4Si_8O_{10}\cdot nH_2O$.

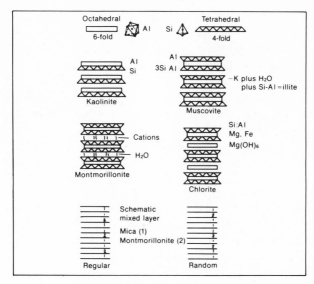

Fig. 8—15 Terminology and structure of clay minerals *(after Pettijohn, Potter, and Siever)*

Potassium ions fit between the hexagonal rings of the silica tetrahedra. When they do so, they bind the layers together so they cannot swell. These clay minerals are called illite. Some iron and magnesium is also present in illite. Mixed layer clays consist of stacked layers alternating between smectite and illite. The chlorites are similar to illite but contain iron and magnesium. Other types of clay minerals also occur, but nowhere in abundance.

Smectite as an authigenic clay mineral is the most harmful in a reservoir sandstone. It partly fills the pores with fluffy material whose microporosity is unavailable to store hydrocarbons. This microporosity is filled with immovable water, so that at a water saturation of 75% of the total pore volume the water will not flow and the rock may produce clean oil or gas. Smectite swells and the flakes disperse when contacted by fresh water, so the use of fresh water in a waterflood operation may damage the sand. The dispersed clay flakes are carried by the motion of the water to the throats of the pores where they pile up and clog them.

The swelling and dispersion of smectite can be inhibited by using salt water or potassium-chloride solutions. The damage can be partly remedied by treatment with hydrochloric and hydrofluoric acids.

Illite is related to smectite. When deposited in the pores, it often has a very open honeycomb or fibrous structure that causes large loss of effective porosity. It does not swell as much with fresh water, but the particles do migrate to pore throats, resulting in a loss of permeability. The damage can be prevented by the use of salt water or potassium-chloride solutions and can be partly corrected by acids.

It was formerly believed that the permeability damage was due mainly to the swelling of smectite, but it seems that kaolinite is also a harmful clay mineral because the flakes are so delicately attached and easily displaced. They are large and effective in blocking the pores. Kaolinite is less soluble in acids. However, the migration of fines may be prevented by treatment with certain polymers such as polyhydroxy aluminum immediately after the completion of the well.

Chlorite contains large amounts of iron and magnesium. It dissolves readily in hydrochloric acid, but it forms a gelatinous, flocculent precipitate of iron hydroxide that is extremely harmful to the permeability. Many wells when treated with acid have decreased instead of increased their rate of production.

When clays are surrounded by water with dissolved salts, some of the metallic ions attach themselves to the shales. These are called *exchangeable bases* because one type of ion may exchange with another. The most common naturally occurring ions are sodium and calcium. Potassium and ammonium will displace calcium. The amount of exchangeable bases is expressed as *cation exchange capacity* (CEC), measured as milliequivalents per gram. Smectite has a much greater CEC than any of the other clay minerals because the cations can be adsorbed between the tetrahedral layers. In the case of others, they attach only on the broken edges.

Some of the properties of the clay minerals depend on the nature of the exchangeable base. Calcium tends to flocculate smectite while sodium deflocculates it and makes it fluffier. Salt tends to flocculate clay suspensions in drilling mud. The small particles come together and settle out.

The exchangeable cations on the clay surfaces can move just like the dissolved ions in the pore water. Consequently, the conductivity of a sandstone containing clay minerals depends on both the conductivity of the ions in the pore water and the conductivity of the ions in the clay minerals. C_o is the conductivity of the rock fully saturated with water $(1/R_o)$ and C_w is the conductivity of the water $(1/R_w)$. The excess conductivity is due to the exchangeable cations in the clays. Waxman and Smits suggested the following formula to take this into account:

$$\frac{1}{R_t} = \frac{1}{F} \left[\frac{1}{R_w} + BQ_v \right] \qquad (8\text{--}1)$$

where:

B $\quad=$ constant

Q_v $\quad=$ cation exchange concentration (CEC) in milliequivalents of Na per cubic centimeter of pore space

R_w $=$ resistivity of the water

R_t $\;=$ resistivity of the rock

F $\quad=$ formation factor of Archie

The effect of the conductivity of the clay minerals is to make the resistivity of the rock R_t lower than it really should be for a known R_w. In calculating the water saturation by the formula (5–10):

$$S_w{}^2 = \frac{FR_w}{R_t} \qquad (5\text{–}10)$$

if R_t is too low, S_w will come out too high and the oil saturation will come out too low.

Effect of drilling mud on sandstones. When drilling overbalanced, a filter cake develops on the face of a permeable sandstone. In order to minimize invasion of mud and mud filtrate, steps are taken to decrease the permeability of the filter cake. Most of the solids in the mud remain in the filter cake and do not enter the sand pores. However, a small amount of fine particles does enter and tends to plug the pores in the vicinity of the well bore.

The filtrate from the drilling mud enters the sand and often invades it for a distance of several feet. If the sand contains swelling clays, they may originally have been flocculated in the salty, connate pore water. When this is displaced by fresh water, swelling occurs. Certain chemicals put in the drilling mud to reduce its water loss will deflocculate the clays in the pores. Even if there is no swelling, the flow of the water may pick up and move the delicate authigenic clay particles, causing blocking.

If there is oil in the rock, the water may form stable emulsions or asphalt films that block the pores. If the chemicals in the drilling mud react with the compounds in the pore water to form precipitates, these will also cause blocking. The surfactants in the mud may cause the oil to be displaced so completely that the permeability of the sand to oil is greatly decreased.

It is difficult—in some cases impossible—to correct the damage done to sand by drilling-mud filtrate. The only cure is to fracture the sand hydraulically farther back from the hole than the zone of invasion. In many soft formations fracturing is not practical.

Where the pressure in the pore fluids is low, as it is in many low-permeability gas sands, the weight of the drilling fluid is kept as low as possible. Sometimes nitrogen is bubbled through water, and sometimes efforts are made to drill with gas and cement the pipe, never letting liquid touch the face of the sand.

Effects of drilling mud on shales. The water in drilling mud has bad effects on the shales. First, it causes many shales to swell, which results in the walls of the hole sloughing. Large chunks of shale cave in and must be drilled up and circulated out. In the worst cases, they may stick the drillpipe. Second, the water disperses the compacted particles of clay in the shales. Cuttings disintegrate on the way up the hole. The dispersed fragments are so small they go through the shale-shaker screens and increase the solids content of the mud.

O'Brien and Chenevert (1973) classify the problem shales as shown in Table 8.1. The smectite (montmorillonite) shales tend to swell and disperse. Increased amounts of illite and chlorite decrease the dispersive tendencies, but even hard shales sometimes cave badly. Analysis of the shale by X-ray and other methods helps to diagnose the problems.

TABLE 8.I

CLASSIFICATION OF PROBLEM SHALES

Class	Characteristics	Clay content
1	Soft, high dispersion	High in montmorillonite, some illite
2	Soft, fairly high dispersion	Fairly high in montmorillonite, high in illite
3	Medium hard, moderate dispersion, sloughing tendencies	High in interlayered clays, high in illite, chlorite
4	Hard, little dispersion, sloughing tendencies	Moderate illite, moderate chlorite
5	Very hard, brittle, no significant dispersion, caving tendencies	High in illite, moderate chlorite

Laboratory and field tests have shown that polymers in the drilling mud help control dispersion, but not the swelling and sloughing. Potassium chloride in the mud at a concentration of 3–5 percent helps both dispersion and sloughing. Apparently, the potassium ions fit in the cavities in the silica tetrahedral layer, which binds the plates together.

REFERENCES

Almon, W.R., 1976, Pore space reduction in Cretaceous sandstone through chemical precipitation of clay minerals: Jour. Sedimentary Petrol. v. 46, p. 89–96.

Almon, W.R., and D.K. Davies, 1978, Clay technology and well stimulation, Trans. Gulf Coast Ass'n of Geological Societies, vol. 28, p. 1–6.

Almon, W.R., and A.L. Schulz, 1979, Electric log detection of diagenetically altered reservoirs and diagenetic traps, Transactions Gulf Coast Ass'n of Geological Societies, vol. 29, pp. 1–10.

Folk, Robert L., 1974, Petrology of sedimentary rocks: Austin, Texas, Hemphill Publishing Company, 182 p.

Friedman, G.M., and John E. Sanders, 1978, Principles of sedimentology, John Wiley and Sons, New York, 792 p.

Hayes, John B., 1979, Sandstone diagenesis—the hole truth, in Aspects of diagenesis, SEPM Special pub. 26, p. 127–139.

Holub, R.W., G.P. Maly, R.P. Noel, and R.M. Weinbrant, 1974, Scanning electron microscope pictures of reservoir rocks reveal ways to increase production: SPE Pap. 4787, p. 187–196.

Hower, Wayne F., 1974, Influence of clays on the production of hydrocarbons: SPE Pap. 4785, p. 165–176.

Kicke, E.M., and D.J. Hartmann, 1973, Scanning electron microscope application to formation evaluation: Gulf Coast Assoc. of Geol. Soc. Trans., 23rd Annual Convention, p. 60–67.

Krumbein, W.C., and G.D. Monk, 1942, Permeability as a function of the size parameters of unconsolidated sand: Am. Inst. Min. Met. Eng. Tech. Pub. 1492, 11 p.

——, and L.L. Sloss, 1963, Stratigraphy and sedimentation, 2nd. ed.: San Francisco, Freeman, 660 p.

Krynine, P.D., 1943, The megascopic study and field classification of sedimentary rocks: Jour. Geol. v. 56, p. 130–165.

McLaughlin, H.C., E.A. Elphinstone, and Bobby Hall, 1976, Aqueous polymers for treating clays in oil and gas producing formations, Soc. Petr. Engrs, SPE Paper 6008.

O'Brien, D.R., and M.E. Chenevert, 1973, Stabilizing sensitive shales with inhibited potassium-based drilling fluids: Jour. Pet. Tech., September, p. 1089–1100.

Pettijohn, F.J., 1975, Sedimentary rocks, 3rd. ed.: New York, Harper and Row, 628 p.

——, P.E. Potter, and R. Siever, 1972, Sand and sandstone: New York, Springer, 618 p.

Pittman, E.D., 1972, Diagenesis of quartz in sandstones as revealed by scanning electron microscopy: Jour. Sedimentary Petrol., v. 43, no. 3, p. 507–519.

Pittman, E.D., and J.B. Thomas, 1979, Some applications of scanning electron microscopy to the study of reservoir rock. Jour. Petr. Techn., November 1979, p. 1375–1380.

Powers, M.C., 1953, A new roundness scale for sedimentary particles: Jour. Sedimentary Petrol., v. 23, p. 117–119.

Swanson, B.F., 1979, Visualizing pores and non-wetting phase in porous rock, Jour. Petr. Techn., January 1979, pp. 10–18.

Swanson, R.G., 1981, Sample examination manual, American Association of Petroleum Geologists, Tulsa.

Wai, Thit, 1975, Reservoir properties from cores compared with well logs, Taglu gas field, Canada: M.S. Thesis, Univ. of Tulsa, 103 p.

Waxman, M.H., and L.J.M. Smits, 1967, Electrical conductivities in oil-bearing shaly sands: Soc. Pet. Eng. Jour., June 1967.

Williams, H., F.J. Turner, and Charles M. Gilbert, 1954, Petrography: San Francisco, W.H. Freeman, 406 p.

9

Geology of Carbonate Reservoirs

1. OIL IN CARBONATE ROCKS

An abundant sedimentary rock that often contains oil is limestone. Sometimes the limestone contains substantial amounts of magnesium, replacing calcium, and it then becomes *dolomite*. It has become customary in the oil business to call both limestone and dolomite *carbonates* to avoid making a distinction.

Oil was first discovered in carbonate rocks in Ontario in the 1850s and later (in the early 1900s) in the Tampico region of Mexico. In the 1920s, the carbonates of West Texas became important. By the 1930s and 1940s, the great oil fields of Iran and Saudi Arabia were found in the Asmari limestones of Miocene age and the Jurassic limestones, respectively. It has been estimated that about half the world's oil reserves are in carbonates, although there are numerically fewer carbonate than sandstone reservoirs outside the Middle East.

Carbonates differ in many respects from sandstones. They are mostly formed from the remains of animals (shellfish) and plants (algae); they are therefore found in nearly the same place where they originated and were not transported and then deposited like sandstones. Calcium carbonate can easily be dissolved by water solutions, so that solution and recrystallization of the carbonates after their deposition (*diagenesis*) is very common. This solution forms some of the cavities that can store oil. Limestones are much more brittle than sandstones, and as a result of folding or faulting they may break, leaving open *fractures* that serve as routes of fluid flow.

2. MODERN CARBONATE DEPOSITIONAL ENVIRONMENTS

Calcium carbonate is precipitated from sea water by many types of organisms. Molluscs such as oysters and clams make their shells of calcium carbonate, but many other animals and plants also form shells which are called *exoskeletons*. Some animals, notably the corals, live in large colonies and form sturdy buildups or reefs. These are attacked by waves and fish, producing fine calcareous mud that washes down the sides of the reefs. Corals grow only in water which is *warm, clear*, and *shallow*. Consequently, they are found predominantly in the tropics and grow especially well on the eastern shores of islands or continents where the trade winds blow waves and currents shoreward.

In the South Pacific, there are many ring-shaped islands composed only of coral and shell detritus. Their origin was first explained by Charles Darwin. He suggested that there was once a volcano above sea level at the site of the island (Figure 9–1). Coral reefs formed in the shallow water around its shores. The volcano became extinct, and because of crustal subsidence, began to sink slowly beneath the sea. If the sinking was slow enough, the growth of the reef-building corals was able to keep up with the rising sealevel. Finally a ring-shaped island called an *atoll* developed.

The study of modern carbonate environments by petroleum geologists has proved as rewarding as the study of modern deltas. In the U.S., the most accessible area of carbonate deposition is the Bahama Islands. Many other studies of modern carbonate environments have been made elsewhere, especially in the Caribbean (Cuba, Belize, and Mexico) and the Persian Gulf area.

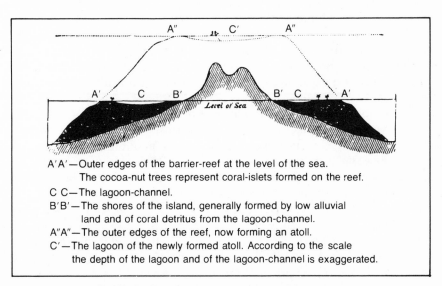

A′A′—Outer edges of the barrier-reef at the level of the sea.
  The cocoa-nut trees represent coral-islets formed on the reef.
C C—The lagoon-channel.
B′B′—The shores of the island, generally formed by low alluvial
  land and of coral detritus from the lagoon-channel.
A″A″—The outer edges of the reef, now forming an atoll.
C′—The lagoon of the newly formed atoll. According to the scale
  the depth of the lagoon and of the lagoon-channel is exaggerated.

Fig. 9–1 Growth of coral as volcanic island sinks

It is a marvellous experience for anyone, especially a petroleum geologist, to snorkel or scuba-dive in a reef environment. The fantastic shapes of the corals and the motions of spectacularly colored fish are visible in the clear water. The islands are also beautiful from the air. The deep water is dark purplish blue, while the lagoons and banks are various shades of light greenish blue. The underwater dunes of skeletal sands are clearly seen, even from high-flying jets.

The Bahamas. The Bahama Islands stretch in a chain 1,600 km long from near Florida to near Puerto Rico (Figures 9–2 and 9–3). The bathymetric pattern is very much the same in most of the islands (Figure 9–4). To the northeast, the sea bottom slopes steeply (as much as 40°) down to great depths (3,000 m and more). The living reef emerges a few feet above the sea level where the waves continually break against it and wash over it at high tide (Figure 9–4).

Back (west) of the reef about a kilometer, there is usually an island that consists of carbonates of various types formed when the sea level was higher. They were exposed and cemented into rock when the sea level was lower than now, during the ice age.

Between the island and the reef is a *lagoon* about 1 or 2 km wide where the water is up to 10 m deep. The twice daily lunar tides cause strong currents to flow in and out of the lagoon through gaps in the reef. This winnows the reef detritus, removing the fine lime mud, piling up the *skeletal sands* into underwater dunes, and forming round *oolite* grains (also called ooids or ooliths).

Behind the island, to the west, is a wide stretch of shallow water (0–10 m) called a *bank*. Near the island there is little movement of the water so that lime mud accumulates. The mud is full of organic life. Algae grow abundantly in the shallow water. Worms live in the mud, ingesting it and excreting it in oval *fecal pellets* a millimeter or so in diameter.

The lunar tides twice a day raise the water level on the bank and then lower it again; consequently there are strong tidal currents near the western edge of the bank. The mud is swept out, and the sand-size grains are continually rolled backward and forward to form oolites.

Another modern pattern is found along the shores of the Persian Gulf, which has an arid climate (Figure 9–5). Adjacent to the reef is a skeletal facies as in the Bahamas; back of it, in the lagoon, is a zone where oolites form. Landward of it where the water gets so shallow that waves cannot winnow the sediment, a pellet mud forms. This zone grades landward into a salt flat, some of which is above high tide. This salt flat is called a *sabkha*, an Arabic word. This is defined by Friedman and Sanders (1978) as a "surface of deflation in an arid

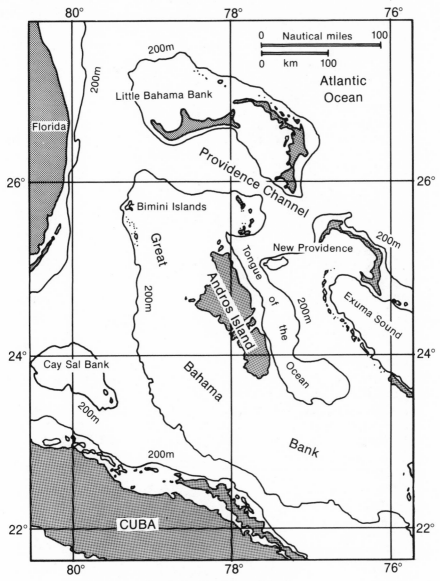

Fig. 9—2 The Bahama Islands

environment, formed by the removal of dry, loose particles down to the level of the ground water or to the zone of capillary concentration." On it, blue-green algae grow abundantly and form mats. Sea water is

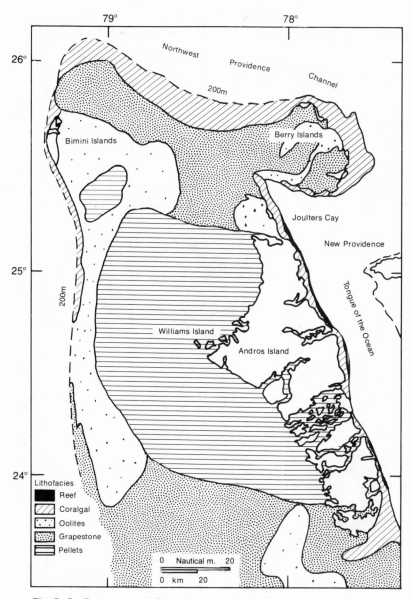

Fig. 9—3 Facies map of Great Bahama Bank *(after Friedman and Sanders)*

drawn by capillary forces to the surface where it evaporates. Anhydrite, gypsum, and sometimes salt form as crusts. As these minerals

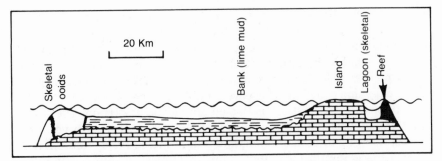

Fig. 9–4 Profile across typical Bahama bank *(after Friedman and Sanders)*

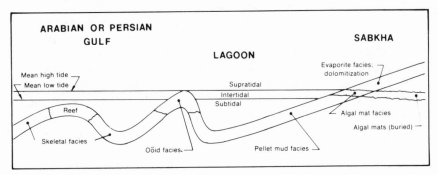

Fig. 9–5 Diagram of reef complex in the Persian Gulf *(after Friedman)*

crystallize out, the water becomes enriched in magnesium, and it sometimes converts the lime muds of the sabkha to dolomite.

Modern carbonate depositional environments have been studied in many other places around the world. Mostly they show a similar pattern—a shallow-water shelf bounded by a reef that drops farther offshore.

3. NAMING CARBONATE ROCK TYPES

The interpretation of carbonate facies to define zones of good reservoir properties has been held back by the lack of a generally accepted system for naming the different rock types. Two different geologists describing the same core might very well use different words.

In 1913 A.W. Grabau proposed the words *calcilutite* for consolidated lime mud, *calcarenite* for limestones whose particles are sand size, and *calcirudite* for limestones with pebble size grains. These terms are simple, descriptive, and still widely used.

Beginning about 25 years ago, it became obvious to oil company geologists that some uniformity in the terms used to describe carbonates was desirable. Several major companies developed classifications about the same time, and these were published together as Memoir 1 of the American Association of Petroleum Geologists (Ham, 1962). The most widely used have been those of Robert Folk of the University of Texas, Robert Dunham of Shell, and Leighton and Pendexter of Exxon. The latter is similar to Folk's. Folk introduced a number of rather unfortunate words, some of which are now abandoned. Perhaps the most practical and easily understood classification is a modification of Folk's suggested by G.M. Friedman.

Folk said that a carbonate rock consists of three textural components: grains, matrix, and cement. The cement is clear calcite that filled or partially filled the pores after the original deposition. There are several different kinds of grains, of which four are the most important. These are (1) shell fragments, called "bio"; (2) fragments of previously deposited limestone, called "intraclasts"; (3) small round pellets, the excreta of worms and other small burrowing organisms; and (4) ooliths, spheres formed by rolling and coating lime particles along the bottom.

The matrix is lime of clay-particle size (lime mud). It is called micrite. The clear secondary calcite cement is called sparite.

Thus, a rock consisting mainly of clear secondary calcite with intraclast grains would be called "intrasparite." A rock consisting mainly of micrite (lime mud) with grains consisting of broken shell fragments would be called "biomicrite." Biomicrite and pelmicrite are the most common limestone types. These eight types are shown diagrammatically in Figure 9–6.

Besides these eight combinations, there are some limestones consisting only of micrite and some consisting of the remains of upstanding reef-building organisms. So there are ten types of limestones in all.

The Dunham classification (Table 9–I) is based on the relative abundance of grains, that is, particles more than 0.020 mm in diameter, and lime mud. If there is a large amount of lime mud, the grains are mud supported; if a small amount, they are grain supported. Dunham also proposed some new words that at first sight seem unfortunate. However, Shell uses the classification worldwide and has taught it to geologists of many foreign companies. The words *mudstone, wackestone, packstone,* and *grainstone* are now widely used and convey an impression of the origin and reservoir properties of the rock.

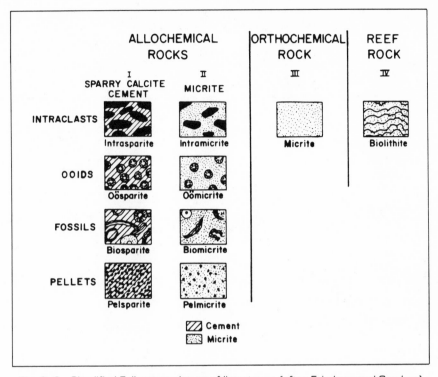

Fig. 9—6 Simplified Folk nomenclature of limestones *(after Friedman and Sanders)*

TABLE 9—I

DUNHAM CLASSIFICATION OF LIMESTONES (AFTER R. G. SWANSON)

Depositional Texture Recognizable					Depositional Texture Not Recognizable
Original Components Not Bound Together During Deposition				Original components were bound together during deposition, as shown by intergrown skeletal matter, lamination contrary to gravity, or sediment—floored cavities that are roofed over by organic or questionably organic matter and are too large to be interstices.	(Subdivide according to classifications designed to bear on physical texture or diagenesis.)
Contains mud (particles of clay and fine silt size, less than 20 microns)			Grain-supported		
Mud-supported		Grain-supported			
Less than IO percent grains	More than IO percent grains	More than* IO percent mud	Less than* IO percent mud		
Mudstone	Wackestone	Packstone	Grainstone	Boundstone	Crystalline Carbonate

*Modification of original Dunham classification by changing percent mud from 1 to 10%

TABLE 9–II

LEIGHTON AND PENDEXTER CLASSIFICATION OF LIMESTONES

GRAIN MICRITE RATIO (a)	% GRAINS (b)	GRAIN TYPE (c)					Organic Frame-Builders	No Organic Frame-Builders
		Detrital Grains	Skeletal Grains	Pellets	Lumps	Coated Grains		
9:1 ~90%		Detrital Ls.	Skeletal Ls.	Pellet Ls.	Lump Ls.	Oolitic Ls. Pisolitic Ls. Algal encr. Ls.	Coralline Ls. Algal Ls. Etc.	
1:1 ~50%		Detrital-Micritic Ls.	Skeletal-Micritic Ls.	Pellet-Micritic Ls.	Lump-Micritic Ls.	Oolitic-(Pisolitic-Etc.) Micritic Ls.	Coralline-Micritic Ls. Algal-Micritic Ls. Etc.	
1:9 ~10%		Micritic-Detrital Ls.	Micritic-Skeletal Ls.	Micritic-Pellet Ls.	Micritic-Lump Ls.	Micritic-Oolitic (Pisolitic Etc.) Ls.	Micritic-Coralline Ls. Micritic-Algal Ls. Etc.	Caliche Travertine Tufa
		← Micritic Limestone →						

The Leighton and Pendexter classification (Table 9–II) was developed by the Esso geologists in Tulsa and Calgary. Basically, it is similar to Folk's but did not propose new words. Like Dunham's, it was based primarily on the relative proportion of grains and lime muds. Secondarily, the type of grains is considered significant. It names five types of grains (Figure 9–7): *detrital* (Folk's "intraclasts"); *skeletal* (Folk's "bio"); *pellets* (Folk's "pellets"); *lumps* (irregular clumps not recognized by Folk); and *coated grains* (Folk's "ooliths").

4. ANCIENT CARBONATE DEPOSITIONAL ENVIRONMENTS

With the knowledge gained from the study of modern carbonate environments, we are better able to interpret what we see in ancient limestones and dolomites.

Many, if not most, ancient carbonates were deposited simultaneously in three different macroenvironments—shelf, slope, and basin (Figure 9–8).

GRAIN TYPES IN LIMESTONES

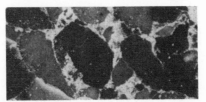

1. ROCK FRAGMENTS

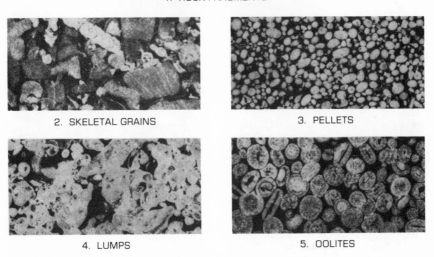

2. SKELETAL GRAINS   3. PELLETS

4. LUMPS   5. OOLITES

Fig. 9—7 Types of grains in limestones *(after Leighton and Pendexter)*

Shelf. The shelf environment consists of broad, shallow seas, mostly less than 100 ft of water. Currents are weak, so generally lime mud has been deposited. Usually there is enough current or wave motion to keep the water oxygenated. Scattered isolated coral heads or larger *patch reefs* are common. Sometimes in mud banks, oxygen is used up and organic matter is preserved.

After a rise in sea level, the shelf sediments build outward and upward so a regressive *shoaling-upward* sequence is formed (James, 1978). This consists of fine-lime muds below, coarser material affected by tidal currents above, and *supratidal* sediments made up of algal mats and sometimes evaporites. If later there is a sinking of the crust or another rise in sea level, the water deepens and the shoaling-upward sequence starts over. The great Jurassic oil fields of Saudi

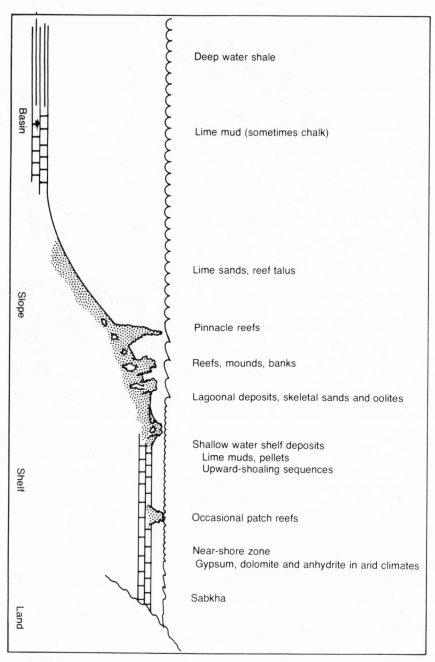

Fig. 9—8 Carbonate depositional environments and characteristic rock types

Arabia consist of shoaling-upward carbonates deposited in a regressive situation, each sequence being terminated by a bed of anhydrite formed when the sea almost dried up.

The very widespread carbonate units of the Williston Basin (Figure 9–9) were deposited mainly in a shallow shelf regressive environment.

In some places the shelf is narrow, perhaps only a few kilometers wide. In these cases a back-reef or lagoonal environment occurs. If the region is arid, the lime mud of the tidal zone may extend landward into supratidal flats (sabkhas).

Shelf edge. The edge of the shelf is a zone where the conditions are highly favorable for carbonate-forming organisms. Currents in the open sea bring nutrients. Often organisms build reefs right up to the surface of the sea. If the sea level rises slowly, the reef-forming organisms are able to build up the reef so that it remains at sea level. There is a balance between the upward growth of the frame-building organisms and their destruction by rasping, burrowing, and grazing organisms. These provide a continual stream of lime mud and sand which is swept into the shelf areas and may provide the bulk of the sediment deposited there.

Very diverse animals build reefs. Modern reefs are formed mainly by colonial corals, on the surface of which calcareous algae may form coatings. Sponges, molluscs, and bryozoa may also contribute. Loose algal flakes and shells may form *banks* or *mounds* that have considerable relief, even when corals or other frame-building organisms are absent. Algal mats collect lime mud and form layered structures called *stromatolites*.

Most reefs may be divided into three parts: the reef core, the reef flank, and the interreef (Figure 9–10). The reef core consists of remains of the reef-building organisms, but these are often partly destroyed and packed down and their interstices are filled with cryptocrystalline cement. Consequently the reef core is massive, unbedded, and its porosity is largely secondary.

The reef flanks consist of bedded lime sands and lime conglomerates derived from the reef. They dip away from the core and become thinner with distance.

The interreef areas consist of normal shelf sediments, some of which have been described.

Reefs, like the regressive sequences of the shelf, are often stacked vertically as the sea level rises relative to the sea bottom. Periods of rapid growth of the reef are often separated by periods of exposure to the atmosphere. These give rise to cemented layers that are permeability barriers.

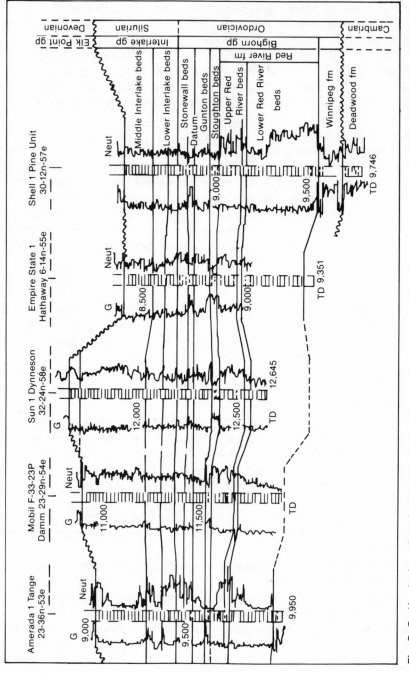

Fig. 9–9 Key beds and marker horizons in the Orodovician and Silurian of the Williston basin. Individual beds identified by sample studies and gamma-neutron responses can be traced over an area of a quarter of a million square miles *(after Porter and Fuller)*

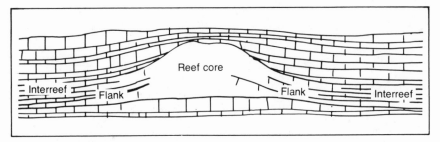

Fig. 9–10 Component parts of a reef *(after James)*

Colonial corals have been important reef-building organisms since the Cenozoic. During the Mesozoic, the principal reef-forming organisms were rudists. During the Paleozoic, the calcareous alga Ivanovia formed mounds at the shelf edge, and these are important oil reservoirs at Putnam, Oklahoma, and Aneth, Utah. Stromatoporoids and rugose corals were reefbuilders during the Paleozoic.

Reefs commonly occur in well-defined, sometimes continuous lines. These are often called *barrier reefs.* An important barrier reef containing the Redwater, Leduc, and Rimbey oil fields occurs in the Devonian of Alberta (Figure 9–11). This barrier is one part of a complex pattern of a basin surrounded by broad shelves. Sometimes the reefs are scattered randomly over the shelves as they are at Swan Hills and Judy Creek. Other times, the reefs form tall pinnacles rising from the slope, as they do in the Silurian of Michigan (Figure 9–12).

Slope deposits. The material on the slope consists of lime sands and blocks that have been broken off the reef by waves and deposited in strata with an initial dip. They are called *reef talus* and sometimes form excellent reservoirs. Some reefs are ring shaped (atolls), and flank material is found in the central lagoon.

Basin deposits. The material in the basin is fine grained, usually lime mud. Normally it does not have sufficient permeability to produce hydrocarbons. In a few places, chalk has accumulated, formed from the tiny shells of algae called *coccoliths.* Curiously, these deposits, as at Pine Island, Louisiana, and Ekofisk in the Norwegian North Sea, are of nearly the same age—upper Cretaceous. They have considerable porosity but very low permeability. The basinal carbonates often grade laterally into shale. In the case of epicontinental basins, it often happens that there is little circulation of the water in the deeper parts of the basins. Organic matter is preserved because not enough oxygen is brought in to destroy it. Occasionally such deposits become highly organic and may become source rocks of hydrocarbons.

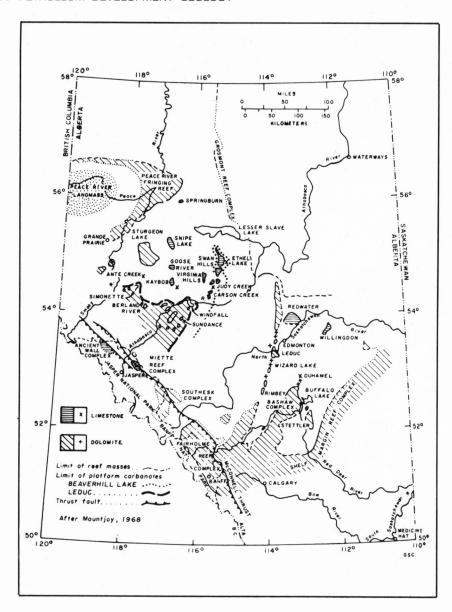

Fig. 9–11 Patterns of platforms and reef trends of Late Devonian age in Alberta (after Toomey et al.)

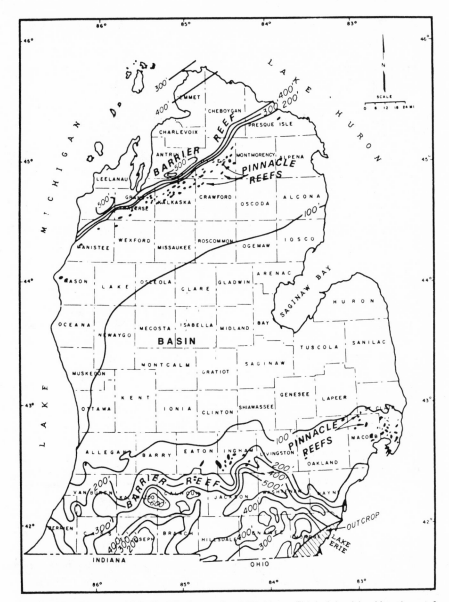

Fig. 9–12 Pattern of Middle Silurian reefs of Michigan. Thick dolomitized barrier-reef complexes at the basin margins grade into thinner limestones in the basin. Oil-bearing pinnacle reefs occur on the slope facies (after Mesolella et al.)

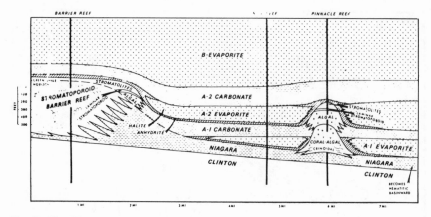

Fig. 9–13 Diagrammatic cross section of Silurian reefs, northern Michigan Basin (after Mesolella et al.)

5. DIAGENESIS OF CARBONATES

Lithification.   Calcium carbonate ($CaCO_3$) is slightly soluble in water, but calcium bicarbonate ($CaHCO_3$) is very soluble. When carbon dioxide gas (CO_2) is dissolved in water, it forms carbonic acid (H_2CO_3), which changes calcium carbonate to calcium bicarbonate. The reactions are complicated but may be summarized in the following equation:

$$CaCO_3 + H_2O + CO_2 \rightleftharpoons Ca^{++} + 2\ HCO_3^- \qquad (9\text{–}1)$$

The $CaCO_3$ is crystalline, the Ca^{++} and $2\ HCO_3^-$ are ions in solution. The reaction is reversible, so that crystalline calcium carbonate may be either dissolved or precipitated, depending on conditions in the water solution. Of these, the most important is pH. When carbon dioxide dissolves in water, it makes the water more acid, that is, it lowers the pH. Carbon dioxide is dissolved out of the air and is also produced by bacteria that decompose organic matter and by animals in their respiration, It is taken out of the water by plants such as algae that fix it to form organic carbon compounds. Slight changes in pH (and also in Eh, which is oxidation-redaction potential) thus cause solutions to dissolve or precipitate calcium carbonate.

The porosity and permeability of carbonate rocks, like those of sands, are controlled by the currents and waves in the original depositional environments. However, the original texture is vastly altered by the solution and reprecipitation of calcium carbonate after burial.

Recently deposited lime muds consist mostly of aragonite (which is a different and more unstable crystallographic form of calcium carbonate) and high-magnesian calcite. Consolidated limestones consist of low-magnesian calcite and sometimes dolomite. Profound changes take place soon after burial.

When originally deposited, lime muds have a porosity of 50 percent or more, but when they are consolidated into limestone their porosity is generally less than two percent. Shales lose porosity by a compaction process that involves flattening. However, limestones are formed from lime mud by recrystallization, and the pores are filled by precipitation of calcite, apparently brought in from elsewhere, because no compaction has occurred. Oolites and fossils are not squashed and flattened. Where did the calcite come from that filled the pores? The only possible source is adjacent layers of lime mud. Irregular, thin, black bands called *stylolites* are believed to be the insoluble remains of a bed of lime mud that was almost totally dissolved. This process is very poorly understood.

When recently deposited carbonate sediments are exposed to the atmosphere, fresh meteoric water enters the pores and percolates downward. At first it dissolves the aragonite, but as it goes deeper it becomes saturated and reprecipitates calcite, filling the pores. Because most carbonates were deposited in shallow water, exposure to the atmosphere occurred often.

Consolidated limestones show abundant evidence of solution and reprecipitation. Micritic skeletal (wackestone) limestones often have the original shells dissolved out, leaving cavities (molds). Irregular channels and cavities (vugs), formed by solution permeate some limestones. These, like fractures, are usually lined and are sometimes filled with clear crystalline calcite.

Dolomitization. Limestones are often partially or completely changed to dolomite. Dolomite has the composition $CaMgCO_3$ and it is crystallographically similar to calcite. However, it has greater density, less solubility in water, less ductility, and more brittleness. Obviously, waters enriched in magnesium permeated the calcium carbonate deposits sometime after their burial. They laid down an atom of magnesium and picked up one of calcium. Usually the dolomitization involved a recrystallization. Figure 9–14 is a diagram showing how a pelletal-skeletal micritic limestone (wackestone) is converted to crystalline dolomite. First, dolomite crystals (rhombs) form from the micrite in a random manner. Later, the nondolomitized micrite is dissolved, leaving intercrystalline porosity. Some of the Arabian fields produce from beautiful granular crystalline dolomite, which resembles granulated sugar.

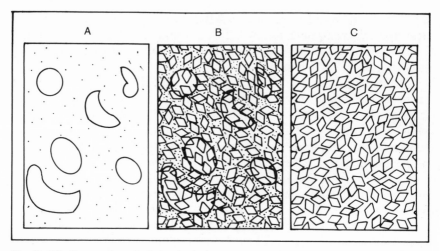

Fig. 9—14 Conversion of skeletal micrite to sugary dolomite. At first, tiny rhombic dolomite crystals form in the micrite. Later, the micrite is dissolved, leaving intercrystalline porosity (after Friedman and Sanders)

Recent studies of shorelines in arid areas have suggested the origin of dolomitizing solutions. The evaporation of sea water in lagoons and sabkhas makes supersaline brines that precipitate calcium carbonate and gypsum, and this enriches them in magnesium. On the coasts of the Red Sea, Bonaire, and Australia, these brines are converting recently deposited calcium carbonate to dolomite. It has been suggested by Friedman (1980) that the enormous epicontinental seas of the past that deposited shallow-water carbonate (e.g., Williston Basin) were subject to periods of excessive evaporation. Sometimes the evaporation was so intense that gypsum and salt were deposited. The brines would then permeate and dolomitize the carbonates. Sometimes fresh water from occasionally heavy rainfall would flood the salt flats. Mats of algae grew on the bottom of the shallow supersaline seas.

This theory is very plausible. Dolomites are always found in the shelf environment, never in the deep basin. They are commonly associated with evaporites (salt and especially gypsum and anhydrite) and with stromatolites, which are layers of algal mats.

Dolomitization may be enhanced by fracturing. In the Scipio field of Michigan, fractures are associated with a fault. The limestone is dolomitized only in the vicinity of the fractures. Increased permeability seems to increase dolomitization, probably because of the dolomitizing solutions. The Leduc reef of Alberta contains water that is in hydraulic communication with water in other reefs and it is dolomitized. The Golden Spike reef, only a few miles away, contains no water and is not dolomitized.

There is no doubt that dolomites generally are more porous and permeable than limestones. It is not obvious, however, whether the dolomitization enhanced the permeability or the permeability enhanced the dolomitization.

6. TYPES OF PORES IN CARBONATES

The interstitial pores in carbonates basically resemble those in sandstones. They are the open cavities between the grains, usually more or less clogged by mud or precipitated substances, or opened by water solution. However, carbonates differ from sandstones by being *soluble* and *brittle*. Because they are soluble, they often have large cavities called channels or vugs. Because they are brittle, they often have fractures that, if they are open or enlarged by solution, are also large openings. Carbonates therefore often—but by no means always— have a secondary porosity that may be greater than the primary interstitial, or *matrix*, porosity. The fractures may contribute only slightly to the porosity but vastly increase the permeability. The behavior of fluids in carbonates containing only interstitial porosity resembles that in sandstones, but in vuggy or fractured carbonates it is entirely different.

Choquette and Pray (1970) classified carbonate pores into 15 types (Figure 9–15). Only a few of these are important; some are curiosities.

Interparticle (intergranular) porosity. When the pores between ooliths, or sand-sized skeletal grains (Figure 9–16), are not filled by lime mud or secondary calcite, the rock has excellent porosity and permeability. Oolites that have been very prolific occur in the Mississippian St. Genevieve of Illinois and the Jurassic Smackover of the Florida panhandle. Skeletal sands have excellent permeability and are important in the Permian San Andres of West Texas. The pores between flakes of phylloid algae provide the porosity in the Pennsylvanian carbonates at Aneth, Utah. Interparticle porosity is often enhanced or destroyed by exposure to the atmosphere. Such exposure and changes in water depth leave reservoirs with stratified zones of low porosity and permeability, which form barriers to vertical flow of fluids.

Intercrystalline porosity. Dolomitizing solutions sometimes convert the rock to crystals of dolomite rhombs, about the size and shape of sugar grains (Figure 9–17). Such rocks are important reservoirs in the

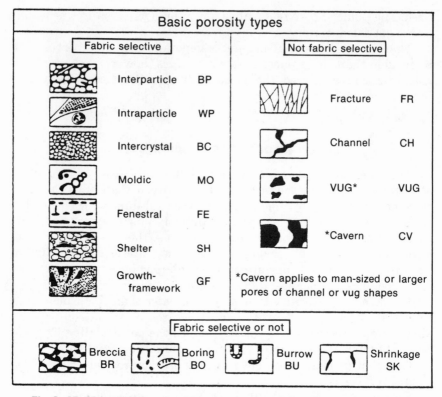

Fig. 9–15 Classification of carbonate pore types *(after Choquette and Pray)*

Jurassic Arab formation of Abqaiq and Ghawar in Arabia, and also in the Permian of West Texas.

Channel and vug porosity. Carbonate rocks often contain small tunnels or channels (Figure 9–18). These range from less than a millimeter in diameter (pinpoint porosity) to several millimeters or even centimeters (vuggy porosity). The channel walls are usually coated with secondary calcite crystals. Limestones with channel or vuggy porosity often have excellent permeability but have porosities of 10 percent or less. Reef cores often have vuggy porosity, while the adjacent reef flanks have interparticle porosity. Very low pinpoint porosity characterizes the Mississippi Lime of the Kingfisher, Oklahoma, area, while large vugs are common in the Mississippian Madison fields of the U.S. Rockies.

Fracture porosity. When sandstones are sharply folded, the beds

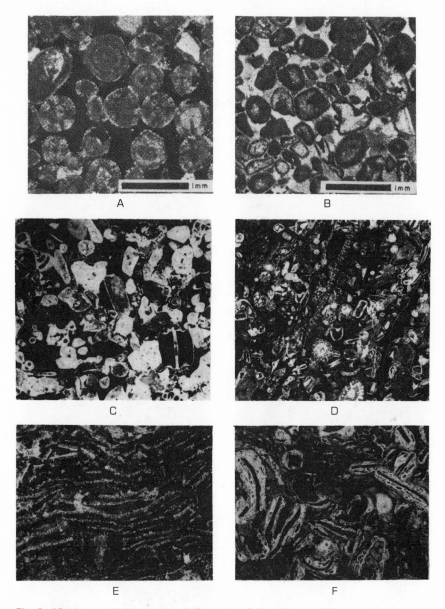

Fig. 9–16 Interparticle porosity. A,B—oolitic. C,D—skeletal, E,F.—phylloid algae *(after Choquette & Pray and Leighton & Pendexter)*

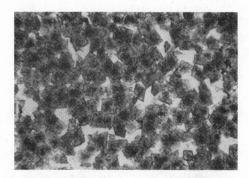

Fig. 9–17 Intercrystalline porosity in sugary dolomite *(after Powers)*

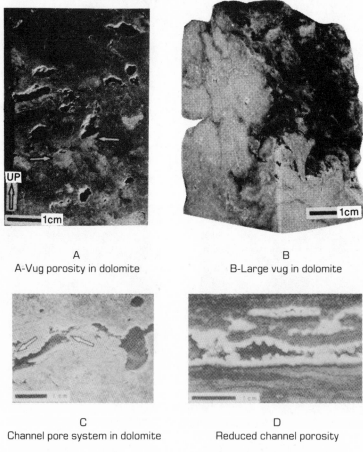

A
A-Vug porosity in dolomite

B
B-Large vug in dolomite

C
Channel pore system in dolomite

D
Reduced channel porosity

Fig. 9–18 Channel and vug porosity *(after Choquette and Pray)*

slide over each other; but when carbonates are folded or faulted, they develop fractures (Figure 9–19). If the stress environment is compressional, as it often is along faults and sharp folds, the fractures may be tightly closed. In this case the porosity due to the fractures will be negligible, and even the permeability will not be very great. Oil fields in this type of fracture porosity may be noncommercial. Much more prolific are fractures resulting from tension. Over an anticlinal fold, tensional fractures are perpendicular to the bedding and either parallel to the axis of the fold or, more commonly, at right angles to it (Figure 9–20).

Fractures usually have a volume of less than 1 percent of the rock, while the matrix may have a porosity of 5 to 10 percent. Consequently, fractures contribute much more to the permeability of a reservoir than to its porosity.

Most hard rocks, including both sandstones and limestones, exhibit *joints.* Joints are vertical fractures that occur in sets. The direction of the fractures is the same over regions hundreds of miles across. They are not affected by local structures. Their origin is quite unknown. Most joints terminate vertically at shale beds, but sometimes the same joint will extend through a shale bed and cut another hard bed above or below. They are spaced from a few centimeters to several meters apart: the thicker the bed, the wider the spacing. Joints appear to be usually closed at depth. The weight of the overburden causes a slight elastic compression vertically and expansion horizontally, which keeps them closed. Joints are planes of weakness and open easily if the pressure of the injected fluids exceeds a certain critical value.

Fractures are also caused by drilling. Induced fractures can be recognized because they have no mineralized coatings on their surfaces. They usually are located centered along the axis of the core.

Intraparticle porosity. Often the carbonate grains, skeletal and oolitic, have a microporosity of their own (Figure 9–21). This porosity is too fine to admit hydrocarbons, but it is measured by any method of core analysis. These measurements may give a very erroneous value for the porosity available for storage of oil or gas. Inasmuch as the micropores are full of water, resistivity curves may indicate high water saturation even when the large pores contain oil.

Moldic porosity. Sometimes the grains are dissolved, leaving the micrite (Figure 9–22). If the rock is mostly lime mud with occasional grains (mudstone or wackestone), moldic porosity is not important for hydrocarbons because the large pores are connected to each other only through the fine pores of the micrite. If the grains are sufficiently abundant to touch each other (packstone), then the cavities left by

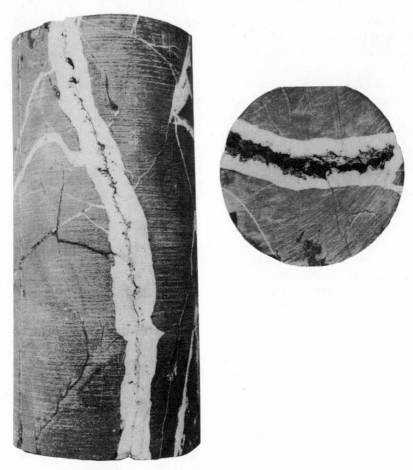

Fig. 9—19 Fracture porosity *(after Davidson and Snowdon)*

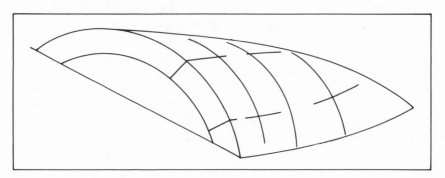

Fig. 9—20 Tension-type fractures produced by folding

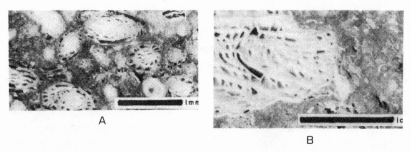

Fig. 9—21 Intraparticle porosity. A—fusilinids; B—horn coral *(after Choquette and Pray)*

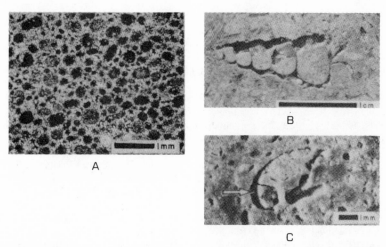

Fig. 9—22 Moldic porosity: A—pellets, B—gastropod shell, C—pelecypod shell dissolved away from micrite matrix *(after Choquette and Pray)*

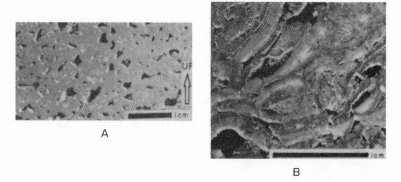

Fig. 9—23 Fenestral (A) and shelter (B) porosity *(after Choquette and Pray)*

dissolved grains become available for hydrocarbons. Such reservoirs are called *oomoldic*; they are not common.

Fenestral, shelter, growth framework, boring, burrow, and shrinkage porosity. It often happens that small cavities are left open as lime mud settles to the bottom. These are usually unimportant types. Fenestral (also called bird's-eye) pores are caused by algal organic matter rotting or by gas bubbles (Figure 9–23). Occasionally, the spaces between the skeletons of the growth framework-forming organisms remain open. However, they are usually filled with lime mud or crypto-crystalline calcite. While reefs are usually porous zones, the porosity is more often due to secondary channels and vugs than growth framework.

REFERENCES

Bathurst, R.G.C., 1976, Carbonate sediments and their diagenesis, 2 ed., Elsevier, New York, 658 pp.

Choquette, P.W., and L.C. Pray, 1970, Geologic nomenclature and classification of porosity in sedimentary carbonates, AAPG Bulletin vol. 54, pp. 207–250.

Dunham, Robert J., 1962, Classification of carbonate rocks according to depositional texture, *in* W.E. Ham, ed., Classification of carbonate rocks, AAPG Memoir I, pp. 122–192.

Folk, Robert L., 1962, Spectral subdivision of limestone types, AAPG Memoir I, pp. 62–84.

Friedman, G.M., 1980, Review of depositional environments in evaporite deposits and the role of evaporites in hydrocarbon accumulation, Bull. Centre Recherche Explor., Prod. Elf-Aquitaine, vol. 4, no. 1, pp. 589–608.

——, and J.E. Sanders, 1978, Principles of sedimentology, John Wiley, New York, 792 pp.

James, N.P., 1977, Facies models 7: introduction to carbonate facies models, Geoscience Canada vol. 4, no. 3, pp. 123–125.

——, 1977, Facies models 8: shallowing upward sequences in carbonates, Geoscience Canada, vol. 4, no. 3, pp. 126–136.

——, 1978, Facies models 10: Reefs, Geoscience Canada, vol. 5, no. 1, pp. 16–26.

Leighton, M.W., and C. Pendexter, 1962, Carbonate rock types, *in* W.E. Ham, ed., Classification of carbonate rocks, AAPG Memoir I, pp. 33–61.

Mesollela, K.J., et al., 1974, Cycle deposition of Silurian carbonates and evaporites in Michigan Basin, AAPG Bulletin vol. 58, pp. 34–62.

Pittman, E.D., 1971, Microporosity in carbonate rocks, AAPG Bulletin vol. 55, pp. 1873–1881.

Powers, Robert, 1962, Arabian upper Jurassic carbonate reservoir rocks, *in* W.E. Ham, ed., Classification of carbonate rocks, AAPG Memoir I, pp. 122–192.

Scholle, P.A., 1978, A color illustrated guide to carbonate rock constituents, features, cements, and porosities, AAPG Memoir 27, 241 pp.

Swanson, R.G., 1981, Sample examination manual, AAPG, Tulsa.

Toomey, P.F., et al., 1970, "Upper Devonian (Frasnian) algae and foraminifera from the Ancient Wall Complex, Jasper National Park, Alberta, Canada, Canadian Journal of Earth Sciences, Vol. 7, pp. 946–981.

Wilson, James L., 1975, Carbonate facies in geologic history, Springer-Verlag, New York, 471 pp.

10

Oil Fields in Carbonate Reservoirs

---

1. OIL FIELDS IN REEFS

Horseshoe Atoll, West Texas.   The Horseshoe Atoll consists of a line of reefs more than 150 miles long. During Pennsylvanian and Permian time, a large platform formed in the middle of the Midland Basin (Figure 10–1). Along the southern side of the platform, a topographically elevated reef formed.

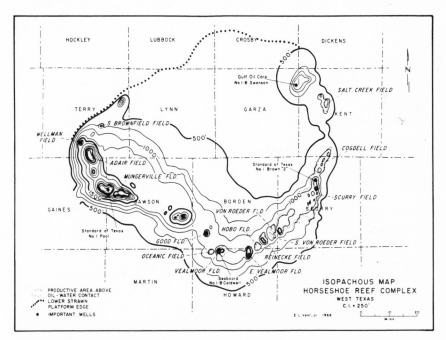

Fig. 10–1   Isopach map of Pennsylvanian reef complex, West Texas (after Vest)

The reef core consists of skeletal sand-sized grains of animals and algae. There were no frame-building organisms, and the reef core is stratified and not massive. As the basin subsided, the reef grew upward and maintained its top near sea level. The flanks and basin continued to subside, so eventually the reef stood several hundred feet higher than the basin. The total thickness of the reef is more than 1500 ft (Figure 10–2).

Frequently, the surface of the reef was exposed to the atmosphere. Meteoric water in places enhanced the porosity, which averages 10%. This periodic exposure also gave rise to frequent beds of micrite and shale (Figure 10–3). These are barriers to vertical flow.

After deposition, the whole area was tilted to the west. This caused oil to accumulate mainly in the southeast quadrant. Here the mounds are filled to the spill point. The recoverable oil from the whole complex of fields is 2.5 billion barrels. The field was waterflooded and is now (1981) subjected to a large-scale enhanced-recovery operation using carbon dioxide.

Judy Creek field, Alberta. The Judy Creek reef is part of the Swan Hills reef complex, shown near the center of Figure 9–11. It is near the western edge of a rather complex basin of Devonian age. The main organic reef consists of fragments of stromatoporoids, algae, brachiopods, cup corals, Amphipora, and crinoids. The reef was atoll shaped with steep outer flanks and an inner lagoon (Figure 10–4). The

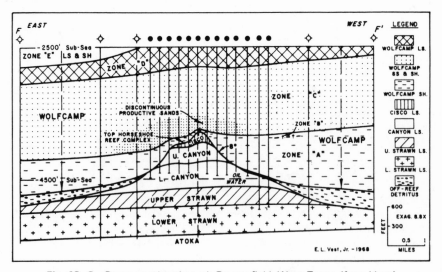

Fig. 10–2 Cross section through Scurry field, West Texas *(from Vest)*

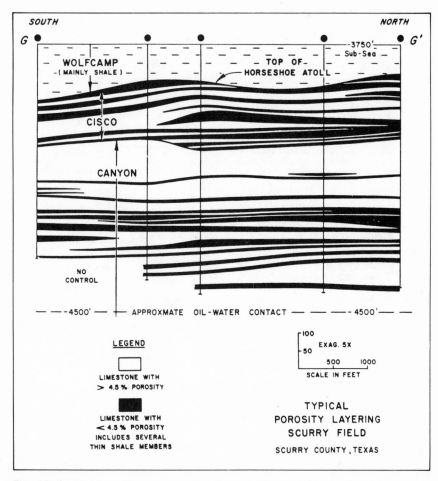

Fig. 10–3 North-south cross section through Scurry field, showing impermeable layers *(after Vest)*

best porosity is in the organic reef and reef detritus; in the reef interior the porosity is patchy and discontinuous. About 75 percent of the reef is filled with oil; there are 830 million bbl oil in place, of which 47% is recoverable.

There are many other important oil fields in reefs. Among these are the Redwater-Leduc-Rimbey Trend, the Swan Hills complex, and Rainbow-Zama reefs of Alberta; the Eocene reef fields of the Sirte Basin of Libya; the Cretaceous Golden Lane of Mexico; and the small but prolific reefs in the Silurian of northern Michigan.

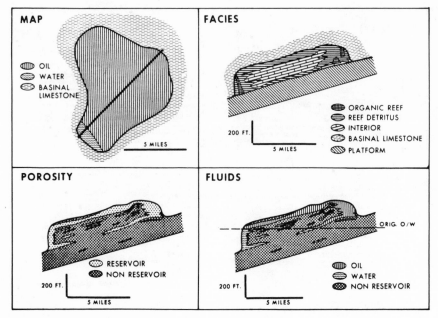

Fig. 10—4 Judy Creek oil field, illustrating lithofacies, porosity, and fluid distribution (after Jardine et al.)

2. OIL FIELDS IN SHELF CARBONATES

Huge areas are covered by shelf carbonates, like those of the Williston Basin. Most shelf carbonates are micritic and lack permeability. In certain zones, probably where the water was shallow and agitated by waves, skeletal sands and oolites developed. The water became shallower with time because the carbonate production built up to near sea level. Such regressive sequences are cyclic.

Ghawar field, Saudi Arabia. The Ghawar field is the largest oil field in the world. Its recoverable reserves have never been published, but they are certainly more than 70 billion barrels. The oil is contained in shelf carbonates of the upward-shoaling type. The Jurassic Arab formation consists of four main cycles of deposition and dessication named A, B, C, and D. Each zone starts at the base with normal marine limestone and is capped by anhydrite. The zone with the greatest permeability is the D. It consists of clean, current-washed pellet, oolitic, and skeletal grains. It appears to have been formed at a shelf margin. To the west, carbonates of a lagoonal facies predominate, while to the east deeper water rocks (mostly micrite) are found. The other great

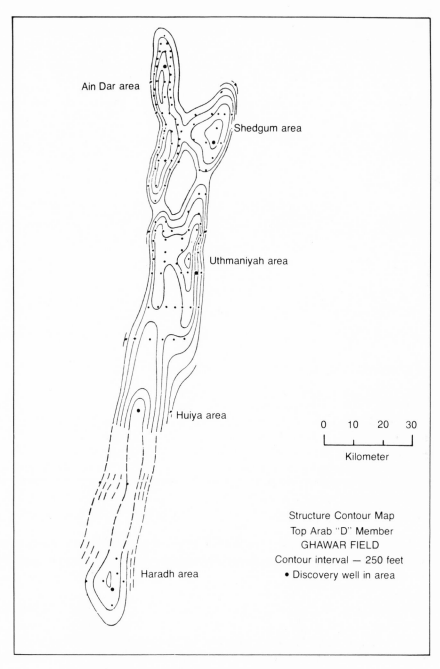

Fig. 10—5 Structure map of Ghawar field, Saudi Arabia

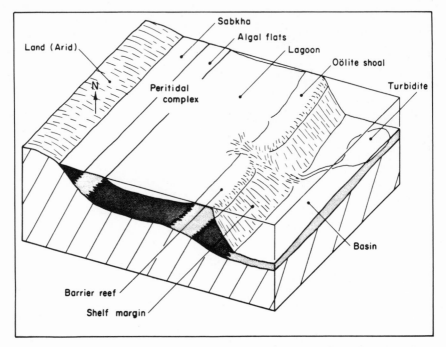

Fig. 10–6 Depositional environments in New Mexico and West Texas during the Permian *(after Friedman and Sanders, Silver and Todd)*

fields of Saudi Arabia—Abqaiq, Fadhili, Kursaniyeh, and Safania—line up with Ghawar in a north-south trend.

The structure of Ghawar (Figure 10–5) is a large, gentle, rather symmetrical anticline, striking north-south. It shows well on gravity maps and is probably caused by some basement dislocation. There are 1200 ft of relief with dips of 10–12 degrees on the flanks. It is productive over a length of 225 km. The oil-water contact is tilted to the east, and there is said to be a thick asphalt layer above it.

Wasson San Andres field, West Texas. On the north side of the Permian Basin of West Texas, there was a line of reefs during the Permian. Turbidities, lime muds, and shales accumulated in the deep basin to the south. North of the reefs there was a wide lagoon where salt and anhydrite were deposited intermittently (Figure 10–6). The whole area was extremely arid because there was probably no Atlantic Ocean nor a Gulf of Mexico at that time. The reefs migrated southward with time, building outward on the reef talus (Figure 10–7). The lagoonal deposits constitute the San Andres and Grayburg dolomites,

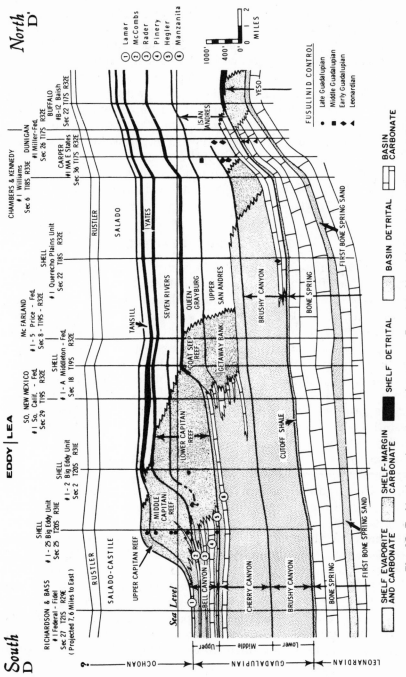

Fig. 10—7 North-south cross section across Permian, New Mexico *(after Silver and Todd)*

which are permeable and productive in many large fields. The reef and the adjacent back-reef deposits are the most permeable. They consist of skeletal sand-sized particles, oolites, and sucrose dolomite. The lagoonal sediments are micrites with little porosity. Anhydrite occurs in nodules in the reefs and also fills the pores in some units.

The Wasson field is an anticline capped by dense dolomite. There was an original gas cap above and an oil-water contact below. The aquifer must have been limited because the water never advanced into the field. Permeable strata alternate with impermeable ones, so there is limited vertical continuity. Lateral continuity is also limited, and the original 40-acre spacing (400 m between wells) was too wide to drain all of them. For water injection, it was decided to space the wells 880 ft (270 m) apart.

3.   FRACTURE-TYPE FIELDS

When carbonate strata are bent sharply, as over an anticlinal structure, they fracture. In a compressional tectonic environment like an overthrust belt, the fractures are tightly closed and provide little porosity. When the strata are bent in a tensional environment, as over the edge of a fault block in the basement, wide-open gash fractures develop. Even then, however, the fractures seldom provide a porosity greater than 1 percent of the rock volume. The best fracture-type pools also have interstitial porosity in the solid blocks between the fissures. This is called *matrix porosity*. The principal effect of the fractures is to provide passages for the flow of fluids. Even if the matrix has a permeability of 10 md or less, the wells will produce thousands of barrels per day if there are abundant fractures.

Gach Saran, Iran.   A series of anticlines at the west foot of the Zagros mountains in Iran constitute the most prolific oil fields in the world. The largest of these is the Gach Saran structure that produces oil from the Miocene Asmari limestone (Figure 10–8). It is bent over a sharp anticline which has produced a system of fractures. There are tensional-type faults parallel to the axis of the structure, and a system of fractures parallel with and at right angles to the axis (Figure 10–9).

The Asmari has substantial matrix porosity. In one cored well, 282 feet (13% of the total thickness of the Asmari) has an average porosity of 11.1% and a permeability that ranges from 0.5 to 100 md. Interstitial water (presumably in the micropores) is generally below 30%. Another 7% of the total thickness (302 ft) has a permeability less than 1 md and interstitial water up to 50%. The rest of the Asmari (1461 ft, 71.5%) has a porosity of 2% and very low permeability.

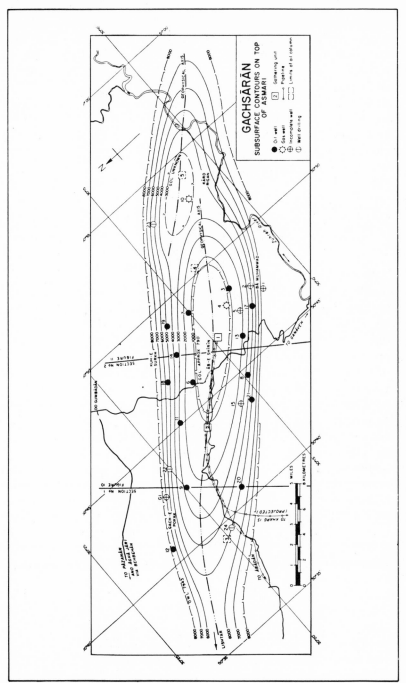

Fig. 10—8 Map of Gach Saran field, Iran *(after Crichton and Slinger)*

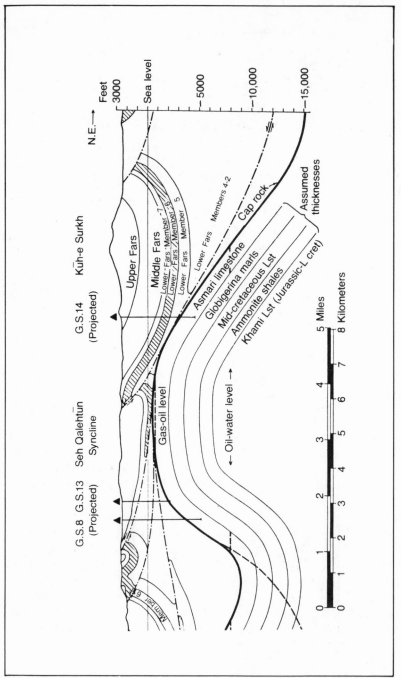

Fig. 10—9 Cross section of Gach Saran field *(after Crichton and Slinger)*

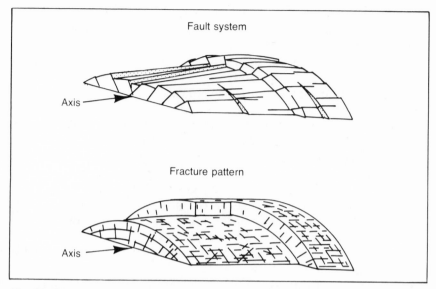

Fig. 10—10 Pattern of faults and fractures in Asmari limestone fields of Iran *(after Andresen)*

Fractures can be recognized in cores and from the behavior of the wells. Most wells are very prolific, producing more than 20,000 b/d. Flowmeter measurements show that most of the huge production comes from a few short intervals in each well. Pressure tests showed that shutting in one well caused immediate pressure buildups in wells 1 km or more away.

The fractures result from the folding. They are mostly at right angles to the bedding and are oriented parallel and perpendicular to the axis of the structure. They do not extend as far as the adjacent structure, Agha Jari (about 50 km distant), because the pressure there has been drawn down much more than at Gach Saran.

Beaver River gas field, Canada. The Beaver River gas field is a sharply folded and faulted anticline. Initial production from the discovery well was 85 million cfd. The Middle Devonian carbonate was deposited on a broad shelf in subtidal to supratidal environments. The rock is dolomitized. In the upper 350 ft of the reservoir, there is intercrystalline porosity, pinpoint vugs, and partially filled vugs. Below this zone, the matrix porosity is low—less than 2 percent. The water saturations in the matrix could not be determined from logs but were estimated from capillary pressure curves to be 50 to 80 percent.

The rock is intensely fractured. It was difficult to estimate the volume of the fractures or vugs. However, the overall average porosity was estimated to be 2.7%, and the field was estimated to contain 1.5 trillion cu ft of gas.

From 1971 to 1973, the field produced about 200 million cu ft per day, and then the wells suddenly started to produce water. At first it was thought that water had coned upward into the wells that had produced heavily. It was found, however, that the gas-water contact had risen uniformly. The field was shut in after having produced only 178 billion cubic feet, slightly over one-tenth the original estimates.

The fractured formation outcrops in the mountains 85 miles away, with a potentiometric surface of 1500 ft above sea level. This amounts to a hydraulic head of 13,000 ft at the gas field. Apparently, the water advanced rapidly into the field as the gas was withdrawn, moving through the fractures and maintaining a high pressure. Because the pressure was never drawn down, the gas was never able to bleed out of the low-permeability matrix and vugs before they were shut off by the water in the fractures. It also seems possible that there might have been more interstitial water in the matrix porosity than had been estimated, and that the porosity in the fractures was overestimated.

Other pools producing from fractured carbonates include Mara (Venezuela), Roosevelt (Utah), West Edmond (Oklahoma), and several pools in the Ellenburger dolomite of West Texas. Except for the latter, the fractures provide a porosity of less than 1 percent, and most of the hydrocarbon is in the matrix.

Oil was discovered in highly fractured limestone in two pools in the Magdalena Valley of Colombia around 1950. The rate of oil production was excellent initially, but it declined rapidly and the fields were soon abandoned. Evidently the fractures provided excellent permeability but little porosity.

4. OIL FIELDS IN DEEP-WATER CARBONATES

Most carbonates were deposited in shallow water. This is because organisms forming calcite live only in shallow water where there is plenty of light. However, two kinds of deep-water carbonates have sufficient porosity to be oil reservoirs. One type is formed near the foot of an upstanding reef by successive slides of reef talus. The other type is chalk formed by very small algae or foraminifera that lived on the surface but whose tests settled on the bottom.

Poza Rica, Mexico. The Golden Lane fields of Mexico were developed between 1904 and 1920. They must have had immense vuggy and even cavernous porosity, because the wells were very prolific. Production was from the Middle Cretaceous El Abra formation, which

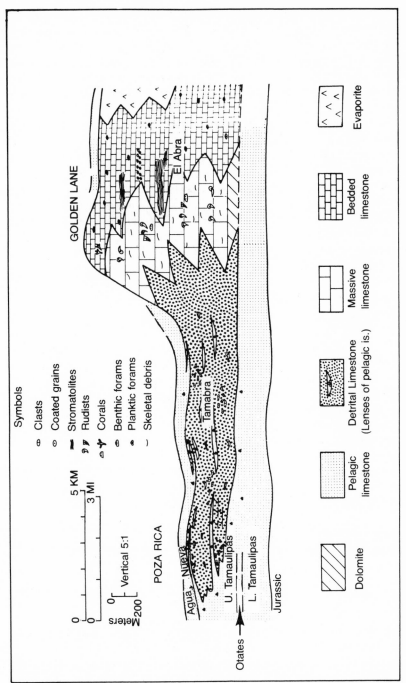

Fig. 10–11 Cross section from Golden Lane reef to Poza Rica field, Mexico *(after Enos)*

formed a well-defined reef. The reef itself consists of many different lithologies formed by rudists, corals, and stromatolites. Oolites are present and also evaporites in the back-reef or lagoonal facies. The reef must have stood up 1,000 m above the basin to the west, with a slope of 30 degrees.

West of the reef is a detrital limestone formed by successive slides of reef talus material that extends 20 km into the basin. It is called the Tamabra formation because it lies between the reef, El Abra, and the basin facies, called the Tamaulipas formation (Figure 10–11). It forms the reservoir rock of the Poza Rica pool, which contained more that 2 billion bbl of recoverable oil. The most productive lithology is skeletal fragment grainstone and packstone. Interparticle porosity ranges up to 25 percent. Wackestones and breccias are also common. The Poza Rica field provided most of the oil production of Mexico from 1938 until 1978 when new Cretaceous reefs were discovered in Chiapas and the Gulf of Campeche.

Ekofisk, Norwegian North Sea. During the Upper Cretaceous there was a great eustatic rise in sea level. Much of northwestern Europe, including the North Sea, England, Belgium, and Holland, was submerged under a quiet sea. Its depth is uncertain but was probably between 200 and 400 m. In this sea microscopic algae called *coccoliths* flourished enormously. Their tiny ring-shaped shells, 1–20 μm in diameter, were made of calcite. These settled to the bottom, forming an enormous thickness of chalk (as much as 1,500 m) in the southern North Sea between England, Denmark, and Norway. Over wide areas, however, its thickness is between 300 and 600 m. The chalk is an exceedingly pure carbonate rock, soft, white, and porous, that contains very little terrigenous material. Locally, the chalk may contain small amounts of shell debris. Nodules of chert are common. The chalk underlies much of southern England and constitutes the famous white cliffs of Dover. In the central North Sea the deposition of chalk continues into the lowermost Tertiary, called the Danian Stage. *Creta*, Latin for chalk, gave its name to the Cretaceous system. Around the sea's margins, the chalk becomes shaly (marl).

At about the same time, large areas of chalk were deposited in the U.S. The Selma and Austin chalks of the Gulf Coast Plain and the Niobrara of Nebraska are also soft, pure, white carbonates formed of coccoliths.

When first deposited on the sea bottom, the porosity of the shells is 70%. With increasing depth of burial, porosity is lost rapidly. Some of the loss is due to mechanical compaction and some to chemical cementation. At a depth of burial of 1,000 m, the porosity of most chalk is reduced to about 35%, at 2000 m to 15%, and at 3 km practically to zero. The permeability of chalk is typically very low because the particle size is so fine. It decreases from about 10 md at 40% porosity to 0.1 md at 10%. Figure 10–12 shows how increasing depth of burial results in decreasing porosity and increasing cementation.

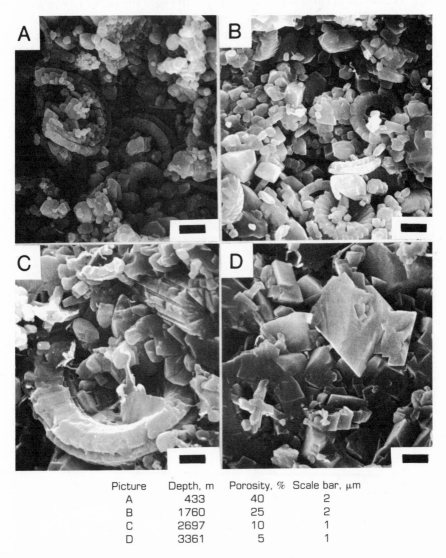

Picture	Depth, m	Porosity, %	Scale bar, μm
A	433	40	2
B	1760	25	2
C	2697	10	1
D	3361	5	1

Fig. 10–12 Sequence of scanning electron photomicrographs showing texture of chalk and increasing recrystallization with depth of burial *(after Scholle)*

The Ekofisk field was discovered by Phillips Petroleum Company in Norwegian waters in December 1969. Production from the chalk was quite unexpected. The chalk is a good seismic reflector, and Phillips had found a series of nearly circular domes, caused by pushups from the Zechstein salt (Figure 10–13). There were 200 m of saturation in the upper part of the chalk sequence. In this field the porosity was 30%, although its depth was 3,500 m. There has been considerable speculatin on why the porosity was so high at this depth. The overlying Tertiary clays and shales are undercompacted and have abnormally high pore pressure, apparently because the section was so impermeable that the pore water could not be expelled. The presence of the abnormally high pore pressure in the chalk would support part of the weight of the overburden so that the compaction pressure would be less.

The permeability of the chalk, however, is extremely low: 0.2 to 10 md. The wells produce several thousand barrels per day. At Ekofisk there is a well-developed set of fractures that acts as channels to drain

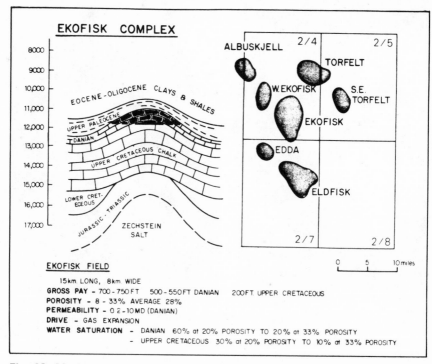

Fig. 10–13 Map and diagrammatic cross section of Ekofisk complex, Norwegian North Sea *(after Pergrum et al.)*

the highly porous chalk. The fractures are probably tension types developed over the dome caused by the rising salt. Nearby fields are less prolific, and apparently the fractures are necessary for commercial production.

REFERENCES

Andresen, K.H., R.I. Baker, and J. Raoofi, 1963, Development of methods for analysis of Iranian Asmari reservoirs, 6th World Petroleum Congress, part II, paper 14, pp. 13–25. 13–25.

Davidson, D.A., and D.M. Snowden, 1978, Beaver River Middle Devonian carbonate: performance review of a high-relief, fractured gas reservoir with water inflow, Jour. Pet. Techn., December, pp. 1672–1678.

Enos, Paul, 1977, Tamabra limestone of the Poza Rica Trend, *in* Deep-water sedimentary environments, SEPM Spec. Pub. 25.

Ghauri, W.K., 1979, Production technology experience in a large carbonate waterflood, Denver Unit, Wasson San Andres field, West Texas, SPE paper 8406.

Jardine, D., et al., 1977, Distribution and continuity of carbonate reservoirs, Jour. Petr. Techn. July, pp. 873–885.

McQuillen, Henry, 1974, Fracture patterns on Kuh-e Asmari anticline, Southwest Iran, AAPG Bull. vol. 58, no. 2, pp. 236–246.

Pergrum, R.M., G Rees, and D. Naylor, 1975, Geology of the Northwest European continental shelf: vol, 2. North Sea, Graham Trotman Dudley Ltd., London, 225 pp.

Powers, R.W., 1962, Arabian Upper Jurassic carbonate reservoir rocks, *in* Classification of Carbonate Rocks, AAPG Memoir 1, pp. 127–192.

Scholle, Peter A., 1977, Chalk diagenesis and its relation to petroleum exploration, AAPG Bull. vol. 61, no. 7, pp. 982–1009.

Slinger, F.P.C., and J.G. Crichton, 1959, The geology and development of the Gach Saran field, southwest Iran, Proceedings, Fifth World Congress, pp. 349–376.

Thralls, W.H., and R.C. Hessom, 1956, Geology and oil resources of eastern Saudi Arabia, 20th International Geological Congress, Mexico, vol. 2, pp. 9–32.

Vest, E.L., Jr., 1970, Oil fields of Pennsylvanian-Permian Horseshoe Atoll, West Texas, *in* Geology of Giant Petroleum Fields, AAPG Memoir 14, pp. 185–203.

11

Oil and Gas

1. CRUDE OIL

Crude oil consists mostly of hydrocarbons, that is, organic compounds comprised only of hydrogen and carbon. There is a wide variety of hydrocarbons of different molecular weight and molecular type. Most are liquid at reservoir temperatures and pressures. Some are gases, which occur both free and dissolved in the liquid crude oil. Also, some high molecular weight heavy molecules which are normally solid can occur dissolved in crude oil or emulsified with it in a colloidal form. These high molecular weight compounds often contain small amounts of nitrogen, oxygen, and sulfur, and therefore they are not hydrocarbons. They are commonly referred to as NSO compounds and seldom constitute more than a few percent of the total volume of crude oil, except in very heavy oils.

Carbon and hydrogen go together to form hydrocarbons in a large variety of patterns, and the number of atoms in the molecule increases from one carbon per molecule to several hundred carbons per molecule. The result is that there are hundreds of different hydrocarbons which occur naturally in crude oil. Consequently, in order to characterize and describe crude oil, these different chemical compounds must be classified. They are classified on the one hand by *molecular weight* and on the other by *molecular type*.

Molecular type is defined as the pattern with which the carbon and hydrogen atoms are attached to each other. The simplest molecular type is the series called the straight chain or normal paraffin (Figure 11–1). One atom of carbon surrounded by four atoms of hydrogen is methane, which is common natural gas. A straight chain consisting of two carbon atoms connected to each other, each carrying three hydrogen atoms, is called ethane, and it has the chemical formula C_2H_6. It is also a gas at normal temperatures and pressures but has a slightly

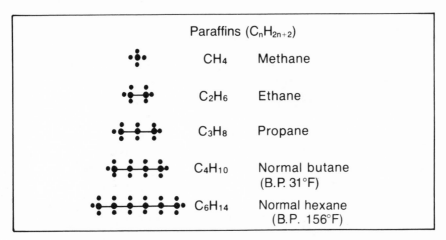

Fig. 11—1 Structure of light paraffins *(after Hunt)*

higher boiling point and compressibility. A molecule of three carbon atoms attached to each other with the two end ones carrying three hydrogens and the middle one carrying two hydrogens is called propane, C_3H_8. It is also a gas at normal temperatures and pressures but can be liquefied at high pressures and normal temperatures and is one of the principal constituents of LPG (liquefied petroleum gas). Four carbon atoms with the chemical formula C_4H_{10} is called butane, and it also is a gas which can be liquefied at normal temperatures. Six carbon atoms in a chain is called normal hexane. It is a liquid at normal temperatures although it is easily volatilized. It is usually included with some heavier molecules in gasoline.

It will be noted that as the number of carbon atoms increases, the density and boiling point also increase. There are two hydrogen atoms for each carbon, except that the two end carbons have three hydrogen atoms each. Consequently, the type formula for the normal paraffins is C_nH_{2n+2}.

The carbon atoms can also be attached to each other in branched instead of straight chains, and this series is called the iso-paraffins. Three carbons can only be attached to each other in a straight line, but when there are four carbon atoms they can be attached together in a branched form. Consequently the lightest branched paraffin is isobutane. If there are six carbon atoms in the molecule, they can be put together in four different ways, as shown in Figure 11–2. They all have the same chemical formula, but they have slightly different boiling points.

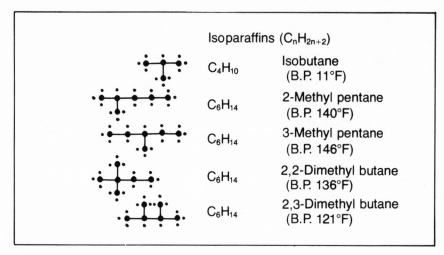

Fig. 11—2 Structure of iso- or branched-chain paraffins *(after Hunt)*

The paraffins, either straight or branched, can be hooked around in the form of a ring. This series is called the cycloparaffins or naphthenes. These are shown in Figure 11–3. The cycloparaffins lack the two hydrogens at each end of the chain; therefore, each carbon has two hydrogens and the type formula is C_nH_{2n}.

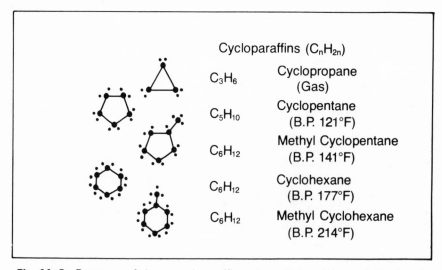

Fig. 11—3 Structure of ring or cycloparaffins, also called naphthenes *(after Hunt)*

The aromatic hydrocarbons have a basic structure consisting of a ring of six carbon atoms each with one hydrogen attached to it (Figure 11–4). Various types of hydrocarbon chains can be hung onto the aromatic ring and several rings can be attached together. The lowest molecular weight aromatic is benzene, and these rings are frequently called benzene rings.

There are several other homologous series or patterns by which carbon and hydrogen can be put together, such as the olefins, but they occur in crude oil only in very small quantities.

As the number of carbon atoms in the molecule increases, the molecular weight, boiling point, and density increase. This is shown in Table 11–I, which lists the first eight carbon numbers in the paraffin series with their boiling points and density (specific gravity).

A refinery separates crude oil into different useful liquids by means of their boiling points. Table 11–II shows the amounts of the different fractions separated by a refinery out of a typical crude oil. Gasoline may be around 30%, kerosene 10%, gas oil 15%, lubricating oil 20%, and residuum 24%. The residuum can be sold for heavy fuel oil or, if there is greater need for gasoline, can be cracked to form lower molecular weight compounds, some of which fall in the gasoline range.

An average crude oil might have its molecular types divided as shown in the table. However, there is much variety in crude oil. Some consist almost entirely of low molecular weight paraffins with very little lubricating oil and residuum, while others consist of high molecu-

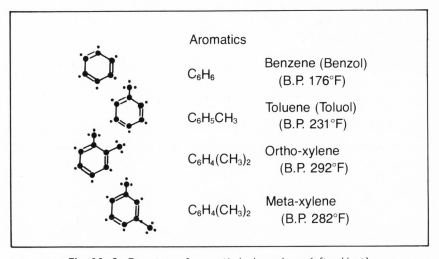

Fig. 11–4 Structure of aromatic hydrocarbons *(after Hunt)*

TABLE 11–I

PHYSICAL PROPERTIES OF LOW MOLECULAR WEIGHT HYDROCARBONS

Name	Boiling point °F	Specific gravity 60/60°
Normal paraffins		
Methane	−258.7	
Ethane	−127.5	
Propane	− 43.7	0.508
Butane	31.1	0.584
Pentane	96.9	0.631
Hexane	155.7	0.664
Heptane	209.2	0.688
Octane	258.2	0.707
Cycloparaffins or naphthenes		
Cyclopropane	− 27	
Cyclobutane	55	
Cyclopentane	121	0.750
Cyclohexane	177	0.783
Cycloheptane	244	0.810
Cyclooctane	300	0.830
Aromatics		
Benzene	176	0.885
Toluene	231	0.872
O-xylene	292	0.885

TABLE 11–II

COMPOSITION OF A TYPICAL CRUDE OIL (AFTER HUNT)

Molecular size	Wt. %	Molecular type	Wt. %
Gasoline (C_4–C_{10})	31	Paraffins	30
Kerosene (C_{11}–C_{12})	10	Naphthenes	49
Gas oil (C_{13}–C_{20})	15	Aromatics	15
Lubricating oil (C_{20}–C_{40})	20	Asphaltics	6
Residuum (C_{40+})	24		100
	100		

lar weight napthenes and aromatics with much residuum and very little gasoline.

Table 11–III shows a report of a Hempel distillation of crude oil which serves to characterize it in a preliminary way.

About a liter of crude oil is placed in a distillation flask and heated to 50° C. At this point, 2.0% of the oil (in this example) boils over and is collected and measured in a small graduate. The specific gravity of this fraction was 0.630. The temperature is then raised to 75° C, at which

TABLE 11—III
HEMPEL ANALYSIS OF CRUDE OIL BY BOILING POINT

Sample 41339

WARREN COUNTY, SOUTHWEST TWP., GOODWILL HILL FIELD
Third Stray sand; 695-720 feet
General characteristics
Specific gravity, 0.802; A.P.I. gravity, 44.9°; Sulfur, percent, less than 0.10; Saybolt
Universal viscosity at 100° F., 38 sec.; Color, dark green.

Distillation, Bureau of Mines Hempel method
Distillation at atmospheric pressure, 742 mm.　　　　First drop, 30° C. (86° F.)

Fraction no.	Cut, °C.	at °F.	%	Sum %	Specific gravity, 60/60° F.	°API, 60° F.	C.I.	S.U. visc., 100° F.	Cloud test, °F.
1	50	122	2.0	2.0	0.630	93.1	—		
2	75	167	2.6	4.6	.658	83.6	1.8		
3	100	212	5.2	9.8	.700	70.6	12		
4	125	257	7.8	17.6	.728	62.9	16		
5	150	302	6.8	24.4	.746	58.2	17		
6	175	347	6.7	31.1	.762	54.2	18		
7	200	392	5.8	36.9	.774	51.3	17		
8	225	437	5.5	42.4	.786	48.5	17		
9	250	482	5.7	48.1	.798	45.8	18		
10	275	527	6.2	54.3	.811	43.0	19		
Distillation continued at 40 mm.									
11	200	392	2.2	56.5	0.826	39.8	22	39	15
12	225	437	5.9	62.4	.832	38.6	21	44	30
13	250	482	5.1	67.5	.842	36.6	23	53	50
14	275	527	4.8	72.3	.851	34.8	24	72	65
15	300	572	5.0	77.3	.858	33.4	24	105	80
Residuum			21.1	98.4	.897	26.3			

Carbon residue of residuum, 2.0 percent; carbon residue of crude, 0.4 percent.

Approximate summary

	%	Specific gravity	°API	Viscosity
Light gasoline	9.8	0.675	78.1	
Total gasoline and naphtha	36.9	0.731	62.1	
Kerosene distillate	17.4	0.799	45.6	
Gas oil.	8.8	0.832	38.6	
Nonviscous lubricating distillate	10.9	0.839—0.857	37.2—33.6	50—100
Medium lubricating distillate. . . .	3.3	0.857—0.862	33.6—32.7	100—200
Viscous lubricating distillate. . . .	—	—	—	Above 200
Residuum.	21.1	0.897	26.3	
Distillation loss.	1.6			

point an additional 2.6% of the oil has come over; it has a density of 0.658. The temperature is raised by 25° increments, and the amount of the oil that distills over at each temperature is recorded. When the temperature reaches 275° C, the heating is halted because at higher temperatures the large hydrocarbon molecules would crack, forming smaller molecules that were not originally present in the crude oil. However, distillation can be continued at a vacuum of 40 millimeters of mercury. The temperature is reduced to 200° and distillation is continued to 300°, at which point the oil remaining in the flask is called residuum. In this case, it was 21.1% of the original volume of oil. The second part of the table combines these data into a summary of the familiar products of a refinery: gasoline and naphtha, kerosene, gas oil, and distillates of various boiling points.

A more complete breakdown of the various fractions of a crude oil is the *true boiling* analysis. This gives a better appraisal than the Hempel of the various products that can be obtained in the refinery.

In the 1920s the American Petroleum Institute started a research project on the composition and properties of petroleum which has been continued ever since. In the early years, the only way to distinguish and separate the various hydrocarbons occurring in petroleum was by their boiling point, and very elaborate distillation apparatus was used. Recently, the gas chromatograph has been developed. With it much better separation of the various hydrocarbon compounds can be made. As of 1967, 234 different compounds have been identified in crude oil (Table 11–IV). These amount to only about half the total volume. The other half consists of higher molecular weight compounds which are more complicated and more difficult to determine.

Analyses of crude oil based on gas chromatograph and mass spectograph may be used to compare two samples of oil and see whether they are from the same or different reservoirs. This has proven very useful in field development. Even oils from different fault blocks in the same field show slight differences.

The most important physical property of crude oil is the specific gravity, which may be defined as the ratio of the density of the oil to the density of water, both taken at the same temperature and pressure. In the United States, specific gravity is expressed as *API gravity*, which is defined as follows

$$°API = \frac{141.5}{\text{specific gravity } 60/60} - 131.5$$

The expression 60/60 means that the density of both the oil and the

TABLE 11–IV

DISTRIBUTION, BY CLASS AND CARBON NUMBER, OF THE COMPOUNDS
ISOLATED IN CRUDE OIL (FROM MAIR)

Carbon No.	4	5	6	7	8	9	10	11	12	13	14	15	16	17	18	Total
Branched paraffins	1	1	4	6	15	7	5	..	..	..	1	1	..	..	..	41
Alkyl cyclopentanes	..	1	1	5	13	2	..	..	..	..	..	..	..	..	..	22
Alkyl cyclohexanes	..	..	1	1	8	3	1	..	..	..	..	..	..	..	..	14
Alkyl cycloheptanes	..	..	..	1	..	..	..	..	..	..	..	..	..	..	..	1
Bicycloparaffins	..	..	..	..	3	3	5	1	..	..	..	..	..	..	..	12
Tricycloparaffins...........	..	..	..	..	..	..	1	..	..	..	..	..	..	..	..	1
Alkylbenzenes...............	..	..	1	1	4	8	22	4	..	..	..	..	..	..	..	40
Aromatic cycloparaffins...	..	..	..	..	..	1	4	3	..	1	..	..	..	2	1	12
Fluorenes	..	..	..	..	..	..	..	..	..	..	1	2	3	1	..	7
Dinuclear aromatics.......	..	..	..	..	..	..	1	2	12	15	5	1	1	..	..	37
Trinuclear aromatics......	..	..	..	..	..	..	..	..	..	1	4	1	1	..	..	7
Tetranuclear aromatics ...	..	..	..	..	..	..	..	..	..	..	..	1	1	1	..	3
Sulfur compounds.........	..	..	..	..	..	..	..	..	..	1	1	..	1	1	..	4
Total.................	1	2	7	14	43	24	39	10	12	17	9	8	7	6	2	201
Normal paraffins C_1 to C_{33}																33
Grand total.............																234

water are taken at 60° F. Outside the U.S., the specific gravity of the oil
is expressed as density, i.e., grams per cubic centimeter at a temperature of 20° C.

Viscosity is expressed in Saybolt Universal Seconds (SUS) at
100° F. This is an arbitrary number determined by the time for 60 cc of
liquid to flow by gravity from a hole in a special vessel. It is also expressed in centipoises.

The sulfur content is important because it affects the refining process. Recent environmental requirements in the U.S. and elsewhere
require the removal of sulfur before the refined products can be mar-
keted.

Table 11–V gives the API gravity, sulfur content, and viscosity of
some important types of crude oil.

2. NATURAL GAS

Crude oil at depth always contains some methane dissolved in it.
However, natural gas frequently occurs without any liquid petroleum,
and it is then called *nonassociated* gas. The origin of nonassociated gas
is not obvious; it seems probable that there may be certain types of
source rocks that generate gas but not liquid hydrocarbons.

TABLE 11–V
CHARACTERISTICS OF SOME TYPICAL CRUDE OILS (FROM BUTHOD)

Name, state, and country	Gravity, °API	Sulfur, %	Viscosity SSU @ 100° F
Smackover, Ark., USA	20.5	2.30	270
Kern River, Cal., USA	10.7	1.23	6,000 +
Kettleman, Cal., USA	37.5	0.32	
Loudon, Ill., USA	38.8	0.26	45
Rodessa, La., USA	42.8	0.28	
Oklahoma City, Ok., USA	37.3	0.11	
Bradford, Pa., USA	42.4	0.09	40
East Texas, USA	38.4	0.33	40
Leduc, Alberta, Canada	40.4	0.29	37.8
Boscan, Venezuela	9.5	5.25	
Poza Rica, Mexico	30.7	1.67	67.9
La Rosa, Venezuela	25.3	1.76	
Kirkuk, Iraq	36.6	1.93	42
Abqaiq, Saudi Arabia	36.5	1.36	
Seria, Brunei, Malaysia	36.0	0.05	

In some deep fields, both methane and a considerable amount of light hydrocarbons (up to C_8 or even higher) occur in the gas phase. These are called *gas-condensate* reservoirs. They may be the product of maturation and partial destruction of crude oil.

Nonassociated gas containing no condensable hydrocarbons is called *dry*. If it contains more than 3 gallons per thousand cubic feet (3 gal/Mcf), it is called *wet* gas. These are removed before it is put in the lines for sale, and sold as LPG (liquefied petroleum gas). Table 11–VI gives the composition of various natural gases.

Gas also frequently contains other gases than methane. Small amounts of nitrogen and carbon dioxide are common, and they sometimes amount to so much that the gas will not burn. Hydrogen sulfide is a common constituent of natural gas. It is extremely poisonous, and many fatalities have occurred among drilling crews bringing in *sour* gas wells. The sulfur can be removed and sold. Helium is a rare but valuable constituent of certain gas fields.

3. PHASE BEHAVIOR OF OIL AND GAS

The hydrocarbons are all mutually soluble. In natural oil reservoirs, the light gaseous hydrocarbons may be all in solution in the oil or there may be two separate phases—the gas phase above and the oil phase below.

TABLE 11–VI

AVERAGE COMPOSITION OF VARIOUS COMMERCIAL NATURAL GASES
(FROM LEVORSEN)

Pool and location	SG (air = 1.0)	Methane	Ethane C_2H_6	Propane C_3H_8	Butane C_4H_{10}	Pentane and heavier	CO_2
United States							
Panhandle-Amarillo		91.3	3.2	1.7	0.9	0.56	0.1
Hugoton, Kansas		74.3	5.8	3.5	1.5	0.6+	
Carthage field, Texas	0.616	92.54	4.7	1.3	0.8	0.6	
Velma, Oklahoma		82.41	6.34	4.91	2.16	1.18	
Canada							
Turner Valley, Alta.		92.6	4.1	2.5	0.7	0.13	

A single pure light hydrocarbon (like propane or butane) can exist as either a gas or a liquid, depending on the pressure and temperature, as shown in Figure 11–5. At higher pressures or lower temperatures, the material is liquid, as in the upper-left part of the diagram. At temperatures and pressures higher than the critical point, the material exists as only a single phase—neither gas nor liquid.

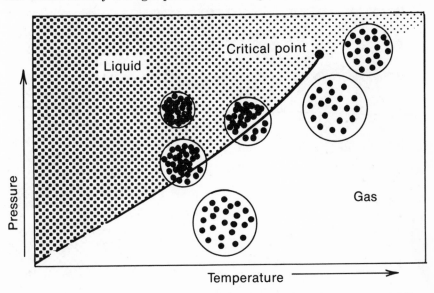

Fig. 11–5 Vapor pressure vs. temperature for a pure hydrocarbon component *(after Clark)*

A crude oil with methane and light hydrocarbons dissolved in it behaves as in Figure 11–6. Suppose oil in the reservoir has a temperature of Tr 1 and a pressure as shown at A'. As it flows through the sand toward the well, the pressure drops but the temperature in the reservoir stays constant. When it reaches the pressure at A, the gas starts to come out of solution, and this is called the *bubble point*. If the temperature were to stay the same and the pressure to decrease to point B, the volume would consist of 75 percent oil and 25 percent gas. If the oil and gas both came up the tubing in the well bore to the tank at the surface called a *separator*, both the temperature and pressure would decrease and the material would follow the dashed line to the point marked Sep. The oil would then go to the stock tank, where the pressure would drop farther, which would cause the evolution of a little more gas.

The gas which is evolved includes methane, ethane, propane, and considerable amounts of the heavier hydrocarbons, which are normally liquid. This results in an appreciable shrinkage of the crude oil. The amount by which an oil decreases in volume from the reservoir to the stock tank is called the *formation volume factor*. It ranges from 1.0 for an oil with few light fractions to 2.0 or more for an oil with large amounts of light (gasoline) fractions.

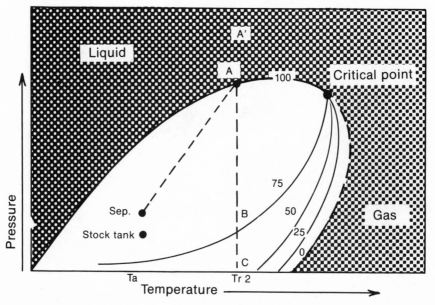

Fig. 11–6 Phase diagram of low-shrinkage oil *(after Clark)*

Some hydrocarbon reservoirs contain predominantly light hydro-carbons at pressures and temperatures higher than critical (Figure 11–7). As pressure in the reservoir decreases from A' to A, the material enters the two-phase part of the curve and liquid starts to condense. This is called retrograde condensation. It continues to point B where a maximum amount of liquid has condensed. Additional loss in pressure results in the liquid vaporizing, and finally it all goes back to the vapor phase again.

Often, gas reservoirs contain considerable amounts of condensable hydrocarbons. The phase behavior of wet gas is shown in Figure 11–8. If the pressure is reduced without changing the temperature, the mate-rial remains all gas. On its trip to the surface, however, the tempera-ture is reduced, and a substantial amount of liquid is formed.

It is important to determine the formation volume factor of a crude oil, especially in the case of reservoirs containing much methane and light hydrocarbons. The gas and oil are collected at the wellhead and put in a pressure vessel with a glass window. The volume inside the vessel can be reduced by injecting mercury at the bottom, and the temperature may also be controlled. The amount of gas and liquid at different temperatures and pressures can be observed by looking in the window or by noting the decrease in compressibility that occurs when

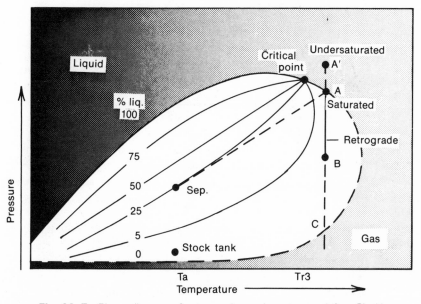

Fig. 11–7 Phase diagram of retrograde condensate gas *(after Clark)*

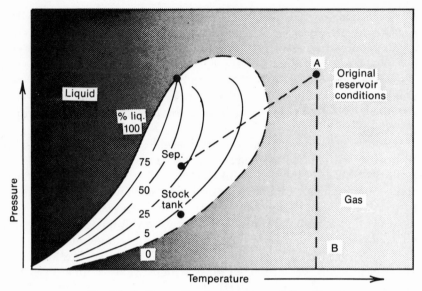

Fig. 11–8 Phase diagram of wet gas *(after Clark)*

all of the gas has gone into solution and only the liquid phase is present. The observations give pressure-volume-temperature relations, called a PVT analysis.

4. ORIGIN OF PETROLEUM

Organic matter in nonreservoir rocks. As soon as any organism dies, its remains are at once attacked by bacteria. These bacteria require oxygen; if plenty is available, as in soils and many continental deposits like red beds, the destruction of the organic matter is complete. However, muds rapidly deposited on the sea bottom have low permeability, so the circulation of water is stopped almost entirely. The aerobic bacteria quickly use up the dissolved oxygen. Anaerobic bacteria, which use oxygen from dissolved sulfates in the water, then take over. However, they disappear in number very rapidly with depth, and their activity practically ceases below 10 or 15 m below the bottom of the ocean. This is probably because they have consumed all of the readily decomposable compounds. By this depth of burial, the organic matter is transformed but not entirely destroyed. The chemical environment (Eh) is reducing rather than oxidizing.

The insoluble organic matter preserved in the shales is called *kerogen*. Kerogen has been divided into three main types (Tissot and

Welte, 1978; Hunt, 1979). Type I is characteristic of oil shales deposited in marine and lacustrine environments in which the organic matter is amorphous and largely algal in origin. It is sometimes called sapropelic or algal. Chemically it consists of normal and branched paraffins with some aromatics. When heated, it produces a large amount of lipids (oil). Type II (also called liptinitic) is predominantly naphthenes and aromatics. It is found in shales where marine organic matter derived from a mixture of phytoplankton, zooplankton, and bacteria have been deposited in a reducing environment. Type III (called humic) contains a high percentage of polycyclic, aromatic hydrocarbons and oxygenated functional groups plus some paraffin waxes. It is primarily derived from terrestrial plants and contains identifiable vegetable debris. It contains little or no long chains. It can generate dry gas but not liquid oil. Some rather hypothetical structures in kerogen are shown in Figure. 11–9.

The changes in chemical composition can be diagrammed, as in Figure 11–10. As the temperature increases, the kerogen loses oxygen. It goes off as CO_2 and water, moving a point to the left along the tracks. The different types of kerogen start out with different amounts of hydrogen: the sapropelic (oily) with a hydrogen-carbon ratio near 2 and the humic (woody) with a hydrogen-carbon ratio less than one. They

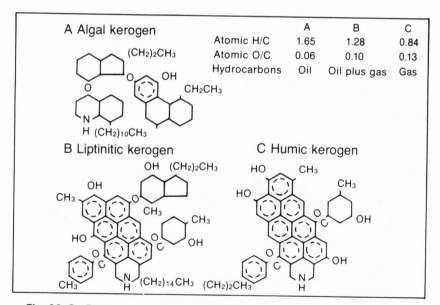

Fig. 11–9 Basic organic structures in different types of kerogen (after Dow)

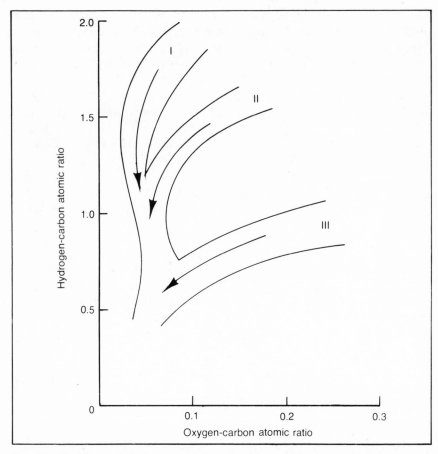

Fig. 11–10 Change in hydrogen-carbon ratio and oxygen-carbon ratio of kerogen of different types during burial *(after Hunt)*

progressively lose hydrogen, which goes off as water and as hydrocarbons. It is now believed that most crude oil is formed from Type II kerogen while it is heated between 60° and 150°C. Past 150°C. the heavier hydrocarbons are cracked to form methane gas. The kerogen becomes rich in carbon and poor in oxygen and hydrogen. Finally they are all gone and the carbon becomes graphite.

As the sediments become more deeply buried, the temperature rises. The average geothermal gradient is about 25°C. for each 1,000 m of burial (15°F. for each 1,000 ft). When the temperature reaches about 85°C., oil starts to form from the kerogen. Time affects the reaction; oil starts to form at 85°C. with an age of 10 million years but at only 40°C.

at 400 million years. The generation of oil breaks loose the long chains of the algal and liptinitic kerogen. Since the hydrogen-carbon ratio of crude oil is about 2 and of methane 4, the hydrogen-carbon ratio of the kerogen decreases markedly. Its color darkens and it becomes shinier.

At temperatures above 150° C (300° F), liquid hydrocarbons decompose, forming methane and solid hydrogen-poor bitumen. Consequently, there is a temperature "window" where oil is generated. Methane can be generated at shallow depths by bacteria and also at great depths from either kerogen or oil. This pattern of hydrocarbon generation is illustrated in Figure 11–11.

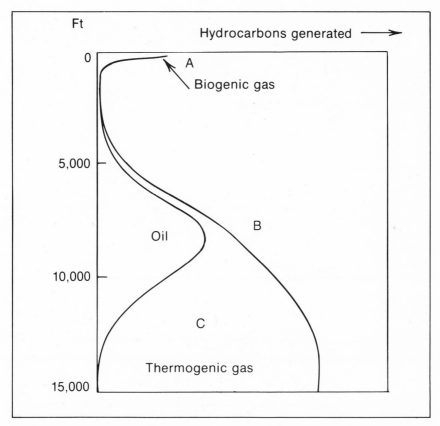

Fig. 11–11 Diagram showing the amount of hydrocarbons generated with increasing depth. Dry gas is generated in the shallow biochemical zone (point A). Oil and wet gas form in the intermediate zone (B), and only dry gas forms in the deep thermocatalytic zone (C) *(after Milner et al.)*

It is important, therefore, when exploring a new area to determine whether the organic matter has been heated enough to produce oil but not enough to convert it to gas. There are various organic indicators that help to determine this.

The normal paraffins extracted from recent sediments mostly have odd numbers of carbon atoms (Figure 11–12). The C_{29} are the most abundant. These are supposed to be waxes from plants, which are mainly odd-numbered. In crude oil, this odd-numbered preference has usually been lost. Thermal cracking breaks the long chains into two or more shorter chains. Breaking a chain with an odd number of carbons results in two shorter chains, one odd and the other even. If a strong odd carbon preference is present in the normal paraffins extracted from the sediment, it is presumed that the kerogen has not been heated high enough to generate petroleum.

Other methods can be used to tell whether the rocks have been heated too high. One way is to extract the solid kerogen from the rock by dissolving the clays and silica by hydrofluoric acid. The kerogen remains, and it can be analyzed chemically or examined with the microscope. If it has been heated too hot, it will be high in carbon and

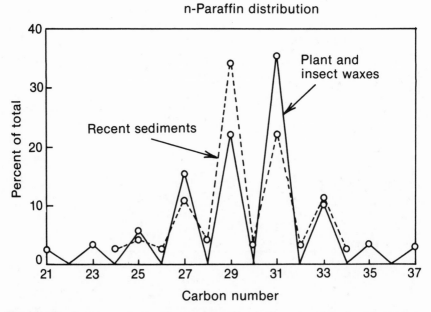

Fig. 11–12 Relative amounts of normal paraffins in Recent sediments showing predominance of odd-carbon chains *(after Hunt)*

low in hydrogen. Under the microscope it will be black and opaque instead of a translucent orange. Another method is to mount the organic matter in plastic, polish it, pick out shiny fragments of plants called vitrinite, and then measure the percent of vertically incident light that is reflected. The more reflective it is, the higher the temperature to which it has been subjected.

When the permeable formation containing the oil pool outcrops, it sometimes permits surface ground water to flow past the oil. When this happens, it makes the oil heavy and asphaltic. Partly it comes about because the light hydrocarbons are more soluble in water than the heavy. Partly it is the effect of bacteria which consume certain fractions of the oil, especially the normal paraffins. This process is called *degradation*.

Degradation has occurred on a large scale in many places. Hundreds of billions of barrels of tar occur in outcropping sands in Canada and Venezuela. In California, there is a large amount of heavy asphaltic crude oil in shallow sands. The oil from the deeper sands is much lighter and contains more gasoline.

5. SOURCE ROCKS OF PETROLEUM

Source rocks of petroleum are rich in preserved organic matter, although reservoir rocks very seldom are. The reason for this is that the properties of a rock which make it a good reservoir, namely porosity and permeability, are derived from a sedimentary environment of considerable energy in the form of waves and currents. These remove the mud particles, leaving the pores open. The same waves and currents that remove the mud particles also remove most of the even lighter and fluffier organic matter. Furthermore, the waves and currents keep the water oxygenated so the organic matter is oxidized before it gets a chance to be buried. Consequently, porous and permeable sandstones and limestones usually contain very little organic matter. Sediments deposited on land above the permanent water table also contain very little organic matter because oxygen is able to remove it all. Such sediments are quite common. They consist usually of light-colored shales and sandstones, red, green, or mottled. Thick sections of these rocks practically never contain petroleum.

Geological formations that contain oil are almost always associated with thick sections of organic rich shales. Figure 11–13 shows a large area in Wyoming and Colorado where the upper Cretaceous is very productive of oil. Analyses of the shale show that the organic content ranges from 1.5 to 3.0.

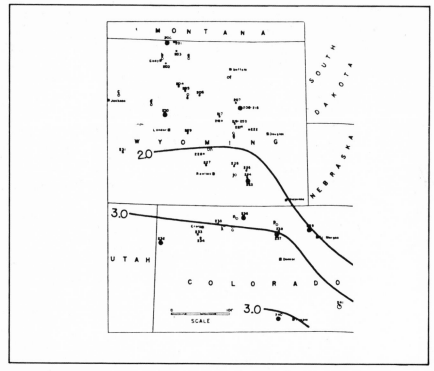

Fig. 11–13 Total organic carbon in Cretaceous shales of Wyoming, Colorado (after Trask and Patnode)

From many studies of this type, it has been determined that the best *source of petroleum is a shale containing 0.5 to 5 or 10% organic carbon.* These shales are usually dark in color, ranging from gray to black. However, things other than organic carbon can give shales a dark color, and it is desirable to make a chemical analysis.

In addition to considerable amounts of organic carbon, source rocks of petroleum still contain some of the petroleum hydrocarbons that were generated during the compaction and burial process. Figure 11–14 shows the relation between total organic carbon and parts per million hydrocarbon in several different lithologic types characteristic of the Pennsylvanian Cherokee shale, which is an excellent source rock. The light-gray shales contain less than 1 percent organic carbon and less than 100 ppm hydrocarbon. The gray shales contain 1–2 percent organic carbon and 100 to 500 ppm hydrocarbon. The black shales are very rich; they contain 5 to 10 percent organic carbon and more than 1,000 ppm hydrocarbon. The coals consist of nearly pure carbon

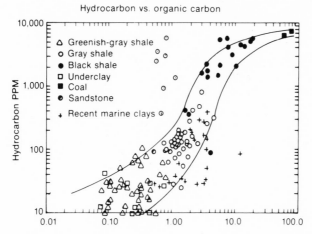

Fig. 11–14 Organic carbon and hydrocarbon content of the Cherokee shale of northeastern Oklahoma *(after Baker)*

and they also contain much hydrocarbon. It is not clear from these data whether the rich character of the Cherokee shale is due to the gray shales or the black shales.

For a rock to be a good source rock, it must contain more than 100 parts per million by weight hydrocarbon. This constitutes a second important characteristic which can be determined by chemical analysis.

A third criterion for a good source rock is the presence of light (C₄–C₇) hydrocarbons. These are routinely detected by mud logging equipment, or the cuttings may be sealed in cans or jars and sent to the laboratory.

The principles discussed in the preceding paragraph are illustrated by Figure 11–15, which is a hydrocarbon log of parts of the Mobil et al. James River, 8–14–34–7W5 in Alberta, Canada, reported by Snowdon and McCrossan (1973). Down to a depth of 5,000 ft (1,500 m), organic carbon is quite abundant, ranging from 1 to 5 percent in most samples. The gas content is low and consists of almost pure methane. This indicates that the rocks are immature and did not generate much oil. Around 8,000 ft (2,500 m) in the Lea Park Formation, the organic carbon averages about 1 percent. The total gas is low, but it is mostly heavier than methane. These are doubtful source rocks. Between the Cardium and Viking zones, the carbon is around 2 percent. The gas is very high and mainly ethane-plus, indicating very rich source rocks. Another zone which is high in carbon and total gas occurs between

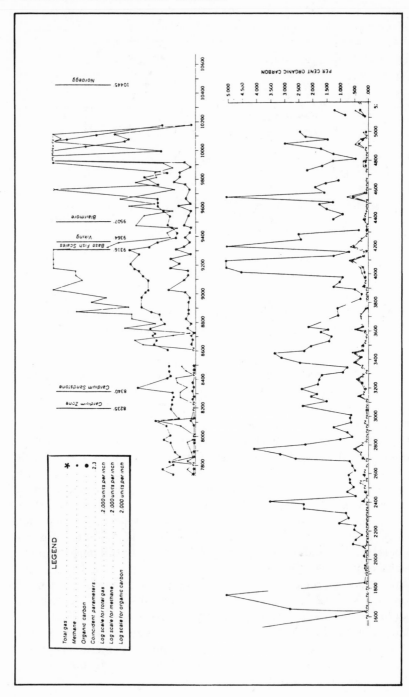

Fig. 11–15 Hydrocarbon log at Mobil James River Well *(after Snowdon and McCrossan, Geological Survey of Canada)*

9,900 and 10,200 ft (3,020 and 3,110 m), but here the gas is mostly methane. This zone may have been too deeply buried.

It would be very useful if it were possible to find some way of fingerprinting source rocks and the oil derived from them to identify the particular source rock from which each accumulation was derived. New analytical methods including the gas chromatograph and mass spectrometer have made this possible, and efforts have been made in some major oil company research laboratories. To date, few results have been published.

A notably successful attempt to relate particular types of oil to particular source rocks was made in the Williston basin by the Amoco research group (Williams, 1974). There are 3 principal types of oil in the Williston basin: Type I oils are found mainly in Ordovician reservoirs, Type II mostly in the Mississippian, and Type III only in the Pennsylvanian. The three types could be distinguished by several different chemical characteristics, all of which are well known and easily measured. Among these, the most significant characteristics turned out to be the relative amounts of straight chain, branched

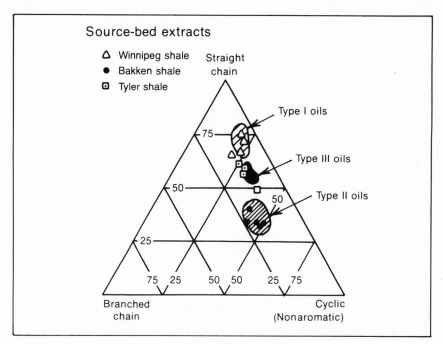

Fig. 11–16 Hydrocarbon-type distribution in C_4-C_7 fraction; comparison of Williston basin basic oil types with source-bed extracts *(after Williams)*

chain, and cyclic hydrocarbons in the C_4–C_7 fraction (Figure 11–16). Types I and III and predominantly straight chain, while Type II contains considerably more cyclic and branched-chain compounds.

Another significant distinction is the distribution of heavy normal paraffins (Figure 11–17). Each curve represents the relative amounts of C_{15+} paraffins. Type I has a very peculiar distribution with hardly any paraffins heavier than C_{19}. There is an odd-number preponderance between C_{15} and C_{19}.

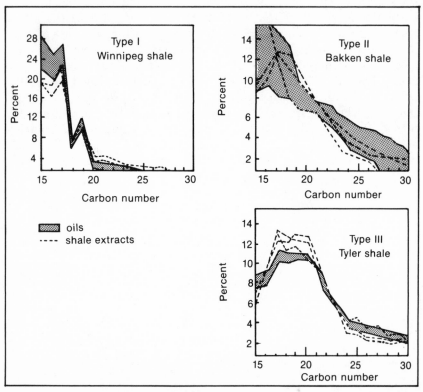

Fig. 11–17 Distribution of C_{15+} paraffins in shale extracts in shale extracts and crude oils *(after Williams)*

6. MIGRATION AND ACCUMULATION OF PETROLEUM

The compaction process. Soft fluffy mud on the sea bottom has a very high porosity—more than 80% water by volume. As it is buried under additional sediments, the water is squeezed out and the clay becomes stiffer. By the time it is buried to 500 m, the porosity has decreased to 30 or 40 percent and additional burial brings about additional compaction of the shales. Numerous studies have been made of the relation between porosity of shale and depth of burial; an average of these is shown in Figure 11–18. It will be noted that a shale buried to a depth of 1,500 m will have a porosity of about 15%. Additional burial has less and less effect on the porosity, so that a depth of burial to 3,000 m (10,000 ft) gives a porosity of about 10 percent.

The reduction of porosity from 10 percent to 5 percent still involves the expulsion of large volumes of water. For every cubic kilometer of sediment, 5×10^7 metric tons of water are removed (300 million barrels).

Primary migration of oil. The physical process of the removal of petroleum from the source bed and its transfer to the reservoir bed is very poorly understood. Most people think it is related to the expulsion of the water. That is, the water moving out of the shales, on its long and possibly tortuous path to the surface, carries small amounts of

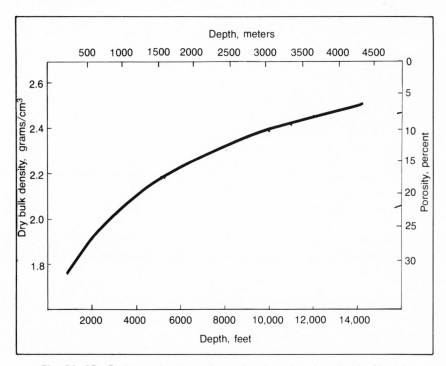

Fig. 11–18 Sediment density and porosity plotted against depth of burial

hydrocarbon which are segregated into oil fields somewhere along the way, possibly by capillary or coalescence effects. The problem is that petroleum-like hydrocarbons are extremely insoluble in water. Gas (methane) is quite soluble; each barrel of water can contain up to 20 or more cubic feet of methane, depending on the temperature and pressure. The light hydrocarbons are soluble to the extent of 1 to 10 parts per million. However, the heavier hydrocarbons are soluble only in parts per billion. The water cannot, therefore, move the oil in solution.

Studies of the motion of oil in reservoir rocks show that unless 20% or more of the pore space is filled with oil, it is kept from moving by capillary forces. Droplets of oil will not move through a porous reservoir, even a very permeable one. The drops hang up at the throats of the pores and cannot be pushed through by even a strong current of water. However, if the porous medium is collapsing, it is possible that some oil might be squeezed out like toothpaste from a tube. It has been suggested (Dickey, 1975) that the oil moves through the pores of the shale in fine thread-like channels. This initial removal of oil from its source rock and its transfer to the reservoir is called *primary migration*, and it is the subject on which additional research is seriously needed.

The problem of the distance of oil and gas migration has been much discussed. The lenticular sands enclosed by the Cherokee shale are full of oil that must have come from the surrounding shales. On the other hand, huge accumulations of oil, like those in Saudi Arabia, Alberta, or Venezuela, have no unusually rich source rocks in the immediate vicinity. The oil must have been collected from a very large volume of ordinary source rocks on the order of several thousand cubic kilometers. The oil must have travelled several hundred kilometers to reach its present position. It follows from this that the lack of obvious source rocks in an area does not mean that oil will not be found.

Segregation of oil. Hydrocarbons are widely dispersed in the source rocks, amounting to only a few hundred parts per million by weight. The process by which they are segregated and become a rich oil field is obviously a very important process which is also poorly understood. It may be that the water squeezed out of shale carries oil through the reservoir rock, in the form of a solution or an extremely fine colloidal emulsion, with oil droplets less than two micrometers in diameter that repel each other by electrostatic charges. These could be screened out of a moving stream of water by a familiar process called capillary filtering. Aircraft pilots in remote areas get their gasoline in drums where it is often contaminated by water. Before pouring it into the tanks, it is customary to wet a chamois skin with gasoline, hang it in a funnel, and pour the contaminated gasoline through it. The chamois skin, being wet with gasoline, will not permit the water droplets to pass through its fine pores and the water will be screened out (Figure 11–19). Conversely, if they had wet the chamois skin with water, water would have gone through the fine-grained filter and oil would have been screened out. Some process similar to this may provide the mechanism for the segregation of oil. If the water containing colloidal or very fine capillary droplets of oil passes through a coarse-grained reservoir rock and then enters a fine-grained rock on its way to the surface, the oil may be screened out at this capillary interface.

Figure 11–20 shows how oil might be trapped in reservoir rocks around the margins of a depositional basin. As the organic muds compact, water is squeezed out of them and makes its way to the surface, mostly following thin silty or sandy beds. If it traverses a permeable but lenticular sandstone, as at A, the oil may be filtered out of the stream of water where it reenters a fine-grained material. Unconformity surfaces often form permeable zones through which the water current will flow. If oil droplets coalesce, they will be filtered out at interfaces between coarse and fine sands. They will collect and saturate any porous reservoir, either above or below the unconformity.

Secondary migration. Once the oil has accumulated in the reservoir, its motions can be better understood because a great deal of study has been given to the flow of oil through porous media. It is reasonable to suppose the oil will collect in the upper part of the sand body where it was prevented from migrating farther, along with the water moving out of the rocks as compaction and consolidation took place. The porous

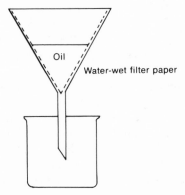

Fig. 11–19 Oil will not pass through water-wet filter paper; it is prevented by capillary pressure

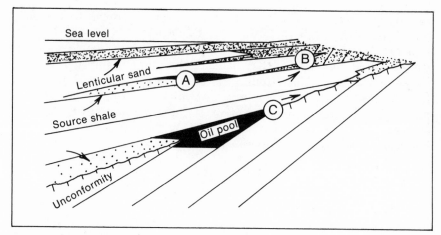

Fig. 11–20 Segregation in wedge belts of porosity *(copyright Gulf Publishing Company, Houston. Used with permission. All rights reserved)*

reservoir rocks are very seldom entirely filed with oil. Usually there is water below the oil in the structurally lower parts of the permeable sand body. The portion of the permeable sand body filled with water is usually called the *water leg* or *aquifer*. The contact between the oil and the water is usually level, just as it would be if there were a glass with oil in the top part and water in the bottom part.

The earth's crust is continually rising and sinking, so the *attitude* (dip) of any reservoir rock does not stay the same throughout geologic time. In sedimentary basins of Tertiary age, the dip usually becomes steeper the deeper the formations because sinking has been continuous and the older beds have had more opportunity to be tilted. In the case of more ancient rocks, such as the Mesozoic and especially the Paleozoic, there may have been a period of mountain building which disturbed the rocks seriously and possibly even tilted them the opposite way.

An example of this is in eastern Oklahoma where the Bartlesville sands of the Tulsa area were originally deposited in a basin whose deepest part lay to the southeast. These are delta-type channel sands. Presumably, the primary migration filled the upper parts of these sands with oil, which collected along the northwest termination of the sands (Figure 11–21). During the Mesozoic period the eastern areas were uplifted and the western areas of Oklahoma subsided, which caused a reversal of the dip. Now the eastern termination of the sand bodies is the higher, and here is where the oil is found. It may be

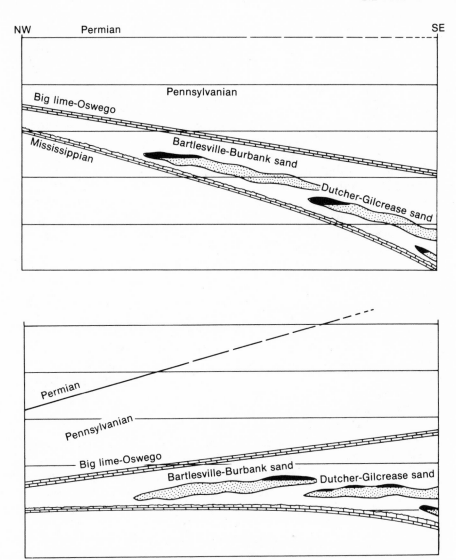

Fig. 11—21 Reversal of original dip in eastern Oklahoma *(after Levorsen)*

supposed that the oil migrated from its original position in the north-west termination of each permeable sand to its present location at the east termination of each permeable sand during the time the tilting of the earth's crust took place.

Oil-stained water sands. Usually the aquifer, down dip from the oil sand, shows no trace of oil in the pores. This is puzzling because it is hard to imagine how the oil could have travelled through the sand without leaving some residual oil behind.

It often happens, especially in the vicinity of an oil field, that the sand does contain oil, although it will produce only water. Apparently the zone once contained oil, which has since moved out, leaving 10 to 20 percent of the pore space still containing residual oil. When this occurs it forms one of the most baffling problems for the production geologist—will the sand produce oil or water? Cores will not tell because they are flushed by drilling mud filtrate. Electric logs are often also difficult to interpret because of the invasion of mud filtrate. Even when a good value for water saturation (S_w) can be obtained from logs, it is still not sure whether the sand will produce oil or water. A high value—75%—may indicate water production. However, at least some of the water is associated with the clays and is immobile. It may all be, in which case the well will produce clean oil even at 75% water saturation. It may be worth the expensive rig time to run a drillstem test or perforate casing and then test. In many, perhaps most cases, the sand will produce only water. It still is worth getting the test so a good pressure measurement and a sample of the water can be obtained. This information will help to decide where to go to find the oil pool. In some cases it may well produce oil or gas. If this happens a few times it will pay for many unsuccessful tests.

Tilted oil-water contacts. It is usually assumed that the oil-water contact is level. In the early development of a pool, step-out wells are drilled to determine the elevation of the oil-water contact with respect to sea level. Then a line is drawn on the structure contour map derived from seismic data, projecting the oil-water contact around the structure. This gives the area of the pool from which preliminary reserve estimates are calculated.

However, it often happens that the oil-water contact is not level. In some giant pools, like Ghawar, Saudi Arabia, the oil-water contact is lower on one side than the other. In Frannie, Wyoming, the oil-water contact is 1,000 ft higher on the northeast than on the southwest (Figure 11–22). In eastern Oklahoma, the oil-water contact is generally tilted to the southwest. In the case of Bradford, Pennsylvania, and southwest Bachaquero, Venezuela, the oil-water contact runs at almost right angles to the structure contour lines.

Consequently, when the water-oil contact is located by the production geologist, he should decide how to project it around the structure. Probably the best way is to know what the tilt is in other pools in the vicinity. It is usually down in the direction of latest tectonic basin

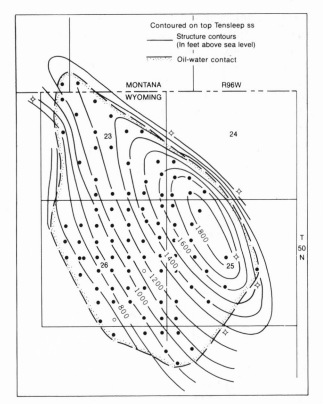

Fig. 11–22 Structure of Frannie field, Wyoming, showing tilt of oil-water contact (after Hubbert)

subsidence—that is, toward the center of the basin. However, in some areas, for example the Big Horn basin of Wyoming, the tilts are in many different directions.

For a while, it was thought that flowing water in the aquifer was the principal cause of tilting (Hubbert, 1953). However, this seems unlikely. Many oil fields in lenticular sands with strong salt water in the aquifer are tilted, and in such cases there could hardly be any flow of the water.

More recently it has been discovered that there is commonly a layer of asphalt at the oil-water contact. At Hawkins, Texas, and Prudhoe Bay, Alaska, the layer is 10 to 30 feet thick. In other cases, it may be much thinner. This layer has been cored in many places. In Saudi Arabia it has been recognized on the electric logs. There is an easily recognizable zone of high saturation of immovable oil immediately above the water (Schlumberger, 1975).

The presence of an asphalt layer (tar mat) can be recognized on resistivity logs. In Figure 11–23, Sand B shows high resistivity on the LLd and LLs logs, but low resistivity on the MicroSFL log. The latter is shallow penetration and shows invasion of drilling mud filtrate. At point b, all three curves show high resistivity, indicating that the drilling mud filtrate was not able to enter the sand. Below it, all three curves show low resistivity, indicating water in the sand.

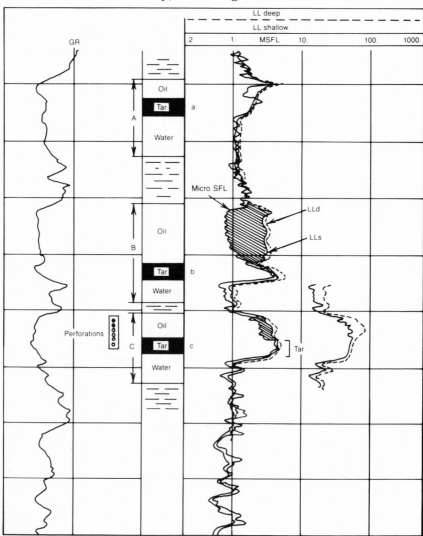

Fig. 11–23 Indication of asphalt layer on electric logs. At points a, b, and c, the shallow penetration curve (MicroSFL) shows no invasion *(after Prins)*

REFERENCES

Bailey, N. J. L., C. R. Evans, and C. W. D. Milner, 1974, Applying geochemistry to the search for oil—examples from western Canada basin: AAPG Bull., v. 58, no. 11, p. 2284–2294.

Baker, D. R., 1962, Organic geochemistry of Cherokee Group in southeastern Kansas and northeastern Oklahoma: AAPG Bull., v. 46, no. 9, p. 1621–1642.

Buthod, Paul, 1962, Properties of crude oils and liquid condensate, *in* Thomas C. Frick, ed., Petroleum production handbook, Reservoir engineering, vol. 2: Soc. Petrol. Eng. of AIME, Dallas, p. 18–23.

Clark, N. J., 1969, Elements of petroleum reservoirs (revised edition): Soc. of Petrol. Eng., Dallas, 250 p.

Dickey, P. A., 1975, Possible primary migration of oil from source rock in oil phase: AAPG Bull., v. 59, no. 2, p. 337–345.

Dickey, P. A., and J. M. Hunt, 1972, Geochemical and hydrogeological methods of prospecting for stratigraphic traps, *in* R. E. King, ed., Stratigraphic oil and gas fields: AAPG Bull., p. 136–167.

Dow, Wallace G., 1974, Application of oil-correlation and source-rock data to exploration in Williston basin: AAPG Bull., v. 58, p. 1253–1262.

————, 1977, Kerogen studies and geological interpretations: Jour. Geochemical Explor., no. 7, p. 79–99.

Evans, C. R., M. A. Rogers, and N. J. L. Bailey, 1971, Evolution and alteration of petroleum in western Canada: Chem. Geology, v. 8, no. 3, p. 147–170.

Hubbert, M. K., 1953, The entrapment of petroleum under hydrodynamic conditions: AAPG Bull., v. 37, p. 1954–2026.

Hobson, G. D., 1954, Some fundamentals of petroleum geology: Oxford, Oxford Univ. Press, 139 p.

Hunt, John M., Origin and migration of hydrocarbons: AAPG Continuing Education Lecture Series.

Hunt, John M., 1979, Petroleum geochemistry and geology, W. H. Freeman, San Francisco, 617 pp.

Levorsen, A. I., 1967, Geology of petroleum, 2nd edition: San Francisco, W. H. Freeman, 724 p.

Mair, Beveridge J., 1964, Hydrocarbons isolated from petroleum: Oil & Gas Jour., Sept. 14, p. 130–134.

McCain, William D., Jr., 1973, Properties of petroleum fluids: Tulsa, PennWell Publishing Co., 325 p.

McIver, R. D., 1967, Composition of kerogen, clue to its role in origin of petroleum: 7th World Petrol. Congress Proc., Mexico, v. 2, p. 25–36.

Milner, C. W. D., M. A. Rogers, and C. R. Evans, 1977, Petroleum transformation in reservoirs: Jour. of Geochemical Explor., no. 7, p. 101–153.

Nelson, W. L., 1958, Petroleum refinery engineering, 4th edition: New York, McGraw Hill.

Phillipi, G. T., 1965, On the depth, time, and mechanism of petroleum generation: Geochim et Cosmochim Acta, v. 29, p. 1021–1049.

Prins, W. J., 1978, Personal communication.

Schlumberger, 1975, Well evaluation conference, Saudi Arabia: Services Techniques, Schlumberger, France, 152 p.

Snowdon, L. R., and R. G. McCrossan, 1973, Identification of petroleum source rocks using hydrocarbon gas and organic content: Geol. Surv. of Canada Pap. 72–36, 12 p.

Tissot, B. P., Califet-Debyser, G. Deroo, and J. L. Dudin, 1971, Origin and evolution of hydrocarbons in early Toarcian shales, Paris basin, France: AAPG Bull., v. 55, p. 2177–2193.

Tissot, B. P. and D. H. Welte, 1978, Petroleum formation and occurrence, Springer-Verlag, 538 pp.

Trask, P. D., and H. W. Patnode, 1942, Source beds of petroleum: AAPG Tulsa, 566 p.

Williams, J. A., 1974, Characterization of oil types in the Williston Basin: AAPG Bull., v. 58, no. 7, 1243–1252.

12

Oil-Field Waters

Normally, oil or gas occupy only a small fraction of the total pore volume in the subsurface; the rest of the space contains water. In fact, all of the upper part of the earth's crust, deeper than a thin layer near the surface, is saturated with water.

Sometimes the water associated with the oil is fresh, but usually it is salty. In general, the deeper and older a rock is, the saltier the water it contains. The origin of oil-field brines is not well understood. Presumably, they are the fossil remains of the sea water in which the sediments were deposited. However, they are always quite different from sea water chemically. Inasmuch as oil must form and migrate in this watery environment, the nature and behavior of subsurface water can tell us much about the origin and accumulation of oil.

1. CHEMICAL ANALYSIS OF OIL-FIELD WATERS

Purpose of chemical analyses. Each water-bearing formation contains water that is different chemically from the waters in the other horizons. Consequently, if an oil or gas well suddenly starts to produce water, a chemical analysis will tell from which horizon it is coming. Identification of the source horizon is obviously the first step in planning to repair the well, and it is the most common reason for making a water analysis. However, the analysis will be recorded as if the water were coming from the oil productive horizon, even though the only reason for making the analysis was that it was not. Very many water analyses in company files are therefore not representative of the horizon with which they are associated.

Quantitative electric log interpretation requires a knowledge of the resistivity of the interstitial water in the formation, R_w. Although this value is often estimated from the self-potential log, the only sure way of ascertaining it is to determine the resistivity of a sample of water known to have been produced from the formation. Such samples of water are also often analyzed chemically. If they were carefully collected, they form the best source of oil-field water information.

Water recovered from drillstem tests is often analyzed, but such analyses may be misleading. Filtrate from the drilling mud penetrates the producing formation, often in large amounts. The water recovered in the drillstem or formation interval tester, at least at first, is contaminated by drilling mud filtrate. The later water may be representative. The drilling mud filtrate should be analyzed also in order to compare it with the water recovered in the test tool. Chromium is present if chrome lignosulfonate mud is used. Nitrates or tritiated water may be added to the drilling mud to serve as tracers.

Analyses may also be made for geological interpretations, based on the dissolved inorganic salts and on dissolved organic matter such as hydrocarbons and other related compounds. Such analyses ought to be made much more commonly than they are.

2. CHEMICAL COMPOSITION OF OIL-FIELD WATERS

Oil-field waters are generally analyzed for 5 major constituents, Ca^{++}, Mg^{++}, Cl^-, SO_4^{--}, and HCO_3^-. A sixth, sodium, is rather difficult to analyze for, so its value is usually determined by difference. New chemical methods are frequently introduced, but most laboratories devoted to oil-field brine analysis use methods standardized by the American Petroleum Institute. Analytical methods will not be discussed.

It is basic to chemistry that each element or radical has its own combining weight; for example, 22.997 grams of sodium (Na^+) will react with 35.457 grams of chloride (Cl^-), no more and no less. The sum of equivalent weights of the positive ions (cations) must equal the sum of the equivalent weights of the negative ions (anions). The equivalent weight of an ion is the actual weight, usually expressed in milligrams per liter, multiplied by the valence and divided by the atomic weight or molecular weight of the ion.

We usually assume that Cl^-, SO_4^{--}, and HCO_3^- constitute practically all (99 percent or more) of the acid radicals. This is a pretty valid assumption; the minor anions such as iodide, bromide, carbonate, etc., are usually rather small in quantity. Ammonium is seldom analyzed for, but is often present. We assume that the positive ions are all Na^+, Ca^{++}, and Mg^{++}, which is often not true at all. Potassium, strontium, and barium may be present in considerable amounts in waters. Recently, they have more commonly been determined separately.

Iodine and bromine are valuable elements that are sometimes present in commercial quantities in oil-field waters. Amoco Production Company in 1976 started producing iodine from water in the Pennsylvanian Morrow sandstone in western Oklahoma. In evaluating a sand containing brine, it is important to determine whether the volume of the aquifer is sufficiently large to provide a driving force for the water.

Iron is important if the water is to be reinjected. Dissolved oxygen is usually absent in produced water. It should be kept out if the water is reinjected because it attacks the pipes and causes a flocculent precipitate.

The reported equivalent weights of the cations (bases) and anions (acids) usually equal each other. This does not mean that the analyses were good; on the contrary, it means that Na^+ (or $Na^+ + K^+$) was determined by difference. There is no way of telling how good the analyses were because all of the errors are taken up in the difference.

Some students (Rittenhouse, 1969) have paid much attention to minor constituents and have drawn geological conclusions based on their patterns. Others (Graf et al., 1965) have made interpretations based on the relative amounts of isotopes of different elements. All of these are interesting and may help in determining the origin of subsurface waters.

Analytical results have usually been expressed as parts per million by weight (ppm). Inasmuch as most chemical methods are volumetric, that is, they determine the amounts of constituents in a measured volume of water, it is preferable to express results in milligrams per liter (mg/l):

$$mg/l \; = \; \frac{ppm}{density}$$

It is advantageous to express the analyses in milliequivalents per liter because equal numbers represent equal combining weights

$$meq/l \; = \; \frac{mg/l \times valence}{molecular\ weight}$$

The total milliequivalents of cations must equal the total milliequivalents of anions. If the total milliequivalents of cations is considered 100 percent, and the total milliequivalents of anions is considered 100 percent, then the value of each ion can be calculated as milliequivalent percent. It is then possible to take into account the relative amounts of the different ions.

TABLE 12–I
CALCULATIONS TO CONVERT MG/L TO MILLIEQUIVALENTS
AND MILLIEQUIVALENT PERCENT

Element	Factor valence mole. weight	Water 1 mg/l	meq	meq%	Water 2 mg/l	meq	meq%
Na^+	0.0435	44,100	1,918	74	3,040	132.2	99
Ca^{++}	0.0499	11,000	549	20	21	1.0	0.5
Mg^{++}	0.0823	1,500	123	6	7	0.6	0.5
Total cations			2,590	100		133.8	100
Cl^-	0.0282	91,800	2,589	100	3,240	91.4	68.3
SO_4^{--}	0.0208	none	—		407	8.5	6.3
HCO_3^-	0.0164	34	0.5		2,065	33.9	25.3
Total anions			2,590	100		133.8	99.9

The data is plotted in the form of bar graphs in Figure 12–1.

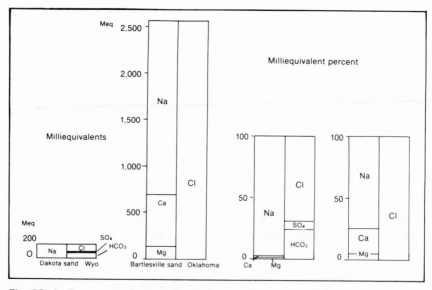

Fig. 12–1 Bar graphs showing chemical composition of two oil-field waters that are quite different. The Dakota sand water is comparatively dilute, contains little Ca or Mg, and considerable SO_4 and HCO_3. The Bartlesville sand water is very concentrated. It contains considerable Ca and Mg but no SO_4 or HCO_3.

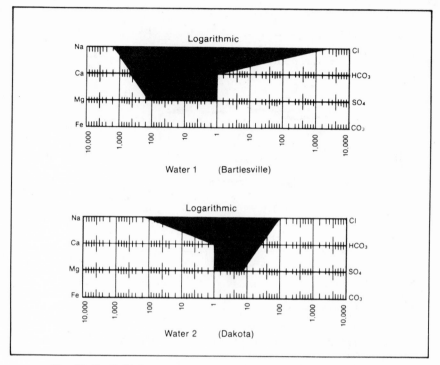

Fig. 12–2 Stiff diagrams used to show water composition on maps.

Table 12–I shows the analyses of two oil-field waters and the calculations to convert the milligrams per liter to milliequivalent percent. Water 1 is typical oil-field brine, probably connate in origin, from the Bartlesville sand of Oklahoma. Water 2 is a typical Rocky Mountain water from the Dakota sand in Wyoming. It is probably of meteoric origin.

For geological purposes, it is often advantageous to show a lot of water analyses on a map. This is facilitated by plotting the chemical composition graphically. The most popular method is the Stiff diagram, shown in Figure 12–2.

The cations are plotted out from the center to the left on three or four lines. The anions are plotted to the right. Usually, milliequivalents are plotted. A logarithmic scale is convenient. There is no zero in a logarithmic scale, so any values less than one meq are plotted as one. The points are connected by straight lines, and the polygon so formed is filled in solid.

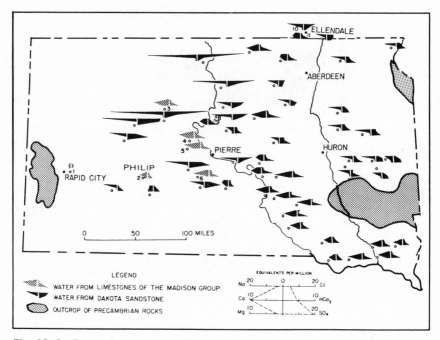

Fig. 12–3 Types of water in the Dakota sandstone and equivalent strata, and limestone of the Madison group in South Dakota and adjacent North Dakota *(after Swenson, courtesy the Geological Society of America)*

An example of how to use the Stiff diagram is shown in Figure 12–3. The water in eastern South Dakota in the Dakota sandstone is different chemically from the water in western South Dakota. It therefore cannot be part of the same hydraulic system (Swenson, 1968).

The Dakota sandstone outcrops on the east flank of the Black Hills at an elevation of 3,200 ft and in the southeastern part of the State at an elevation of 1,200 ft. Wells drilled to the Dakota in the central part of the state flowed water to the surface. It has long been supposed that the water enters at the outcrop in the west and issues on the east (Figure 12–4).

However, the Dakota sandstone is notably lenticular in the western part of the state, although it becomes permeable in the eastern part. It is now believed that the water enters not the Dakota, but the thick and cavernous Madison limestones on the east flank of the Black Hills. This limestone subcrops on the unconformity at the base of the Cretaceous where it is overlain by the Dakota. Along the line of subcrop it feeds a flow of water into the Dakota (Figure 12–5).

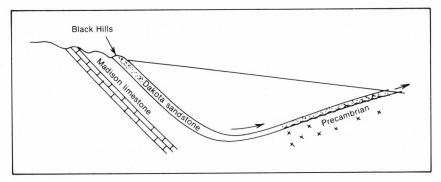

Fig. 12—4 Original concept of Dakota sandstone artesian flow. Water enters at Dakota outcrop on the flank of the Black Hills and discharges in eastern South Dakota

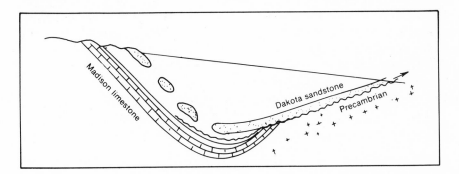

Fig. 12—5 Later concept of Dakota artesian flow. Dakota in western South Dakota is lenticular. Water enters Madison limestone at its outcrop, discharging into the Dakota along its subcrop in central South Dakota

Madison water everywhere is high in bicarbonate, sulfate, and calcium. Dakota water in the western parts of the state is rich in sodium and chloride and low in bicarbonate. In the eastern part it resembles Madison water.

If the Dakota were a prospective oil zone in South Dakota, as it is in Wyoming west of the Black Hills, the map (Figure 12–3) would be exceedingly useful. All of the eastern part of the state is probably flushed and has poor chances of finding an accumulation. The water composition shows that the western, especially the northwestern, part of the state has been protected from flushing and may be considered prospective.

Genetic classification. Based on their history, subsurface waters may be divided into three classes: meteoric, connate, and juvenile.

Meteoric waters. Meteoric waters are those which presently form part of the hydrologic cycle or formed part of it in the recent past, geologically speaking. They include the water in the oceans, the evaporated water in the atmosphere, water in rain and snow, surface water, and ground water in motion. Ground water, under certain geological circumstances, may be circulating through permeable strata in great depths—3,000 meters or more. Chemically, meteoric waters are characterized by low concentration of dissolved solids, usually less than 10,000 mg/l. They usually contain considerable amounts of bicarbonate ion.

Connate waters. Connate waters have been buried in closed hydraulic systems and have not formed part of the hydrologic cycle for considerable amounts of geologic time (White, 1957). The word "connate" means "born with," and it was introduced by Lane (1908) with the idea that they are remains of the original sea in which the sediments were deposited. This is only approximately true because compaction of rocks involves the expulsion of water from the pores and some lateral migration. Chemically, connate waters are usually quite salty, containing 20,000 to 250,000 mg/l total dissolved solids. They are high in chloride, but bicarbonate is scarce or absent. Sulfate is also usually scarce or absent.

The immovable water held by capillary forces in the finer pores of reservoir rock is often called "connate water." This is not a good use of the word because "connate" implies an origin. "Interstitial" is better.

When sediments are buried very deeply, the temperature and pressure increase and recrystallization of the minerals takes place. This chemical process results in the loss of water than originally formed parts of the lattice structure of the hydrous silicates. Recrystallization also results in further loss of porosity. This water must find its way to the surface, although the metamorphosing sediments have long since lost their permeability. In order to escape, it must rupture the rocks by a process similar to "lost circulation" in drilling. As it moves to zones of lower temperature and pressure, silica and other minerals are precipitated, forming veins of quartz. Water in fluid inclusions in quartz crystals in veins is usually salty. It is probable that many, if not most, hydrothermal veins were formed by connate water expelled from sediments by the process of metamorphism.

Juvenile water. Juvenile waters are those which have ascended from the mantle of the earth (the lower part of the crust) and have never taken part in the hydrologic cycle. Some of the water vapor emitted by volcanoes may well have a deep origin. However, juvenile waters are hard to identify with assurance. Much of the water escaping from volcanoes may be ground water which was vaporized by the hot lava. Water from hot springs and geysers is meteoric water which was heated by high-temperature rocks at depth.

Chemical classifications. In the U.S., water analyses have generally been classified according to a system proposed by Palmer in 1911. First, the strong bases (Na^+, K^+) are combined with the strong acids (Cl^-, $SO_4^=$) to form *primary salinity*. Then the strong bases are combined with the weak acid (HCO_3^-) to form *primary alkalinity*. Next, the weak bases (Ca^{++}, Mg^{++}) are combined with the strong acids (Cl^-, $SO_4^=$) to form *secondary salinity*. Finally, the weak acids are combined with the weak bases to form *secondary alkalinity*. In fact, such combinations of ions do not take place, and it is hard to relate the different water classes to their history and geology. This classification is not very useful.

A classification was proposed by Schoeller, a French hydrologist. It is interesting from a chemical point of view but has not been used generally in the U.S.

The Russian hydrologist V. A. Sulin in 1946 proposed a classification which is claimed to have genetic significance. There are four major classes of waters which are divided according to the following table:

TABLE 12–II

MAJOR CLASSES OF WATER BY SULIN CLASSIFICATION

	Types of water (V. A. Sulin)	Ratios of concentrations, expressed as milliequivalent percent		
		$\dfrac{Na}{Cl}$	$\dfrac{Na\text{-}Cl}{SO_4}$	$\dfrac{Cl\text{-}Na}{Mg}$
Meteoric	Sulfate sodium	>1	<1	<0
	Bicarbonate sodium	>1	>1	<0
Connate	Chloride magnesium	<1	<0	<1
	Chloride calcium	<1	<0	>1

The classification is based on the generalization that meteoric waters contain $SO_4^=$ and HCO_3^- but very little Ca^{++} and Mg^{++}. Therefore, the cations are nearly all Na^+, and the milliequivalent percent of the Na^+ must nearly equal the sum of the milliequivalent percent of the anions. Thus,

$$Na^+ = Cl^- + SO_4^{--} + HCO_3^- \quad \text{approximately}$$
$$Na^+ - Cl^- = SO_4^{--} + HCO_3^- \quad \text{approximately}$$

The ratio Na/Cl is more than one. If the ratio $(Na - Cl)/SO_4$ is less than one, the waters are said to belong to the sulfate-sodium class; if it is greater than one, to the bicarbonate-sodium class. Both classes are said to characterize geological zones of large water interchange, that is, artesian circulation.

The typical connate waters contain practically no sulfate or bicarbonate, so almost the only anion is chloride. Consequently,

$$Cl^- = Na^+ + Ca^{++} + Mg^{++} \quad \text{approximately}$$
$$Cl^- - Na^+ = Ca^{++} + Mg^{++} \quad \text{approximately}$$

The ratio of Na/Cl is less than one. If the ratio $Cl - Na/Mg$ is less than one, the water belongs to the chloride-magnesium class; if greater than one, it is of the chloride-calcium class. These waters are said to be

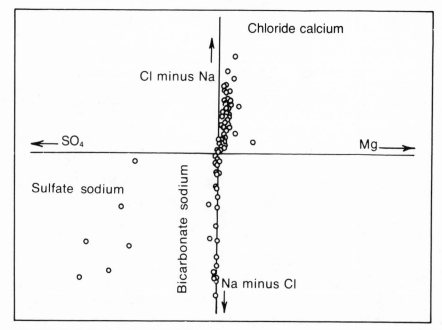

Fig. 12—6 Sulin classification of oil-field waters; values are in milliequivalent percent

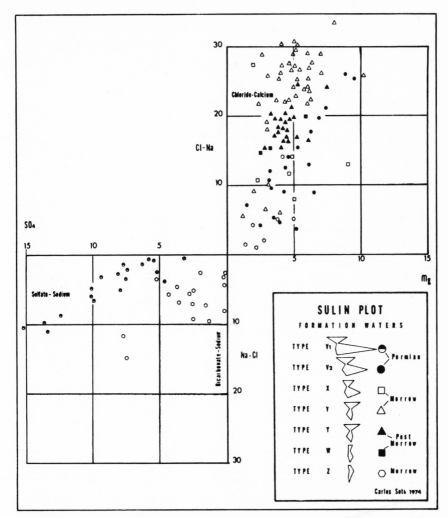

Fig. 12—7 Oil- and gas-field waters of the Texas-Oklahoma Panhandle areas plotted on a Sulin diagram. The Morrow waters are of two types: chloride calcium and bicarbonate sodium. The Permian waters are either chloride calcium or sulfate sodium. The post-Morrow Pennsylvanian waters are all chloride calcium type *(after Dickey and Soto)*

characteristic of zones of small water interchange, that is, stagnant (Kartsev, 1963).

The waters may be grouped by plotting them graphically. The difference Cl − Na (expressed as milliequivalent percent) is plotted up if it is positive and down if it is negative. Then, if the difference is positive, the Mg is plotted to the right of the origin. If it is negative, the SO$_4$ is

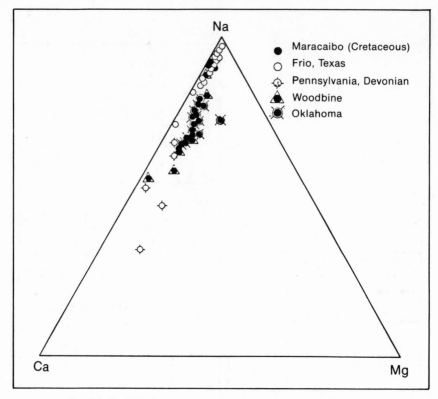

Fig. 12—8   Relative amounts of cations in typical oil-field brines

plotted to the left of the origin. Figure 12–6 is a plot of a number of waters from a variety of formations. Figure 12–7 is a group of waters from the Pennsylvanian and Permian of the western Anadarko basin, Oklahoma, Texas, and Kansas (Dickey and Soto, 1974).

Additional discussion of water classifications may be found in Collins, 1975.

3.   CHEMICAL COMPOSITION OF SUBSURFACE WATER

Chemical composition of connate water.   It was recognized by de Sitter in 1947 and emphasized by Sulin and von Engelhardt (1961) that connate subsurface brines are all similar in chemical composition. The anions are practically all chloride, sulfate is usually scarce or absent, and bicarbonate is always scarce or absent. The cations consist of sodium, calcium, and magnesium, in that order. Calcium is usually 3 to 5 times more than magnesium, expressed as milliequivalents. This

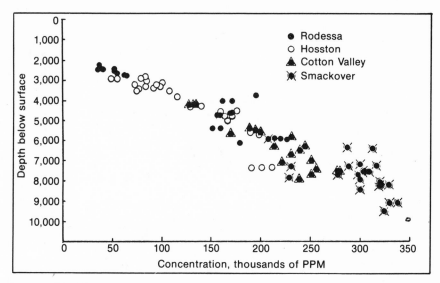

Fig. 12–9 Increase in concentration with depth, southern Arkansas

composition is surprisingly similar in all connate brines, regardless of the mineralogy of the enclosing rocks or whether the depositional environment was in fresh or salt water. Figure 12–8 is a triangular diagram showing the relative amounts of cations in oil-field brines found in rocks of widely differing age.

Connate brines normally increase in concentration with depth. This increase is usually nearly linear and ranges from 25,000 to 100,000 mg/l per thousand ft (80 to 300 mg/l per meter). Figure 12–9 to 12–11 are plots of concentration versus depth for the produced brines of several oil fields (Dickey, 1960).

Connate brines of this composition obviously result from the compaction and lithification of their enclosing sediments. The changes from sea water are drastic: bicarbonate and sulfate are eliminated, the calcium-magnesium ratio is reversed, and the concentration increased.

The chemical processes bringing about this alteration are not understood. It is widely held that the concentration takes place by a process of reverse osmosis (Kharaka and Berry, 1973). As the rocks compact, the waters are forced from sands through shales. The semipermeable character of the shales keeps the salts from reentering them, so they are concentrated in the sands. This explanation does not account for the change in chemical composition nor the linear increase in concentration with depth. It is refuted by the fact that pore water in

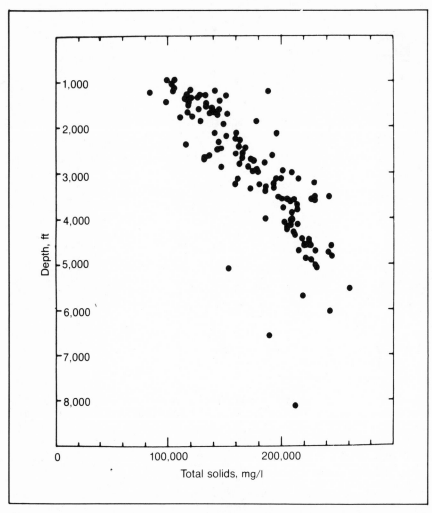

Fig. 12–10 Increase in concentration with depth, sands of Pennsylvanian Cherokee of Oklahoma

shale has quite a different composition from water in the adjacent sands.

Other chemical processes have been suggested. Among these are molecular diffusion of salt from nearby evaporites, thermal effects, and segregation of the heavier salts by gravitational settling. It has been suggested that the sulfate is removed by anaerobic bacteria, but sulfate is abundant in shale waters where the bacteria should have been

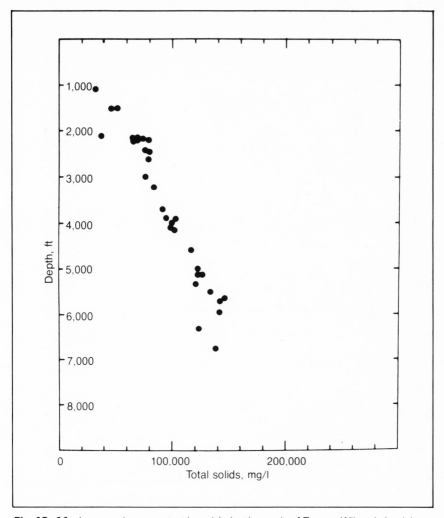

Fig. 12—11 Increase in concentration with depth, sands of Eocene Wilcox in Louisiana

even more active than in sands. Another suggestion is that the magnesium is depleted and calcium is added by dolomitization. This is not plausible either because the calcium-magnesium ratio is about the same regardless of whether the carbonate in the rock is calcite or dolomite.

Another possibility is that as shales compact, more and more of the interstitial water becomes structured. Just as when water freezes, this

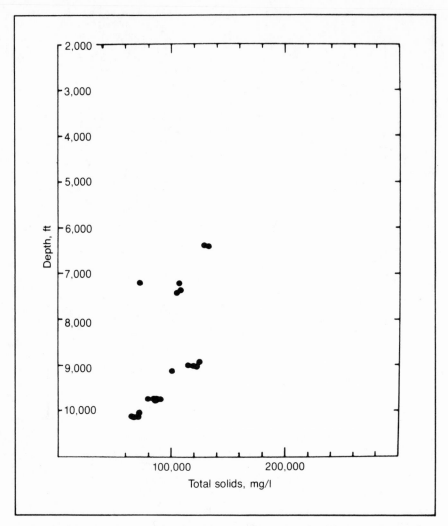

Fig. 12–12 Decrease in concentration with depth, sands of Miocene in Louisiana

structuring expels the salts, concentrating the water in the sands and depleting the water in the shales. Sulfate and bicarbonate are abundant in shale water, and these ions appear to have been held back in the shales because of their ionic size and potential.

Waters from the deep, undercompacted, abnormally pressured zones of the Gulf Coast are less concentrated than the waters in normally pressured zones at equivalent depths. Apparently, whatever

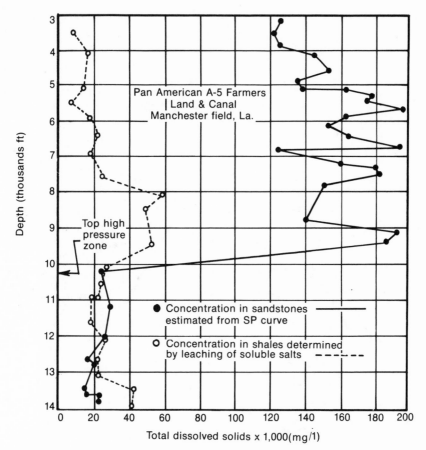

Fig. 12–13 Changes in concentration of interstitial water with depth in sandstones and shales. Total dissolved solids concentration in normally pressured shale water is much less than the adjacent sand, whereas similar concentrations are found in abnormally pressured shale and sandstone waters *(after Schmidt)*

arrested the process of compaction and lithification also arrested the process which concentrated the brines. Concentrations as low as 4,000 mg/l have been reported. This is much less than sea water, although the depositional environment of the abnormally pressured zones is offshore marine. Figure 12–12 shows the change of concentration with depth of waters in Romere Pass field, Louisiana.

Figure 12–13 shows the normal increase in concentration of sand water with depth in Manchester field, Louisiana. The concentration of the water in the sands increases to about 150,000 mg/l at 14,000 ft (4,270 m). Below this depth, the sands are overpressured and the con-

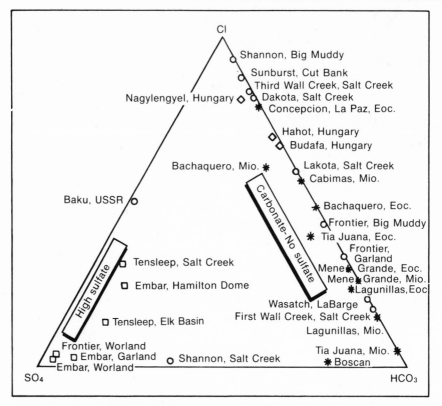

Fig. 12–14 Triangular plot of relative amounts of anions in oil-field waters of meteoric (bicarbonate sodium and sulfate sodium) type

centration of the waters is only about 20,000 mg/l. The shale water at any depth has only about one-third the concentration of the sand water.

The tendency for sands in deep, high-pressure shaly sections to have water of abnormally low salinity is widespread. This tendency must be borne in mind when using electric logs to evaluate potential oil or gas sands. The unusually high value of R_w makes the sand look as if it contained gas, whereas it contains only water. Some very expensive mistakes have been made in offshore Louisiana when wells were completed as gas wells but produced only water.

Chemical composition of meteoric water. Meteoric water enters permeable formations at their outcrops. In plains areas, the relatively fresh meteoric water is unable to displace the heavy connate brines from the deeper parts of the basin, and the transition zone from fresh to salty water usually occurs between 300 and 1,000 ft (100 to 300 m). In mountainous areas, on the other hand, the outcrop on one side of the basin may be much higher than that on the other. When this is the case, meteoric water is able to displace the connate water from the deeper parts of the basin. Usually the displacement is not complete because faults and pinchouts form barriers to fluid flow. Parts of the area will contain meteoric water, parts connate water, and parts mixtures of the two.

Meteoric water is much less salty than connate water, and it has a wider range of chemical composition. Nearly all meteoric waters contain substantial amounts of bicarbonate. Sulfate is commonly absent (bicarbonate-sodium class of Sulin) but often quite abundant (sulfate-sodium class of Sulin). There is no systematic increase in concentration with depth. Concentration is usually less than 10,000 mg/l. Calcium and magnesium are often very scarce in meteoric waters.

Figure 12–14 shows the relative amounts of anions in a number of meteoric waters. Among those from the Rocky Mountains, the Tensleep and Embar (Permian) waters contain sulfate, while the Cretaceous waters do not, even on the same structures (Salt Creek, Wyoming).

The processes which determine the chemical composition of meteoric water have never been adequately studied. It has been assumed that the slowly percolating water dissolves salts from the rocks through which it passes; the farther it goes, the more concentrated it becomes. It also seems likely that the chemical character of the water is determined by the composition of the water in the soils of the intake area. This, in turn, is determined more by the climate than by the mineral composition of the rock. For example, soils in a cold, wet climate are podzols, characterized by a low pH and low Ca and Mg, while soils in a desert climate are characterized by a high pH and abundant Ca and Mg.

In any aquifer where hydrodynamic flow occurs, there will be zones of mixing where the connate water has not been completely flushed out. In these zones, the water concentration will be higher than normal. Also, the chances for finding oil are much better here because they have been protected from flushing.

Figure 12–15 is a diagram of the hydraulic systems commonly found in the Rocky Mountain basins. The Paleozoic rocks, of which the Tensleep sandstone is an example, often have widespread continuity of permeability. They characteristically outcrop high on the mountains and also at low elevations. They are therefore usually thoroughly

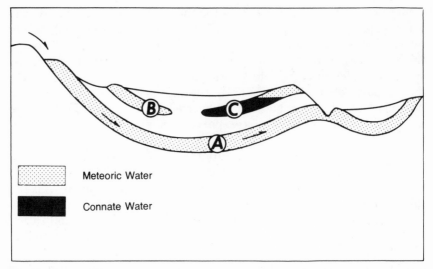

Fig. 12–15 Hydraulic systems in Rocky-Mountain basins *(after Dickey, courtesy Gulf Publishing)*

flushed and contain water of very low concentration. Oil occurs on large anticlines. The sands in the Cretaceous are much more lenticular and outcrop at lower elevations. Consequently, they are not as thoroughly flushed, and the water compositions reflect a wide variety of mixtures of connate and meteoric water.

Table 12–III gives the analysis of six waters from the Denver basin. Five of them are typical meteoric waters with total dissolved solids less than 10,000 mg/l. They contain abundant bicarbonate and some sulfate. Magnesium and calcium are generally low, sometimes absent. However, water B is strikingly different, containing much more dissolved solids. It resembles connate water, except that it contains considerable sulfate and some bicarbonate and it is probably a mixture.

Probably the best way to prospect for oil in country where hydrodynamic flow is common is to look for zones where the total dissolved solids are higher than normal. Increased concentration will show protection from flushing. Chemical analyses are most desirable, but these can be supplemented and extended by estimates of water resistivity from the SP log.

Figure 12–16 shows an area in the eastern Powder River basin where the Dakota sand is productive in an N-S trend of oil pools, probably an old channel. A line of oil fields—Miller Creek, Kummerfeld, Donkey Creek, and Coyote Creek—occupies a narrow sandstone

TABLE 12–III

REPRESENTATIVE J SAND WATERS IN THE NEBRASKA PORTION OF THE
DENVER-JULESBURG BASIN (FROM CRAWFORD)

	A	B	C	D	E	F
Na + K	3,026	26,905	3,428	3,297	2,729	1,672
Ca	33	649	3	240	0	17
Mg	5	120	0	32	0	5
SO_4	437	5,200	265	946	201	876
Cl	2,900	39,000	3,720	4,025	2,621	750
CO_3	0	0	0	0	232	96
HCO_3	2,510	476	2,370	1,513	2,006	1,915
TS	7,686	72,226	8,585	9,285	6,771	4,360
R_w	0.90	0.13	0.80	0.72	1.03	1.50

A—DST sample from wildcat in Sec. 15–12N–52W JL 5263–5268 feet; recovered 1,380 feet water.

B—Produced water at Kugler (Sec. 5–14N–48W) at 4527–4544 feet; producing oilwell K, 115 ppm; Li, 2 ppm.

C—Produced water at a wildcat in Sec. 7–16N–56W at 6712–6720 feet producing oil well.

D—Produced water at Vowers (Sec. 33–17N–54W) at 5894–5899 feet; producing oil well.

E—Produced water at Edwards (Sec. 22–18N–56W) from unit battery; producing oil wells.

F—DST at a wildcat in Sec. 33–23N–55W at 4877–4882 feet; recovered 410 feet water.

body striking north; this sandstone is generally regarded as a channel or barrier-island deposit. Oil fields have been protected from flushing by meteoric water that enters at the outcrop a few km east because there is an impermeable shaly zone directly east of oil fields. Oil is found only in unflushed areas, which are characterized by low resistivity. The water in the flushed area has a resistivity of one ohmmeter, while that in the protected channel is much saltier with a resistivity of 0.2 ohmmeters. The fresh water has gotten past the channel in one or two places so that farther downdip the sand contains fresh water again.

The presence of the barrier is also indicated by the pressure; the potentiometric surface of the fluids in the oil fields is 200 psi (1,400 kPa) less than the potential near the outcrop.

In areas where throughgoing flows of water have occurred, oil is found only where it has been protected from flushing. It was obvious in the case of the Dakota sand pools just described that the resistivity of the water in the flushed areas is much higher than in the unflushed areas where the oil is.

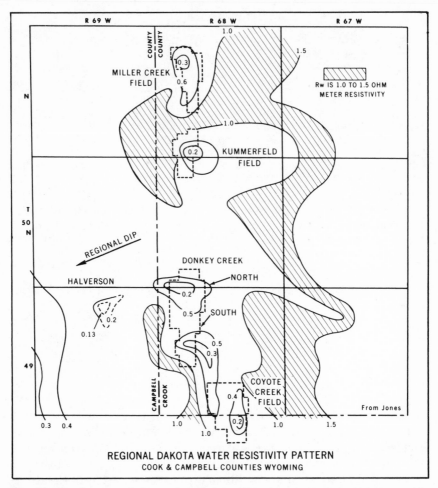

Fig. 12—16 An area of the eastern Powder River basin where the Dakota sand is productive in a N-S trend of oil pools, probably an old channel *(after Jones)*

It has not generally been recognized, although this may be of the greatest importance, that the oil or gas in a tall anticline or faulted structure in a flushed area has also been protected from flushing. In that case the interstitial (connate) water up in the oil or gas reservoir will be chemically very different from the edgewater downdip. It may be much saltier.

The oil or gas content of such a reservoir is calculated from resistivity curves according to the formula:

$$S_w^2 = \frac{FR_w}{R_t} \qquad (5\text{--}10)$$

The value for R_w is often measured on samples of water produced from downdip wells that encountered the oil-water contact. If the true interstitial water is much saltier than the downdip water, its R_w will be much lower. If the R_w value of the aquifer water is put in the formula, the oil or gas volume will be underestimated.

Suppose, as in the case of the Dakota sands, the produced water in the area had an R_w of 1.0 while the interstitial (connate) water had an R_w of 0.2. If F was 16 and R_t was 25 ohms, then an R_w value of 1.0 would give a water saturation of 80 percent, while an R_w value of 0.2 would give a water saturation of 35%. If the inapplicable value for R_w had been used, the sand would have been condemned as water bearing.

This situation has been suggested in a few cases, including the Taglu field of Canada (Wai, 1975), the Dauleb field in Tunisia (Coustau, 1977), and the Bombay High in India.

The Gidgealpa and Moomba gas fields occur near the southwest end of the Cooper basin of Australia (Youngs, 1975). There is evidence of flushing throughout the area—the water increasing in total solids from 3,000 to 15,000 ppm (Figure 12–17). The gas fields are found in areas of higher salinity where faults appear to have protected the formation partly from flushing. The chemical composition of the higher-salinity waters shows more chloride, calcium, and magnesium and less bicarbonate. However, considering the Permian age of the Gidgealpa Formation, the higher-concentration waters are still pretty dilute and are probably mixed with some meteoric water.

Chemical composition should always be used in conjunction with pressure data to work out patterns of fluid flow on the subsurface. Inasmuch as we nearly always have SP logs for every well, resistivity data is apt to be the most complete and available.

The key to finding oil in fresh-water country is to draw maps on water resistivity or concentration. The more concentrated water will indicate zones of protection for flushing, either by a fault or stratigraphic barrier.

The typical chemical composition of meteoric water does not necessarily mean that it is flowing now. In certain localities (Eocene of Lake Maracaibo, Morrow of Anadarko Basin) the waters are definitely not flowing now, but they have the typical composition of meteoric water. Apparently they were flowing during some past geologic time, probably when an unconformity, now buried, was exposed at the surface.

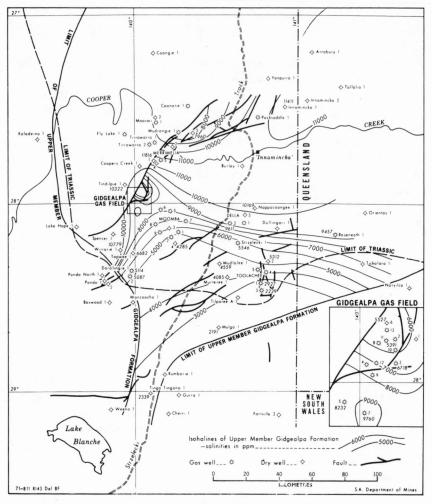

Fig. 12–17 Isohaline map of middle-lower members, Gidgealpa formation (after Youngs)

4. DISSOLVED ORGANIC COMPOUNDS IN FORMATION WATERS

Dissolved gases. In the 1940s, the Humble (now Exxon) research group made an elaborate study of dissolved hydrocarbons in subsurface waters (Buckley et al., 1958). In East Texas, Mississippi, and other places, they took drillstem tests of many important oil and gas sands where they were found to be water-bearing. A sampler was run to the

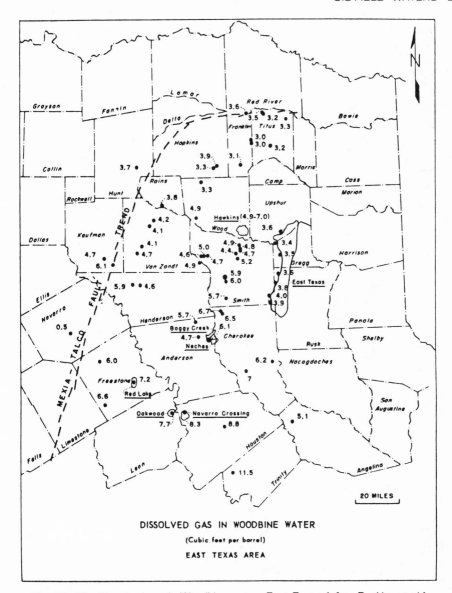

Fig. 12–18 Dissolved gas in Woodbine water, East Texas *(after Buckley et al.)*

bottom of the hole, and a sample of the formation water was recovered at its original pressure. The container was taken to the laboratory where it was opened slowly and the pressure when the first gas started to come off was noted. Then the amount and kind of dissolved gases were determined.

They found that a large amount of methane was dissolved in the water in most of the sands which they sampled. Each sand has its own hydrocarbon distribution pattern which differs from others in the same area. Figure 12–18 shows the content of dissolved gas (mostly methane) in the Woodbine sand of the east Texas basin. The values range rather smoothly from about 3 cu ft of gas per bbl of water in the north to 11 cu ft in the south. At the time the study was being made, water-injection wells were being drilled only one-half mile from the edge of the great East Texas field. Very surprisingly, the gas content of the water was not at all enriched by diffusion from the nearby field; the gas content close to the oil field was even lower than the regional values for the area. A few other fields did show some enrichment close to the oil-water contact.

Gas diffuses through water-saturated rock extremely slowly. Field data are supported by laboratory and theoretical data on the rate of diffusion of gas in water in porous media. Antonov (1963) showed that appreciable increase in gas content occurred 1 km from the edge of the gas pool only after 50 million years (curve 4, Figure 12–19). Kawai

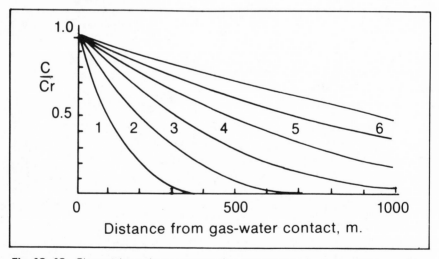

Fig. 12–19 Changes in methane content of water with time at various distances from water-gas or water-oil contact where C = concentration of dissolved gas at gas-water contact, and Cr = concentration of dissolved gas at distance from gas-water contact (after Antonov)

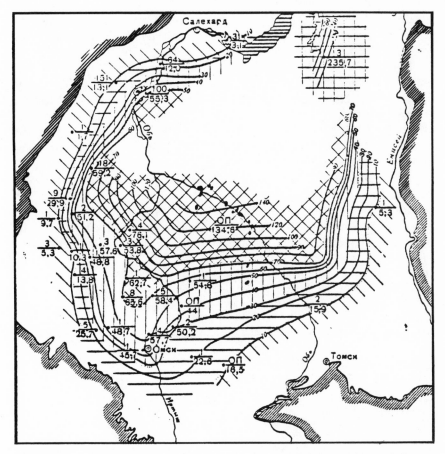

Fig. 12—20 Map of gas saturation and escaping pressure of dissolved gases in formation waters of the Neocomian hydrogeological complex of the West Siberian basin. Contours show total escaping pressure of dissolved gases in atmospheres *(after Zor'kin)*

(1963) made similar calculations, showing that even after 8 million years the concentration of methane 1 km from the gas field would be only about 1 percent of saturation.

This information on slowness of diffusion also has an important bearing on gas and oil content of surface soil samples. It is obvious that methane diffuses through rocks so slowly that anomalously high gas concentrations vertically above fields cannot be expected.

The content of methane in water cannot be used as an indicator of proximity to a gas field. The presence of gas in formation water throughout a large area in the basin may, however, be used as an indicator that gas or oil fields may be present. Conversely, the absence of dissolved gas suggests that no gas or oil will be found at that horizon for a long distance from the sampled well.

Figure 12–20 is a map of the West Siberian basin showing the gas content of the subsurface waters of the Neocomian strata. The huge Taz and Urengoy gas fields were found, after this map was published, in the northeast part of the basin where the highest gas content was in the water (235.7 atm escaping pressure). The big oil fields are in the central part of the basin along the Ob´ River where the escaping pressure exceeds 100 atmospheres.

Other organic compounds in subsurface waters. Hydrocarbons heavier than methane are very insoluble in oil-field waters except for the aromatics: benzene and toluene. The Gulf research people (Zarella et al., 1967) found that benzene is commonly dissolved in oil-field brines. It is much easier to determine than methane because the sam-

TABLE 12–IV
VARIATION OF BENZENE CONTENT OF FORMATION BRINE
WITH DISTANCE FROM PRODUCTION.
(FROM ZARELLA)

Area	Formation	Benzene content of brine, ppm	Distance to production in equivalent zone, miles
1. New Mexico	Pennsylvanian	10·7	0
		6·5	2
2. Saskatchewan	Frobisher-Alida	7·0	0
		4·5	1
		3·4	¾
		2·2	1¾
		1·6	1½
		1·0	5½
3. Alberta	Leduc	6·0—4·8	0
		3·4	½
		2·2	1⅓
		1·8	2¾
		1·6	2¾
4. West Texas	Wolfcamp	2·5	0
		1·2	¾
		1·3	2½
		0·9	5
		0·0	16

ples do not have to be collected under pressure. The water recovered in drillstem or formation testers can be analyzed in a laboratory for its aromatic hydrocarbon content.

Zarella's original paper stated that the benzene concentration decreases away from the oil-water contact, so the analyses can be used to tell how far a dry hole is from a nearby oil field (Table 12–IV). The decrease is not very regular, and the absolute amounts are different at the same distances from different pools. It is doubtful, therefore, if the method will really tell that a wildcat well was a near miss. However, it probably does tell that there are source rocks of petroleum adjacent to the aquifer. That being the case, there may well be an oil field in the vicinity.

Zarella found that barren strata contained no benzene even vertically over a producing field. Thus, in Table 12–V a pool in Saskatchewan produced oil from the Frobisher-Alida formation at 4,645 feet. The Jurassic and younger sands vertically over the oil field contained water with no benzene. This shows that soluble hydrocarbons such as benzene do not migrate vertically across shale beds by molecular diffusion.

TABLE 12–V

VARIATION OF BENZENE CONTENT IN BRINES FROM WELLS
WITH MULTIPLE TESTS. (FROM ZARELLA)

Area	Interval tested, ft	Formation	Benzene content of brine, ppm	Production in equivalent zone
1. Saskatchewan	2950–2980	Viking	0·0	None
	3265–3436	Blairmore	0·0	None
	3415–3436	Blairmore	0·0	None
	3510–3530	Blairmore	0·0	None
	3858–3875	Jurassic	0·0	None
	3993–4002	Jurassic	0·0	None
	4645–4655	Frobisher-Alida	(oil)	0 mile
2. Saskatchewan	2929–2969	Blairmore	0·0	None
	3649–3685	U. Shaunavon	0·0	None
	3739–3760	L. Shaunavon	0·0	None
	4220–4230	Frobisher-Alida	(oil)	0 mile
3. Saskatchewan	4419–4459	Midale	6·8	½ mile
	4490–4502	Frobisher-Alida	trace	

The Russian geochemists also analyze waters for their total soluble organic matter. The waters are shaken with chloroform, which removes the oil-soluble material, and the chloroform is then evaporated. This extract is called *bitumens* or *bitumoids*. It is characterized chemically by treatment with potassium permanganate, which oxides the easily oxidizable compounds like humic acids, and then by potassium iodate, which measures compounds such as hydrocarbons that are more difficult to oxidize. The extract can be titrated with hydrocholoric acid to give the content of naphthenic acid salts. Phenols and amino acids can also be determined.

Many such analyses were made of waters obtained from the first wildcat wells in the West Siberian basin where, more recently, huge oil reservoirs have been discovered (Figure 12–21). The biggest fields are in the area along the Ob´ River where the dissolved organic matter in the water is more than 3 ppm.

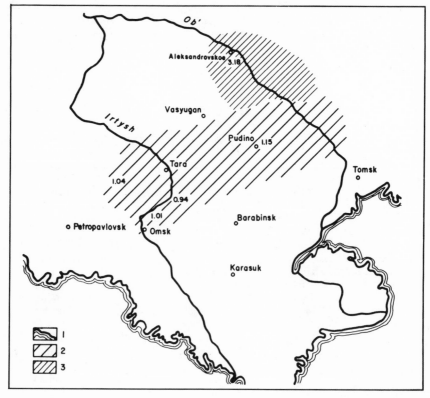

Fig. 12–21 Dissolved organic carbon in waters of West Siberian Lowland *(after Bars and Nosova)*

REFERENCES

American Petroleum Institute, 1968, API recommended practice for analysis of oilfield waters: American Petrol. Inst. RP No. 45, 49 p.

Antonov, P. L., 1963, Dal'nost prodolzhitelnost' diffuzii gazov iz zalezhey v zakonturnye vody (Distance and time of diffusion of gases from fields into edge water): Gazovaya Promyshlennost', no. 9, p. 1–5.

Bars, E. A., and L. N. Nosova, 1962, K voprosu o rastvorennom organicheskom veshchestve v vodakh melovykh i yurskikh otlozhenii sredney chasti Ob'-Irtyshskogo basseyna (The problem of dissolved organic matter in waters of the Cretaceous and Jurassic deposits of the Ob'-Irtysh basin): Geokhimiya Nefti i Neftyanykh Mestorzhdenii, Moscow, Akad. Nauk SSSR, p. 181–198 (whole volume trans. by Israel Program for Scientific Trans., Jerusalem, 1964).

Buckley, S. E., C. R. Hocott, and M. S. Taggart, 1958, Distribution of dissolved hydrocarbons in subsurface waters, in L. G. Weeks, ed., Habitat of oil: Tulsa, Oklahoma, AAPG, p. 850–882.

Collins, A. Gene, 1975, Geochemistry of oilfield waters: New York, Elsevier, 495 p.

Coustau, H., 1977, Formation waters and hydrodynamics, Journal of Geochemical Exploration, 7 (1977) p. 213–241.

Crawford, J. G., 1948, Waters of producing fields in the Rocky Mountain region: AIME Tech. Pub. 2382; Petrol. Technology, v. 11, no. 3, 23 p.

de Sitter, L. U., 1947, Diagenesis of oilfield brines: AAPG Bull., v. 31, no. 11, p. 2030–2040.

Dickey, P. A., 1966, Patterns of chemical composition in deep subsurface waters: AAPG Bull., v. 50, no. 11, p. 2472–2478.

———, 1969, Increasing concentration of subsurface brines with depth: Chemical Geol., v. 4, p. 361–370.

Dickey, P. A., and Carlos Soto R., 1974, Chemical composition of deep subsurface waters of the western Anadarko Basin: Soc. Petrol. Eng. Pap. SPE 5178, 18 p.

Graf, D. L., Irving Friedman, and W. F. Meents, 1965, The origin of saline formation waters: Illinois Geol. Surv. Circ. 393, 32 p.

Hubbert, M. King, 1953, The entrapment of petroleum under hydrodynamic conditions: AAPG Bull., v. 37, p. 1954–2026.

Jones, Bob, 1962, Advancement in exploration technique using log derived data factors: Wyoming Geol. Assoc., 17th Ann. Field Conf. Guidebook, p. 268–272.

Kartsev, A. A., 1963, Gidrogeologiya neftyanikh i gazovykh mestorozhdenii (Hydrogeology of oil and gas pools): Gosudarstvennoye Nauchno-Tekhnicheskoye Izd. Neftyanoy i Gorno-Toplivnoy Literatury, Moscow, 353 p.

Kawai, Kozo, 1963, Some considerations concerning the genesis of natural gas deposits of the dissolved-in-water-type in the Kazusa group, with special reference to diffusion of methane from gas reservoirs: Jour. of the Japanese Assoc. of Petrol. Technologists, v. 28, p. 6–15.

Kharaka, Y. K., and F. A. F. Berry, 1973, Simultaneous flow of water and solutes through geological membranes. Part I experimental investigation: Geochim, Cosmochim, Acta, v. 37, no. 12, p. 2577–2603.

Lane, A. C., 1908, Mine waters and their field assay: Geol. Soc. America Bull., v. 19, p. 502.

McAuliffe, C., 1969, Determination of dissolved hydrocarbons in subsurface brines: Chemical Geol. 4, no. 1/2, p. 225–234.

Palmer, C., 1911, The geochemical interpretation of water analyses: Geol. Soc. America Bull. 479, 31 p.

Rittenhouse, G., R. B. Fulton, III, R. J. Grabowski, and J. L. Bernard, 1969, Minor elements in oil-field waters: Chemical Geol., v. 4, no. 1/2, p. 189–210.

Schmidt, G. W., 1971, Interstitial water composition and geochemistry of Gulf Coast deep shales and sandstones: AAPG Bull., v. 55, no. 2, p. 363.

Schoeller, H., 1962, Les eaux souterraines: Masson et Cie, Paris, 214 p.

Sulin, V. A., 1946, Vody neftyanykh mestorozhdenii v sisteme prirodnikh vod (Waters of oil reservoirs in the system of natural waters): Moscow, Gostoptekhizdat.

Swenson, Frand A., 1968, New theory of recharge to the artesian basin of the Dakotas: Geol. Soc. America Bull., v. 79, p. 163–182.

von Engelhardt, W., 1961, Zum chemismus der Porenlösung der Sedimente: Uppsala Univ. Geol. Inst. Bull., v. 40, p. 187–204.

Wai, Thit, 1975, Reservoir properties from cores compared with well logs: MS Thesis, Univ. of Tulsa, 103 pp.

White, D. E., 1957, Magmatic, connate, and metamorphic waters: Geol. Soc. America Bull., v. 68, no. 12, pt. 1, p. 1659–1682.

Youngs, Bridget C., 1975, The hydrology of the Gidgealpa formation of the western and central Cooper Basin: Rept. of Invest. 43, Geol. Surv. of South Australia, 35 p.

Zarella, W. M., et al, 1967, Analysis and significance of hydrocarbons in subsurface brines: Geochim. et Cosmochim. Acta, no. 13, p. 1155–1166.

Zor'kin, L. M., 1969, O gazakh podzemnykh vod Zapadno—Sibirskogo neftegazonosnogo basseina. (The gases of subsurface waters of the West Siberian Basin): Byul, Moskov. Obshchest. Ispytatelei Prirody, Otd. Geol., v. 44, no. 3, p. 121–125.

13

Subsurface Pressures

All the pores within the rocks that make up the crust of the earth are filled with fluid, occasionally oil or gas, but usually water. All the processes of geology—diagenesis, metamorphism, folding, faulting—take place in a watery environment. The processes that concentrate not only oil and gas but also all the other useful minerals are related to the movements of water.

The behavior of fluids in porous media—either motion or lack of motion—can only be explained if the pressures, or rather potentials, are known. Therefore, the search for and the development of petroleum reservoirs requires the measurement of the pressures of the pore fluids.

The first measurements of bottom-hole pressures in oil reservoirs were made in Oklahoma by engineers of the Amerada Petroleum Corp. (now Amerada-Hess) and the U.S. Bureau of Mines (now DOE) in the 1930s. The Amerada bomb, as now used, is a stainless steel cylinder which is lowered into the well on a wire line. The pressure is recorded on a copper chart, sometimes 4 x 5 inches, which is coated with black. It is inserted inside a cylindrical cup. As the pressure increases, the free end of a Bourdon tube rotates, driving a shaft which is centered in the axis of the cylinder. Attached to the shaft is a stylus that scratches the black coating on the chart. The cup is driven by a clock. The stylus thus makes a complete recording of pressure vs. time.

When the tool is recovered, the chart is removed and examined with a microscope. The deflections are measured to the nearest 0.0001 inch. When properly calibrated, the pressures are accurate to less than one-tenth percent of the full scale deflection. Ranges in pressure are from zero to 250 psi up to zero to 25,000 psi. The clock ranges are from 3 to 72 hours or longer.

Most recently Hewlett-Packard has introduced very sensitive pressure gauges that are lowered to the bottom of the hole on electrical cables and record at the surface. They are particularly useful in measuring transient-pressure buildups.

Pressures in the U.S. have usually been expressed in pounds per square inch (psi). In countries where the metric system has been used, pressures have usually been expressed as kilograms per square centimeter (kg/cm^2). Recently, worldwide scientific and technical organizations have recommended conversion to SI units. SI is the abbreviation for the official Le Systeme International d'Unites, or International System of Units. SI is not identical to the metric or centimeter-gram-seconds (cgs) systems, but it is closely related.

The SI system expresses pressures in *pascals*. It is not strictly correct to use grams (or pounds) to measure pressure. The gram is the unit of mass. Its weight, that is, the pressure one gram will exert resting on a surface, depends on the force of gravity, which is different in different places on the earth. The unit of force is strictly the force that will cause a mass of one gram to accelerate one meter per second per second, and it is called a *newton*. Pressure (and stress) are force per unit area, so the unit of pressure is one newton per square meter (a pascal). This pressure is rather small, so in oil-field work it is common to use *kilopascals* (one thousand pascals), written *kPa*. One pound per square inch is 6.894 kilopascals. Other common measurements of pressure are kg/cm^2 (98.06 kPa) and bar (100 kPa).

The Society of Petroleum Engineers (SPE) of the American Institute of Mining, Metallurgical, and Petroleum Engineers (AIME) has published a complete explanation of the SI system as it applies to petroleum production, with a complete table for the conversion of commonly used units to SI.

In studying the performance of oil and gas reservoirs, the movement of fluids is ascertained by means of pressure measurements. The limits of reservoirs, whether caused by faults or pinch-outs, are marked by pressure discontinuities. In the development of oil and gas fields, different subreservoirs can be distinguished from others by means of their different pressures. Pressure measurements, therefore, are among the most important tools of the development geologist.

1. ORIGIN OF RESERVOIR PRESSURES

Oil and gas fields are often found in permeable reservoir rocks which cover wide areas. The portion of the reservoir rock that is not filled with oil or gas is always filled with water. This water-bearing portion of the reservoir is called the *aquifer* (Figure 13–1).

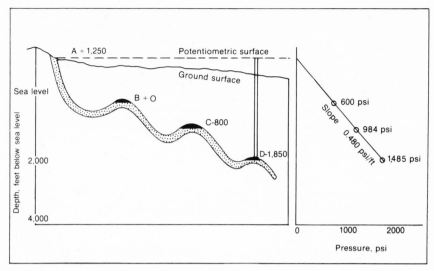

Fig. 13–1 Origin of reservoir pressures

In the cross section, Figure 13–1, the aquifer outcrops at A. There are three oil fields in the same aquifer: B, C, and D. In each field, the pressure will be that necessary to sustain a column of water to the elevation of the outcrop. The surface to which a column of water would rise is called the *piezometric* or *potentiometric* surface.

If the pressure in each field is read in pounds per square inch and plotted against depth, a graph is obtained (Figure 13–1). The points fall on a straight line whose slope is determined by the density of the water. The intercept on the y-axis is determined by the elevation of the outcrop.

Water will flow, if there is a path available, from zones of high potential to zones of low potential, not necessarily from zones of high pressure to zones of low pressure. Pressure in psi (or kPa) must therefore be converted to potential expressed as feet (or meters) of head above or below sea level. This may be done by the following formula:

$$F = \frac{P}{W} - (D\text{-}E) \quad \text{Explained in Figure 13–2.} \qquad (13\text{–}1)$$

E = elevation of Kelly bushing, feet or meters
D = depth of producing sand, feet or meters
S = subsea elevation of producing sand, feet or meters
H = hydraulic head (bottom-hole pressure expressed as feet or meters of water

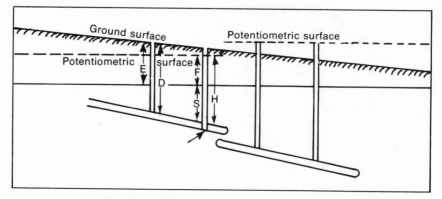

Fig. 13—2 Calculation of potentiometric surface

F = elevation of potentiometric surface, feet or meters above or below sea level

P = bottom-hole pressure, psi or kPa

W = weight of water, psi per foot or kPa per meter

For fresh water, W = 0.434 psi per foot. In metric units it is 9.8 kPa per meter. If the water is salty, the value increases. In the Gulf Coast, normal pressure is 0.465 psi per foot (10.5 kPa per meter).

If there is a hydraulic barrier which separates the aquifer into two portions, each portion will have its own potentiometric surface. If the water is in motion, the potentiometric surface will slope in the direction in which the water is moving.

An example of several oil fields in the same aquifer is the Smackover oil fields of southern Arkansas and northern Louisiana. During early development, the initial reservoir pressure of the oil fields plotted as a straight line vs depth. This indicated interconnection between the pools through the aquifer. Pools discovered later had much lower reservoir pressure indicating that pressure in the entire area had been drawn down by the production of oil and water, Figure 13–3.

Another example is the Leduc system of reefs in Alberta. The initial reservoir pressure of these fields is plotted against depth in Figure 13–4.

Here there are two straight lines, indicating that there are two hydraulic systems. In fact, the map (Figure 13–5) shows that there are two different trends of reefs: the Redwater–Leduc–Rimbey system and the Bashaw system, about 50 miles (80 km) to the east. All of the fields in one system are in hydraulic communication with each other through the aquifer, but the two aquifers are not connected. The Bashaw sys-

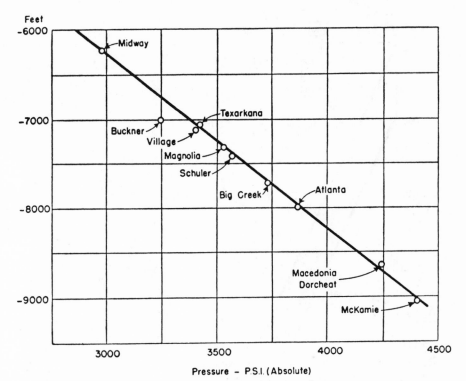

Fig. 13-3 Pressure vs. depth, Smockover fields, South Arkansas *(after Bruce)*

tem has pressures about 500 psi (3,450 kPa) higher than the Leduc system for the same depth. The slopes of the lines are both 0.480 psi per foot (10.87 kPa per m) which corresponds with the density of the water.

Prediction of gas–water contact. Suppose two or more dry holes (A and B, Figure 13–6) have been drilled into the aquifer of a reservoir rock that contains gas. The bottom-hole pressures measured in these wells plotted against depth, as in Figure 13–7, establish the pressure-depth line for the aquifer. If the water is fresh, the slope of the line will be 0.43 psi per foot (9.73 kPa per m); if it is salty, the slope might be 0.46, 0.48, (10.4, 10.8) or even more. The weight of a column of gas is very much less than that of water. It depends on the composition and pressure of gas, but might be as low as 0.03 psi per foot (0.7 kPa per m). If well C encounters a gas reservoir, the estimated gas pressure-depth line can be projected downward almost vertically to where it intersects the pressure-depth line for water, as at D. This point will be at the elevation of the gas-water contact.

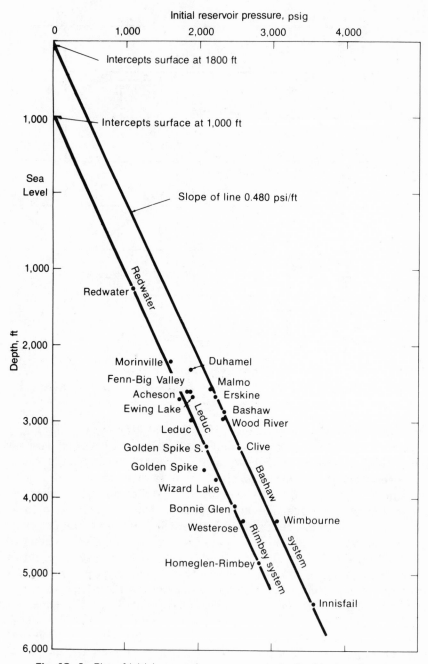

Fig. 13—4 Plot of initial reservoir pressures vs. depth of Leduc oil pools

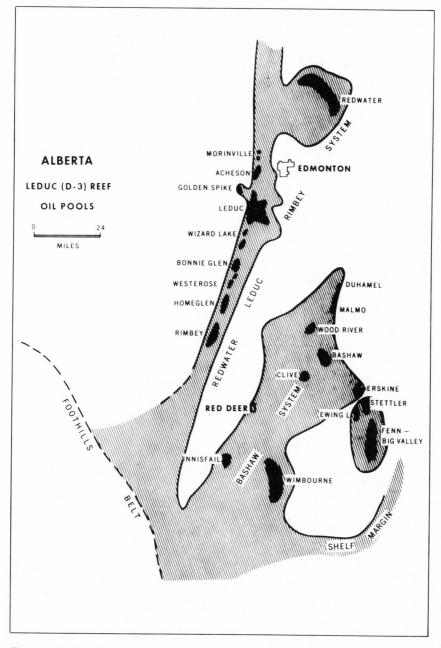

Fig. 13–5 Map showing permeable area in Devonian Leduc formation, Alberta, and principal oil pools

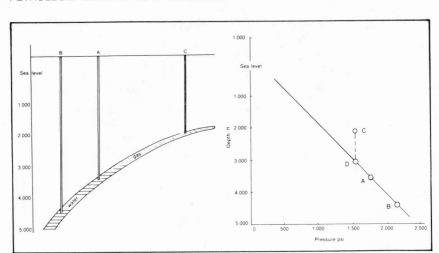

Fig. 13–6 Cross section of a reservoir rock containing gas and water

Fig. 13–7 Plot showing that intersection of pressure-depth line for gas with pressure-depth line in aquifer gives elevation of gas-water contact

It is, of course, necessary that the pressure-depth line for the water be established with assurance, otherwise the prediction of the gas-water contact will be incorrect. In areas where the sands are lenticular and the reservoir pressures vary, the method will not apply. In some areas the pressure-depth line is almost the same over wide areas and depths. For example, in the Texas and Louisiana Gulf Coast it is always close to 0.465 (10.53 kPa per m) in shallow reservoirs. In Nigeria, it is close to 0.434 (9.83). In such areas the pressure-depth line in the aquifer is well established, and the pressure-depth line for gas can be projected downward with satisfactory assurance, although the accuracy of the determination is not high.

2. SUBNORMAL RESERVOIR PRESSURES

Where there is no permeable connection to the outcrop, the origin of the subsurface fluid pressure is obscure. If the shales had vertical permeability, the pressure would be that necessary to sustain a column of water to the ground surface. This value is considered normal and is usually called *hydrostatic pressure*. However, lenticular sands often have pressures much lower than this. This indicates that shales commonly have no permeability to water. If they did, ground water would seep into the lenticular aquifer and the pressure would become normal.

Most of the fields with subnormal pressure are in rather small lenticular stratigraphic-type sands which contain almost only oil or gas and very little water. Notable are many gas fields in the Appalachian basin (Russell, 1972) and the Arkoma basin of Oklahoma and Arkansas. Subnormal pressures are also found in the Morrow of the Oklahoma panhandle and several Cretaceous fields of the Rocky Mountains.

A common situation is that shown in Figure 13–8. The sand body nearest the edge of the basin outcrops. Fresh water is present down to a depth determined by the pattern of meteoric water circulation. At greater depths, it becomes salty and changes chemical composition, losing the bicarbonate typical of meteoric waters and taking on the chloride of connate water. Wherever there is continuity of permeability and little or no water movement, the potentiometric surface will be level or nearly level. Oil is seldom found in this portion of the sand body because it can migrate easily to the surface and be lost. A closed anticline could trap oil.

Following the same sand body downdip, one reaches a point where it is cut by a fault or pinches out. Farther downdip another sand body may appear at the same, or nearly the same, horizon. This sand body is very apt to contain oil at its updip edge where it is banked against the permeability barrier. The fluid potential in this aquifer is often substantially lower than that in the portion of the sand which outcrops.

Still farther downdip the potentials sometimes get even lower. Here it is not unusual to find shoestring sands which have very low potentials, and they often contain almost no water.

Viking sand of Alberta. Figure 13–9 shows a portion of the Alberta

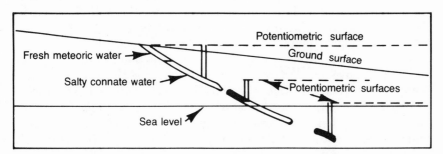

Fig. 13–8 Schematic cross section to show pattern of fluid potentials common in sedimentary basins. Reservoir pressures are normal immediately downdip from the outcrop. However, stratigraphic traps deep in the basin have subnormal pressures (after Dickey and Cox)

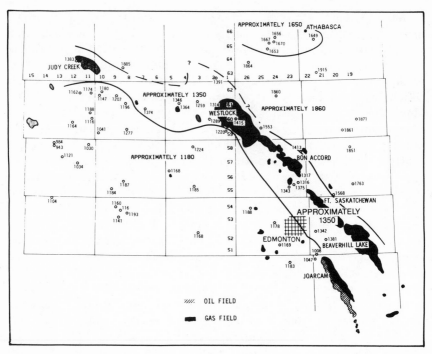

Fig. 13—9 Potentiometric surface of fluids in Viking sands in a portion of the Alberta basin. The gas fields are aligned along a potential discontinuity *(after Dickey and Cox)*

basin in the vicinity of Edmonton. The map covers an area of almost 100 × 150 miles (160 × 250 km). The Viking sands of Cretaceous age produce gas in the Westlock and other fields in the northeast part of the map. The dip of the sands is about 100 ft to the mile (20 m per km) to the southwest, and there are no closed structures. Drillstem tests in the southwest part showed that the Viking sands contain water with a potential of approximately 1,180 ft (360 m) above sea level. The fact that the same potential is found suggests that there is continuity of permeability over the whole area. The elevation of the ground surface is about 2,500 ft (750 m) above sea level. No gas was found in this area. Crossing the map from northwest to southeast is a belt about 20 miles (30 km) wide where the potentials are about 1,350 ft (411 m). This suggests that some sort of permeability barrier exists between this belt and the southwest area. All of the gas fields are found in this belt. Farther northeast the potentials are much higher—about 1,860 ft (560 m). The sand here is at a slightly different stratigraphic horizon. The fluid potentials show that it is not in communication with the sands in the west part of the map.

Potentials determined from drillstem tests can be used in this area to define the gas-rich belt. Readings on noncommercial or dry holes can also be used as well as those on commercial gas wells.

Farther southwest, deeper in the basin, the potentials in the Viking sands get even lower. Figure 13–10 is a schematic cross section showing the potentials in the Ferrier, Leafland, Gilby, and Joarcam fields. The Gilby field has a potential that is only half enough to raise a column of water to the surface. Only about 20 miles (30 km) farther southwest, the Leafland field has a pressure that will sustain a column of water to 600 ft (183 m) higher than the surface. There is a difference of potential of 4,000 ft (1,200 m; 1,750 psi) between the fluids in these two reservoirs. Obviously, they are not connected.

The abnormally high pressures of Ferrier and Leafland may be caused by the proximity to the Canadian Rockies mountain front. Thousands of feet of carbonate rocks have slid over the Cretaceous shales which enclose the Viking sands. This may have increased the pressure of the pore water in the shales, and this pressure might be communicated to Ferrier if the sand body extends far enough west.

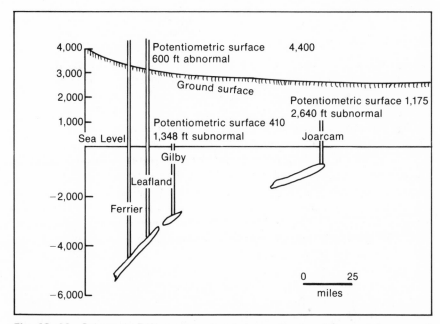

Fig. 13–10 Schematic E-W cross section between Ferrier, Gilby, and Joarcam oil fields. Bars show the height at which a column of fresh water would stand if sustained by the original reservoir pressure. Ferrier is abnormally high and Gilby is abnormally low, although they produce from the same horizon (after Dickey and Cox)

Many of the Cretaceous fields close to the mountain front, for example the Cardium sands at Pembina, also have abnormally high pressures. In contrast, the sands like the Belly River which outcrop on the plains to the east have low pressures.

San Juan Basin, New Mexico. A very similar situation has been described in the Cretaceous Gallup sands of New Mexico by Berry (1959) and later by McNeal (1961). The potentiometric surface is shown in Figure 13–11. The sandstone outcrops around the south and west margins of the basin. Here it is continuous and permeable, but it pinches out downdip toward the center of the basin (Sabins, 1963). Northeast of the pinch-out there are two long, narrow, beach-type sand bodies full of oil which form the Bisti and Gallegos oil fields.

The southwest permeable part of the Gallup sand contains fresh water. The potentiometric contours show that the water enters at the high outcrops and flows in a general peripheral direction, emerging at the low outcrops, as shown in Figure 13–11. No oil is found in this part of the basin. Figure 13–12 is a pressure-depth graph, showing that in this area pressure increases with depth at the appropriate rate for fresh water, that is, 0.43 psi per ft.

In the Bisti field, however, the original reservoir pressure was 570 psi too low for the depth, and at Gallegos 720 psi (3,980 and 5,030 kPa) too low. To show this discontinuity, the potentiometric contours on Figure 13–11 are crowded together, and this crowding indicates a permeability barrier, that is, there is no connection between the Bisti reservoir and the Gallup aquifer. The water at Bisti and Gallegos is salty, which also indicates the lack of any hydraulic connection with the fresh-water area to the southwest.

In the same general area, at a shallower horizon, is the great San Juan gas field. The basin is an assymetrical syncline. The gas occurs deep in the basin on the northeast-dipping southwest flank. Hollenshead and Pritchard (1961) ascribe the trapping of the gas to updip pinch-outs of successive benches of the Mesaverde sandstones. The sea lay to the northeast and the land to the southwest, and the successive benches were laid down as the sea retreated to the northeast. The sand outcrops all around the rim of the basin, and it is hard to see why the gas did not escape, at least along the strike to the northwest and southeast.

Figure 13–12 is a map of the potentiometric surface. As in the case of the Gallup sand, there is widespread continuity of permeability in the southwest half of the basin. The potentiometric contours show a curious ring-shaped permeability barrier which surrounds the gas field. Within the gas field, the potentiometric surface is less than 4,000

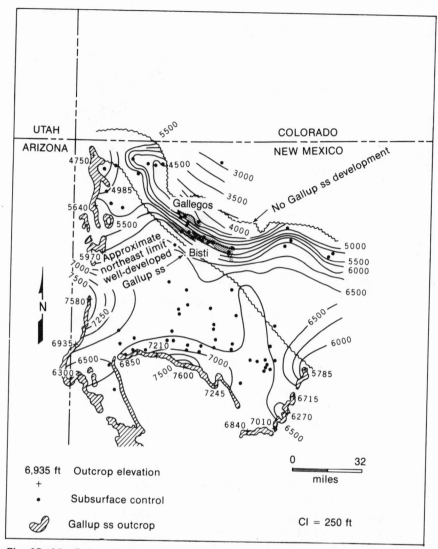

Fig. 13–11 Gallup sandstone potentiometric surface. In the southwest part of the San Juan Basin, the water in the Gallup sand is fresh and flow is mostly peripheral. Potentials are normal. Northeast of the pinch-out line, reservoir pressures are subnormal and the water is salty. Oil is found in stratigraphic traps downdip from the permeability barrier *(after Berry)*

ft (1,200 m) above sea level; outside, it is normal (the elevation of the outcrop is about 7,000 ft (2,100 m) above sea level).

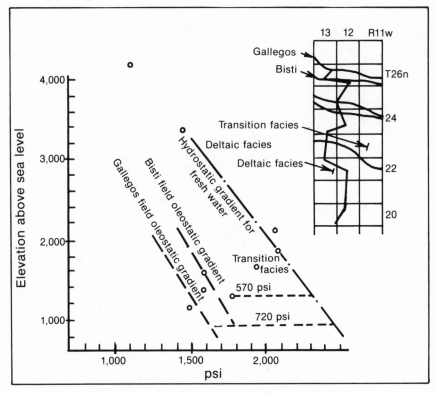

Fig. 13–12 Pressure-depth plot of Gallup sand in San Juan basin. Pressures in the southwest part of the basin (deltaic facies) are normal. Pressures in the Bisti and Gallegos fields are 570 and 720 psi (3,980 and 5,030 kPa), respectively, subnormal *(after McNeal)*

Russell (1972) lists a large number of fields with notably low pressure. Masters (1978) shows that deep down in the Alberta basin nearly all the Mesozoic sands contain gas at abnormally low pressure. Updip they contain water. It is not at all obvious what traps the gas. The most reasonable explanation is the sand bodies containing the gas are small, and there is no regional continuity of permeability.

It is an interesting question why great gas fields like San Juan, Wattenberg, Hugoton, Morrow, Arkoma basin, and the Appalachian gas fields occur low on the flanks of a syncline, with no structural closure and with subnormal pressures. Our current prospecting methods are not directed toward finding this type of accumulation.

Origin of low pressures. Russell (1972) has pointed out that all of the low-pressure reservoirs are in well-consolidated sediments which

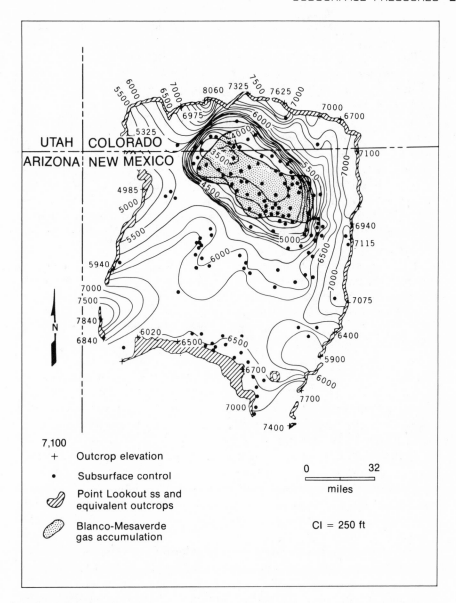

Fig. 13–13 Point Lookout sandstone potentiometric surface. Potentials downdip from the outcrop are normal. Flow is peripheral; in the high outcrops and out at the low ones. The great gas field is near the bottom of the basin and has a very low potential. There is a curious ring-shaped permeability barrier that has prevented the gas from escaping *(after Berry)*

have been uplifted in the recent gelogic past, and which are now undergoing erosion.

Fatt (1958) and McLatchie et al. (1958) have determined that a sandstone reservoir contracts elastically about 7×10^{-6} pore volumes per pore volume per psi (5×10^{-5} pv per pv per kPa) as the internal pressure of the fluids is removed. This is an elastic compression and should not be confused with compaction, which is irreversible. It may be supposed that the removal of overburden will cause an elastic dilation of the sandstone at about the same rate. Shales appear to contract and dilate even more than sandstones. The modulus of compressibility of water is about 3×10^{-6} volumes per volume per psi (4×10^{-7} vol/vol/kPa). Therefore, as overburden is removed, the pore volume dilates but the interstitial water expands only about half as much as necessary to fill the new pore volume thus created. Consequently, its pressure will drop.

If a large pressure drop takes place in the fluids in the shales, it may be supposed that some of the water may be removed from the aquifer. If the reservoir contained gas at normal pressure, removal of water would cause a drop in pressure of the gas.

For the reservoir to remain at subnormal pressure, we must assume that the overlying shales are completely impermeable. Otherwise, the ground water would filter down and pressure up the aquifer. Yet there must be a little permeability, at least locally parallel to the bedding, so that the water can be drawn out of the aquifer into the adjacent shales as their pore pressure drops. This mechanism may account for the fact that most of the low-pressure gas fields seem to contain very little downdip water.

Barker (1978) suggests that as a gas-filled reservoir is buried more deeply, its pressure will increase because of the increase in temperature. However, this pressure increase is less than that resulting from a column of water to the surface. If the gas reservoir is sealed off from communication with the surface, it will appear to have a subnormal pressure.

Use of pressure data in geology. Plans for field development require mainly the knowledge of the extent of the reservoir subdivisions that are hydraulically connected. This can be determined better with pressure measurements than by any other way. Even detailed electric-log correlations may not show permeability barriers. Big, recently discovered fields like Brent and Beryl in the North Sea have been subdivided using the initial pressures in the different sand beds. When an old field is being studied for enhanced recovery, pressure measurements will show the extent of communication.

Suppose electric logs of three wells show a package of three sands in an interval of about 150 ft (50 m) (Figure 13–14). These wells might be exploring for stratigraphic traps and be located a mile or more (2 or more km) apart, or they might be development wells one-quarter mile (400 m) apart. Correlation would indicate that the same three sands, A, B, and C, are present in all three wells. But the sands may not actually connect, and this can be determined only by pressure measurements. Suppose pressure readings are taken at the points indicated with the values indicated. When these values are transferred to a pressure depth plot (Figure 3–15), it appears that the values for sand B fall on a straight line with a slope of 0.48 psi per foot (10.8 kPa per m). This suggests that they are connected. The pressures at 1A and 2A fall on a line with the same slope, but 3A is 40 psi too high and prob-

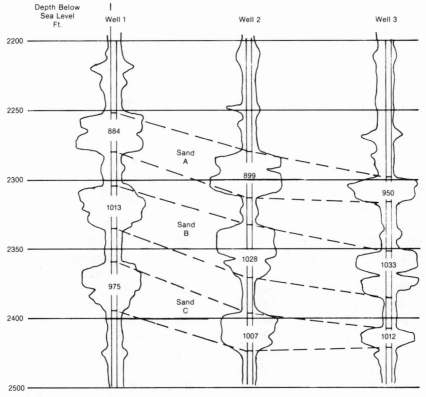

Fig. 13–14 Hypothetical cross section showing three sands that correlate by electric logs. Pressure measurements reveal permeability barriers between the wells.

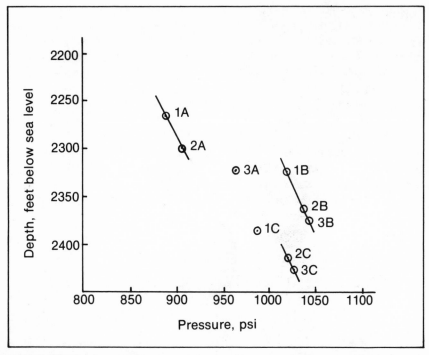

Fig. 13–15 Pressure-depth plot of wells in Fig. 13–14. Sand A in well 3 has too high a pressure, indicating a barrier between wells 2 and 3. Sand C has too low a pressure in well 1, indicating a barrier in this sand between wells 1 and 2.

ably is a separate sand body. Similarly, 2C and 3C are connected with each other, but not with 1C.

A relatively new tool called the *repeat formation tester* (RFT) has greatly facilitated the measurement of pressures (Schultz et al., 1974). The tool contains a cylindrical probe which is driven through the mud cake into the formation. Up to 12 gals of formation fluid can be collected from one or two tests. However, the probe can be retracted and repositioned so that any number of pressure measurements may be made, and the accuracy of pressure measurements is one percent. By the use of this tool, it may be possible to use pressure measurements in subdividing a large reservoir into individual sand units.

In Blaine County, Oklahoma, there is a series of sand bodies deposited in a coastal environment of a transgressive sea that lap out at successively higher positions on the surface of the unconformity between the Mississippian and Pennsylvanian. They terminate abruptly against the unconformity to the northeast and shale-out downdip to the southwest. Each reservoir had its own initial pressure which differed from that of neighboring reservoirs (Figure 13–16). With production of gas, wells in each reservoir had similar and characteristic pressure decline curves which differed from those in neighboring reservoirs (Figure 13–17). Thus, it was possible to tell which reservoir a well was producing from by a drillstem test.

Figure 13–18 is a diagram of an area resembling many basins east of the Rockies and Andes. Horizon A has a high potentiometric surface because it outcrops in the mountains. It may very well have oil at its pinch-out updip on the east flank of the syncline. Sand C outcrops low on the plains and pinches out downdip. It is less prospective. Sand B has no outcrop and may have a very low reservoir pressure. Each sand

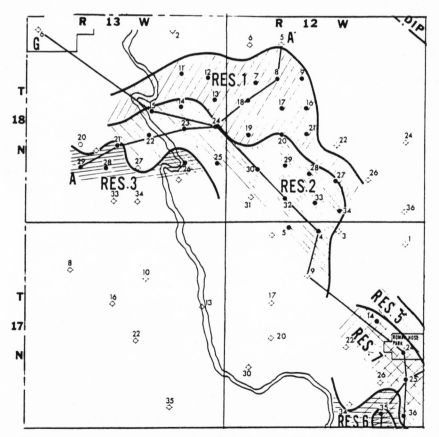

Fig. 13–16 Morrow sand gas fields in Blaine County, Oklahoma *(after Masroua)*

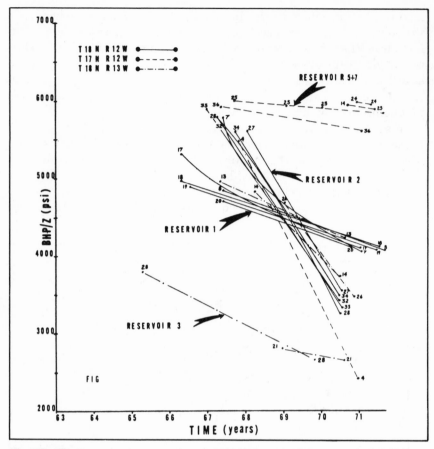

Fig. 13–17 Pressure-decline curves of individual gas wells producing from different reservoirs in T17 and 18N, R12 and 13W, Blaine county, Oklahoma *(after Masroua)*

can be identified by its pressure during a drillstem test, even if it carries only water.

By no means are all, or even the majority, of pressures obtained from drillstem tests usable. The Bourdon gauge must be accurately calibrated. The final shut-in time must be long enough so that the pressure readings can be extrapolated by the Horner method to get the true reservoir pressure. This is explained in Chapter 14.

3. HYDRODYNAMIC (FLOWING) WATER SYSTEMS

Artesian flow. When a formation with widespread permeability outcrops on the flank of a mountain and also at a lower elevation,

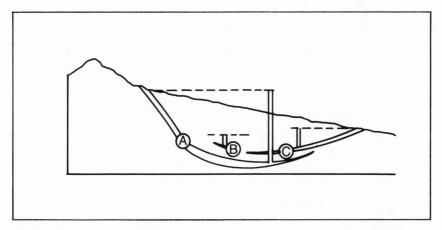

Fig. 13–18 Diagram showing how prospective horizons can be identified by their reservoir pressures

throughgoing flow of water may take place. In Figure 13–19, the water enters at A and discharges at C. Where the ground surface is lower than the potentiometric surface, as at D, a well will be artesian and will flow water to the surface.

As the water flows through the rock, resistance to flow causes a progressive loss of potential. Consequently, the height of the potentiometric surface above sea level declines.

Oil or gas may be trapped in anticlinal structures. If the water in the aquifer is moving, as in Figure 13–20, the potential at A will be

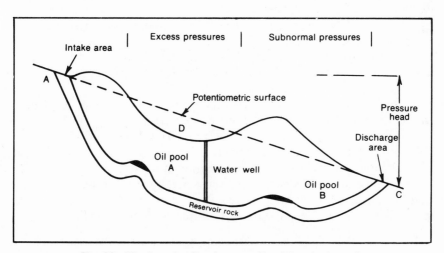

Fig. 13–19 Artesian flow in an aquifer *(after Levorsen)*

higher than at B by the amount ΔZ. This will result in a tilt of the oil-water contact.

The potential in the oil and water is the same along the o-w contact. The potential at A is:

$$P_A = \rho_w h$$

where:

ρ_w = density of water

h = height of potentiometric surface at A

The potential is B is:

$$P_B = (h - \Delta h + \Delta z)\rho_w$$

The oil is not flowing, so the potentiometric surface of the oil is the same at A and at B. Therefore,

$$(h + \Delta z - \Delta h)\rho_w = \rho_w h + \rho_o \Delta z$$
$$\rho_w \Delta z - \rho_o \Delta z = \rho_w \Delta h$$
$$\Delta z = \Delta h \ \frac{\rho_w}{\rho_w - \rho_o}$$
$$\frac{\Delta z}{\Delta x} = \frac{\Delta h}{\Delta x} \ \frac{\rho_w}{\rho_w - \rho_o} \tag{13--2}$$

This equation says that the tilt of the oil-water contact equals the tilt of the potentiometric surface multiplied by the density of water divided by the difference in density between the oil and water.*

Usually in zones of flowing water, the oil is heavy. This is partly because light ends are preferentially soluble and partly because bacterial degradation takes place. In the case of light oil, ρ_o is 0.8 and $\rho_w/(\rho_w - \rho_o)$ is 5. In the case of heavy oil, $\rho_w/(\rho_w - \rho_o)$ might be 10 or more. Often oil-water contacts are tilted, although hydrodynamic flow is not the most common cause.

The density of gas is much less than that of water, so the tilt of a gas-water contact is almost the same as the tilt of the potentiometric surface.

Tilted oil-water contacts are very common in areas of lenticular sands and salt water where there cannot possibly be artesian flow. Among these are Abqaiq, Saudi Arabia; Davenport and Cromwell, Oklahoma; and Bradford, Pennsylvania. These may be ascribed to asphalt layers or to the non-Newtonian character of the oil.

It is hard to demonstrate a clear case of oil-water contact tilt resulting only from hydrodynamic flow. The water is relatively fresh and possibly flowing in Minas, Sumatra; Orito, Colombia; and Sinco, Vene-

*I am indebted to Mr. J.I. Denham of Melbourne, Australia, for this derivation.

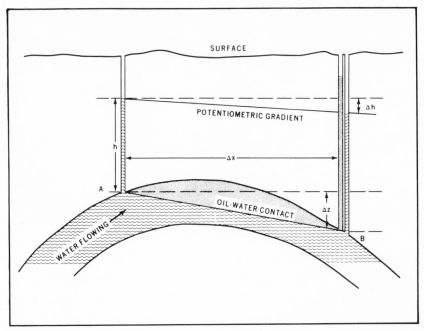

Fig. 13—20 Diagram showing why a flow results in a tilting of the oil-water contact (after Hubbert)

zuela. Tilted oil-water contacts are reported to exist in these fields, but their reservoir behavior has never been published.

Maps of potentiometric surface. Figure 13–21 is a portion of a map of the potentiometric surface of the water of the Tensleep sandstone of the Big Horn basin of Wyoming, recently released by the U.S. Geological Survey. The Tensleep sandstone has widespread permeability, and both the Tensleep and overlying Phosphoria produce oil on anticlines. Among these are Elk Basin, Frannie, Garland, and Byron. A very large anticline called Sheep Mountain has been breached by the Big Horn River. In the gorge are huge flowing springs at the outcrop of the Tensleep and underlying Amsden and Madison formations. The Tensleep outcrops around the margins of the basin, in many places high on the mountains. The water enters at the high outcrops on both sides of the basin and discharges at several outcrops along the Big Horn river. The outcrops on the west side are generally higher than on the east side.

The tilt of the oil-water contact is very great at Frannie, but accord-

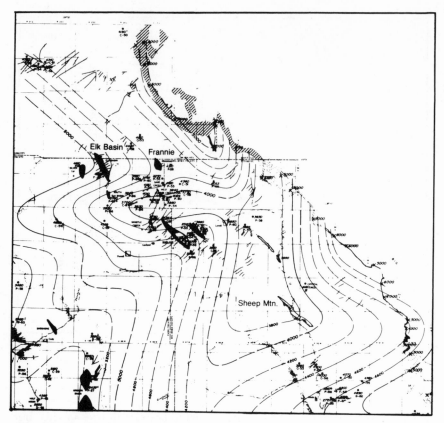

Fig. 13–21 Potentiometric surface of Tensleep sandstone in Big Horn basin, Wyoming *(after Bredehoeft and Bennet)*

ing to this map the potentiometric surface is almost level at that point. Elk Basin has an oil-water contact tilted 50 ft per mile (10 m per km) to the south, although the potentiometric surface tilts 66 ft per mile (12.5 m per km) to the east. There is a strong northerly slope to the potentiometric surface at Garland and Byron, but no corresponding tilt to the oil-water contact.

The Big Horn basin is an area where there ought to be a close relation between oil-water contact and potentiometric surface, but none can be discerned.

Gilman Hill and his colleagues in Petroleum Research Corporation compiled potentiometric maps of the Rocky Mountain intermontane basins during the 1950s. They were based on elevations of the outcrops and on extrapolation of pressure buildups of drillstem tests, using the method introduced by Horner (1951). Many oil companies purchased sets of these maps, and they are still kept up-to-date and used in prospecting.

Careful studies have been made of the hydrology of the Sahara, Figure 13–22 (Chiarelli, 1978). In the northern M'Zab basin, the Hassi Messaoud field is in a zone of high-salinity stagnant water with abnormally high pressure. Farther south in the Illizi basin, Paleozoic strata outcrop around the southern rim. Meteoric water appears to be moving downdip in a northerly direction. Southwest of Rhadames, there is a fault zone which stops the flow, and apparently salty water is ponded against this fault zone (Figure 13–23). The fresh water extends much farther north in the Devonian-Silurian than in the Carboniferous.

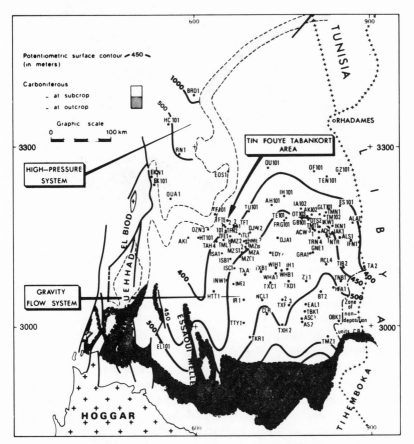

Fig. 13–22 Potentiometric map of eastern Algerian Sahara, showing northward potential decline in the Devonian-Silurian of the Illizi basin *(after Chiarelli)*

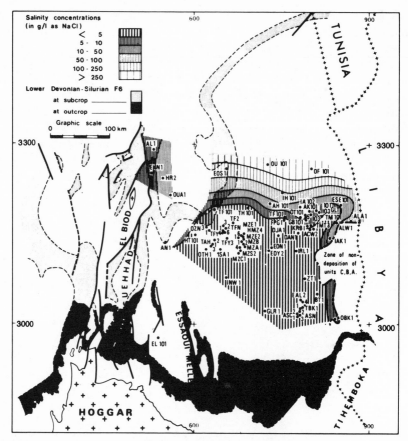

Fig. 13—23 Isosalinity map of the eastern Algerian Sahara, showing the total dissolved solids in the Devonian-Silurian of the Illizi basin *(after Chiarelli)*

Potentiometric maps have also been made by oil companies for various basins in Libya and have been published for certain basins in the Soviet Union. Most of these are on such an extremely small scale that not much can be learned from them.

Regional changes in potential which can be shown on potentiometric maps do not necessarily indicate hydrodynamic flow. If the water in different aquifers has different densities, as is usually the case, then differences in pressure will exist when the water is not flowing, even if the aquifers are interconnected (Bond, 1972).

How to find oil in hydrodynamically active areas. Large areas in the intermontane basins of the Rocky Mountains, Andes, and Alpine-

Himalayan systems contain fresh water that is probably flowing. Besides the Rocky Mountain basins, fresh-water areas include the Llanos of Venezuela and Colombia and the Central Sumatra basin.

Oil has been found in all of these basins, mostly on anticlinal structures of high relief. Many structures have been drilled which contained only fresh water, some with shows of heavy oil. Little information has been released, but in Ecuador the new oil fields are associated with moderately salty water, while many dry holes have been drilled into fresh-water aquifers. Some of the producing anticlines have a major fault parallel to the axis on the mountainward side. Such a fault could protect the area downstream from flushing.

At present the most commonly used method to find oil in hydrodynamically active regions is to draw maps on the water salinity, looking for areas which were protected from flow. These were described in Chapter 12.

Hydrodynamic flow as a trapping mechanism. Gilman A. Hill proposed in 1961 that a downdip flow of water augmented the ability of a trap to hold oil while an updip flow destroyed it. Hill imagined a near, but not quite, pinchout of a sandy body (Figure 13–24). There might be

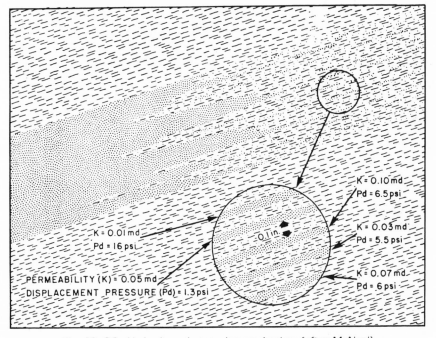

Fig. 13—24 Hydrodynamic trapping mechanism *(after McNeal)*

one thin stringer of sand with a permeability of 0.1 md and a capillary entry pressure of 6 psi. Capillary pressure is the difference in pressure between the oil and the water in the pores. With oil of density 0.8, it amounts to about 0.1 psi per foot (2.3 kPa per meter) of oil column, so 6 psi (40 kPa) capillary pressure would hold an oil column of 60 ft (20 m). Any more than this would leak into the thin sand stringer and out of the trap if the sand got permeable again farther updip. If there is a pressure gradient resulting from a downdip flow of water, it will augment the capillary entry pressure and increase the amount of oil that can collect before it leaks. Conversely, if there is an updip flow, the potentiometric gradient will assist the oil to overcome the capillary entry pressure and drive it through.

This theory has been applied to several reservoirs (McNeal, 1961; Fowler, 1970). However, in all of these cases the facts seem to fit static better than dynamic conditions. Abrupt differences in water salinity coincide with the abrupt pressure differences, indicating no flow across the barrier. Also, it seems farfetched and unlikely geologically for a bed to pinch down to a critical permeability and then open up again. From what we know of fluid flow through porous media, it is impossible for water to flow through a sand fully saturated with oil. It is therefore unlikely that hydrodynamic water flow is ever a trapping mechanism.

4. ABNORMALLY HIGH PRESSURES

Abnormally high pore pressures are especially common in young (Tertiary and Quaternary) sediments but occur in rocks of all ages. They are common at depths below 10,000 ft (3,000 m) but are also found at shallow depths.

Blowouts while drilling are always expensive and may be disastrous. It is thus very important to be aware of the danger of high pressures and the geology of places where they may occur.

Cause of abnormally high pore pressures. High pressures are most common in regressive sedimentary environments where near-shore, shallow-water deposits build out over fine impermeable muds deposited in deep water. Those shales that have permeable connection to the surface compact easily, but the underlying muds cannot compact because their permeability is so low that the pore water cannot be expelled.

The sediments of south Louisiana during the Tertiary were deposited in a regressive sea with continental sandy formations near the shore, mixed sand and shale on the continental shelf, and mostly shale on the outer shelf and slope (Figure 13–25). All of these environments moved seaward as the basin filled. As sedimentation continued, the shales compacted. The pore water was expelled parallel to the bedding along sandy beds which graded laterally into massive sands. In the offshore environment where the sands are lenticular, there was no route of escape for the water. It remained in the pores where it had to sustain part of the weight of the overburden.

Growth faults act as seals to the flow of water parallel to the bedding. In general, the stratigraphic horizon of the first abnormal pressure steps up (gulfward) along each principal line of growth faults (Figure 13–26).

Midland field has many separate reservoirs (Figure 13–27). Most of those below 10,500 ft (3,200 m) have pressures 3,000 to 4,000 psi (20,000 to 30,000 kPa) higher than normal.

Other factors helping to cause high pressures have been suggested. As sedimentation continues, depth of burial increases and temperatures increase. This causes an increase in the volume of the pore water (Barker, 1972). If the system is effectively confined, pore pressures will increase.

The increase in temperature may cause phase changes in the rocks. High pressures have been ascribed to the evolution of gaseous carbon dioxide. The high pressures of the Uinta basin of Utah may have been caused by the thermal cracking of the kerogen in the oil shale.

At temperatures of about 100° C or depths between 8,000 and 10,000 ft (2,500 and 3,000 m), smectite converts to illite. This involves a loss of interlattice water. If there is an increase in specific volume of the water as it comes out of the smectite, or if the smectite loses volume, it could cause an increase in pore pressure. This effect is widely discussed (Burst, 1969) but has not been demonstrated.

Abnormal pressures are very common in Tertiary sediments. All occur in sequences of thick impermeable shales. Examples are offshore Malaysia, Indonesia, Taiwan, the Caspian Sea area, the North Sea, eastern Venezuela and Trinidad, northern Colombia, the Uinta basin of Utah, the Green River basin of Wyoming, Nigeria, and the coastal basins of California (Fertl, 1976).

In Cretaceous sediments, abnormal pressures occur in the Green River basin, the Magdalena Valley of Colombia, and the McKenzie delta of Canada. In Paleozoic sediments, high pressures occur notably in the Morrow of central Oklahoma.

Detection of abnormally high pressures. Geologists of Shell Oil Company (Hubbert and Dickinson) recognized that the shales in abnormally pressured areas were not as compacted as they should be

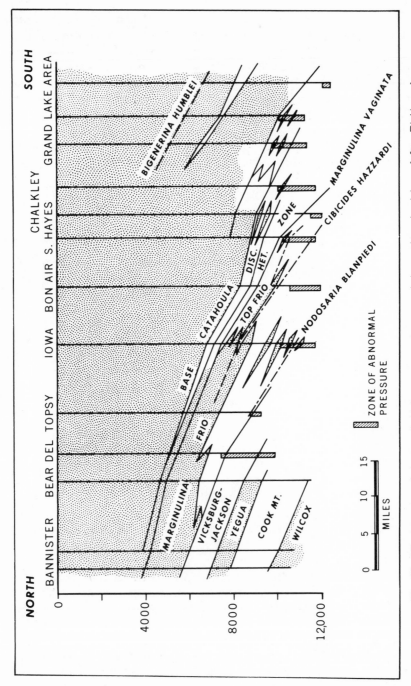

Fig. 13–25 Location of abnormally high pressure across sands in southern Louisiana *(after Dickinson)*

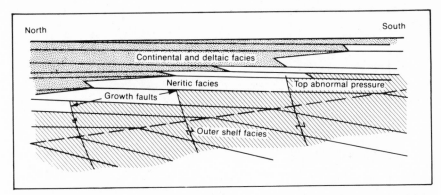

Fig. 13—26 Diagram showing facies patterns and effect of growth faults on high-pressure zones

considering their depth of burial. The water was unable to escape from the pores, where it remained, sustaining part of the weight of the overlying sediments. Thus, the shales in high-pressure areas have higher porosity than normal for their depth of burial. Porosity affects

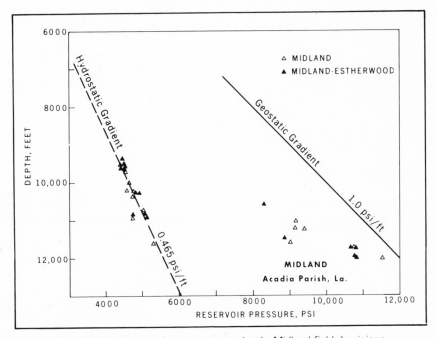

Fig. 13—27 Reservoir pressure vs. depth, Midland field, Louisiana

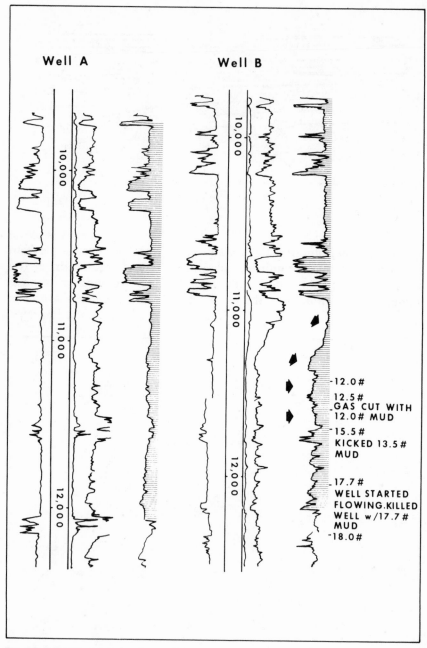

Fig. 13—28 Electric logs of two offshore Louisiana wells only 2,000 ft apart, showing abrupt increase in conductivity of shale in one well *(after Wallace)*

several other attributes of shales which are measured by logging devices, so it is possible to recognize high-pressure zones from the logs.

Resistivity. In well B of Figure 13–28 the induction log (farthest right) shows an abrupt increase in conductivity of shale at about 11,200 ft, as indicated by the arrows. Below this depth, the thin sands produced kicks against heavy mud. Well A, on the other hand, shows no abnormal increase in conductivity. It is separated from well B by a fault. The fault must be impermeable to prevent the pore water from well A to escape laterally.

Hottman and Johnson (1965) showed that the resistivity of shale increases with depth in the Gulf Coast, and the normal curves are shown in Figure 13–29.

If the normal resistivity is divided by the observed resistivity, a parameter is obtained that indicates how far from normal the resistivity is. This parameter is plotted against the amount of pore pressure in Figure 13–30. Pore pressure is expressed as fluid pressure gradient (FPG) in psi per foot of depth. It may also be expressed as mud weight necessary to hold it, shown on the right side of the graph. (1 psi per foot = 22.62 kPa per meter.) An example of the calculation is shown in Figure 13–31a. The normal resistivity of shale in Cameron Parish at

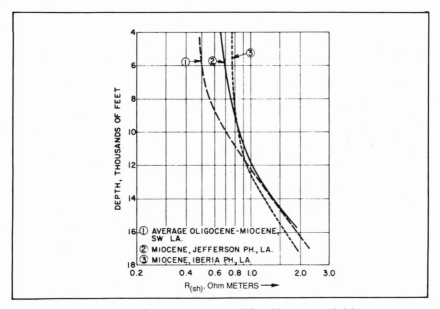

Fig. 13–29 Shale resistivity vs. burial depth *(after Hottman and Johnson, courtesy JPT, copyright SPE)*

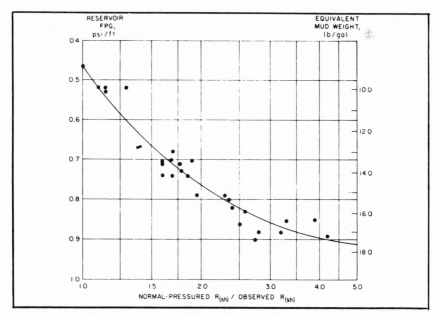

Fig. 13–30 Shale resistivity parameter. resistivity of normally pressured shale divided by observed resistivity of abnormally pressured shale, plotted against depth *(after Hottman and Johnson, courtesy JPT, copyright SPE)*

12,000 ft (3,660 m) is 1.0 ohmmeters, but the actual resistivity observed in well R was 0.4 ohmm. The parameter R_{sh}/R_{obs} is therefore 2.5. Referring to Figure 13–30, the FPG is 2.5 is 0.83 psi per foot (18.7 kPa per meter). At 12,000 ft this is 10,000 psi (69,800 kPa). Actual tests confirmed that this was the pressure at that depth as shown in Figure 13–31b.

Sonic velocity. Normally, the velocity of sound in rock increases with depth. Faust showed that for common shales and sandstones the velocity of seismic waves is proportion to $(ZT)^{1/6}$ where Z is depth of burial and T is geologic time. The velocity of sound depends on the elastic constants of the rock, which are affected by the porosity. The sonic velocity log measures the time in microseconds for a sound wave to travel one foot, usually called Δt. Figure 13–32 shows the normal decrease in travel time Δt with depth in the Tertiary sediments of south Louisiana.

In high-pressure zones, the velocity of sound in shale decreases and the Δt increases. If Δt normal is subtracted from the Δt observed, a parameter is obtained which indicates the degree of abnormality. If this parameter is plotted against fluid pressure gradient (FPG) in psi

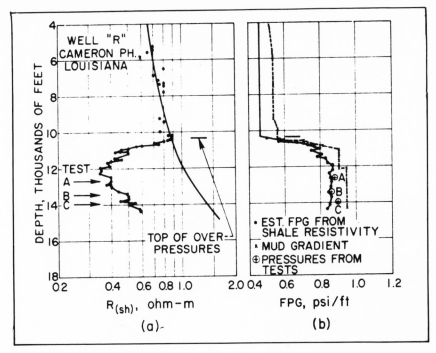

Fig. 13–31 Example of estimating pressures from resistivity log *(after Hottman and Johnson, JPT, copyright SPE)*

per foot (or kg per sq cm per m), an empirical curve is obtained which may be used to estimate FPG, as shown in Figure 13–33.

Normal Δt at 12,000 ft (3,660 m) is 95 microseconds per foot (29 microseconds per meter), but the Δt observed was 130. The parameter is 130 − 95 or 35. From the curve in Figure 13–33, the FPG is estimated at 0.90 psi per foot (20 kPa per m), which agreed well with the observed pressures Figure 13–34.

It is possible to determine the sonic velocity from the surface without drilling a well. Modern methods of seismic prospecting utilize a common depth point array (CDP). With digital methods of processing the data, it is possible to determine the average velocity to the reflecting horizon. If the velocity starts to decrease with increasing depth, this is anomalous, and it may be expected than an undercompacted, high-pressure shale may be present. This technique has been widely used on the U.S. Gulf Coast (Pennebaker, 1968; Reynolds, 1970), the Australian northwest shelf, and the McKenzie delta (Dumont and Purdy, 1976), but very few publications have appeared. Figure 13–35

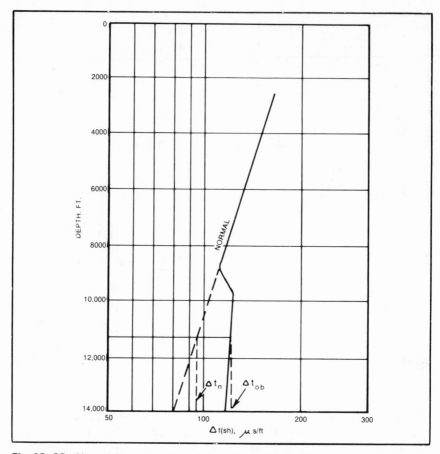

Fig. 13–32 Normal decrease of travel time of sound through shale with depth of burial *(after Hottman and Johnson, courtesy JPT, copyright SPE)*

shows how seismic velocity increases with depth, except below 13,000 ft (left curve). The drilling time curve (center) and the sonic velocity curve (right) also indicated the top of the high-pressure zone at about 13,000 ft when the well was drilled.

Shale density. In zones of abnormal pressure, the porosity of shale is abnormally high for its depth of burial. The density of shales is thus abnormally low. Some advocate measuring density of cuttings continuously. The chips are dropped into a glass cylinder in which there is a mixture of fluids whose density decreases upward. If beads of known density are dropped into the cylinder, the density of the liquid at each level can be ascertained and marked. The density of shale cuttings can

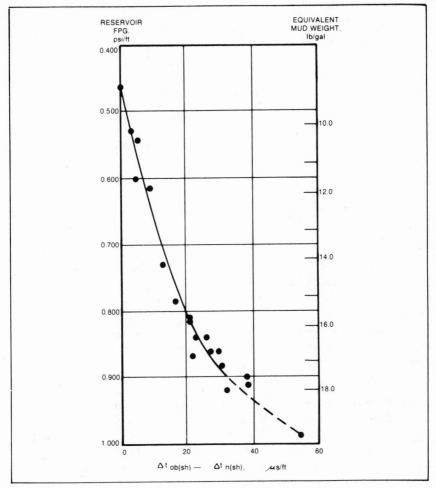

Fig. 13–33 Shale acoustic parameter observed minus normal Δt plotted against FPG
(after Hottman and Johnson, courtesy JPT, copyright SPE)

be determined by noting the level to which they settle. An anomalous
decrease in density indicates a high-pressure zone. If there is a low-
permeability cap rock about the high-pressure zone, there may be an
abrupt increase in density. The shale density method is not very reliable because of the difficulty in obtaining representative cuttings.

Shale density also can be determined by radioactive density logging devices.

Penetration rate.   Normally the rate of penetration when drilling

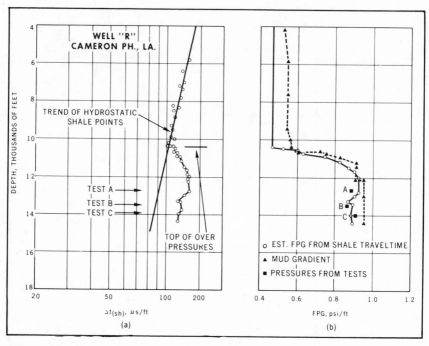

Fig. 13—34 Estimating pressures from sonic log *(after Hottman and Johnson, courtesy JPT, copyright SPE)*

shale decreases with depth. A sudden increase in drilling rate may indicate entry into an abnormally pressured area. Two factors cause the increase in drilling rate: (1) the shales are more porous and softer and (2) if the pore pressure is higher than the mud pressure the spalling of chips cut by the bit is facilitated. Attempts have been made to improve deductions by calculating a *d-exponent* (Jordan and Shirley, 1966; Bolt, 1972), taking into account rate of penetration R in feet per hour, rotary speed N in rpm, weight on bit W in pounds, hole diameter D in inches, as follows:

$$\text{d exponent} = \frac{\log \left(\dfrac{R}{60N} \right)}{\log \left(\dfrac{12W}{10^6 D} \right)} \qquad (13\text{--}3)$$

An example of a decrease in d-exponent indicating abnormally pressured shale is shown in Figures 13—35 and 36.

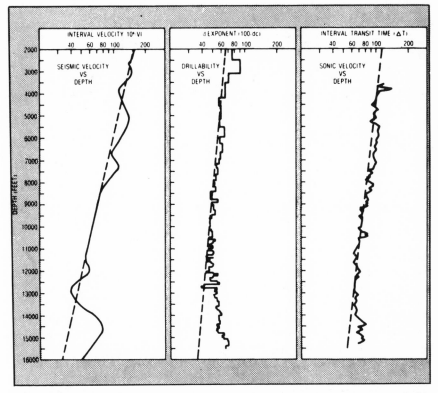

Fig. 13—35 Changes in seismic interval velocity with depth (left curve) compared with drillability (center) and sonic velocity (right), Gulf and Mobil 1—37 Ikhill, McKenzie delta (after Dumont and Purdy, World Oil)

Shale gas on mud logger. If the pore pressure is higher than the mud pressure the shales tend to spall and cave into the hole. If the pore water in the shales is saturated with methane, as is often the case, it comes out of solution on the way up the hole and makes a strong indication in the gas-measuring device in the mud logger. Sometimes this shale gas is enough to expand the mud volume and cause kicks at the surface. Sensitive mud weight indicators at the flow line help early detection of gas in the mud. The caving shales cause increase in torque and drag when pulling drillpipe.

The mud carrying the shale cuttings and the dissolved gas does not reach the surface immediately. In the case of deep wells, the lag time may be an hour or more. In some localities where there is an increase in penetration rate, it is customary to stop drilling ahead but to con-

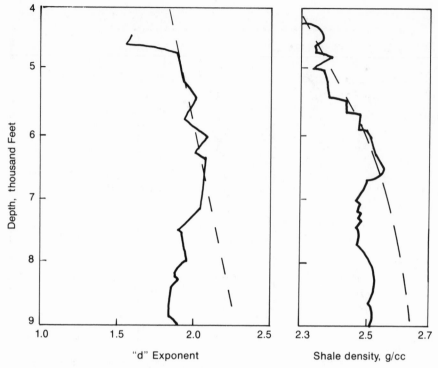

Fig. 13–36 Comparison of shale density and d-exponent vs. depth for a well in Southeast Asia *(after Bolt, courtesy World Oil)*

tinue circulating mud. After the proper lag time, the cuttings and dissolved gas will reach the surface.

Most methods of detecting abnormal pressures during drilling or on well logs depend on the fact that where abnormal pore pressures are found, the associated shales are undercompacted and more porous than normal. There are places in the world where this relationship is not observed (Bradley, 1975; Carstens and Dypvik, 1981). Apparently the porosity of the shales was reduced by diagenetic processes. In such cases the abnormal pressures may be due to temperature changes or to the diagenetic processes themselves. In those places it is difficult or impossible to identify the top of the high-pressure zone by shale resistivity, sonic velocity, shale density, or drilling rate.

Temperature changes. The drilling mud temperature at the surface has been correlated with abnormal pressures. This method seems dubious because abrupt temperature differences cannot exist in the subsurface; even if they did, it is doubtful if they could be detected in drilling mud at the surface. The temperature of the drilling mud returns depends not only on the bottom-hole temperature but also on the rate of penetration and especially the pumping rate. It is suspected that the reported change in temperature of the mud may be an indirect reflection of the abnormal pressure caused by changes in penetration rate, mud velocity, or increased sloughing.

It has also been claimed that the geothermal gradient increases abruptly through high-pressure zones. This conclusion is based on bottom-hole temperatures values indicated by maximum thermometers during logging runs. However, temperature readings from logging devices are always below true, usually by such large amounts as to be worthless. Figure 13–37 shows that a thermometer left on the bottom of

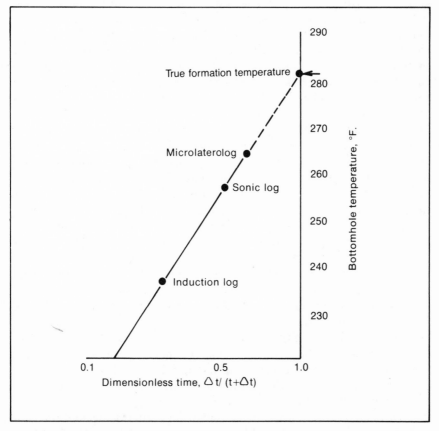

Fig. 13–37 Extrapolation of successive bottomhole temperature measurements to get true value (after Fertl and Timko, courtesy World Oil)

the hole 4 hours after mud circulation stopped recorded 238° F. (114° C.) when the true value was about 280° F. (138° C.). Theoretically, a change in porosity of shale from 20 percent to 40 percent would cause only a small change in geothermal gradient.

The larger gradient changes which have been reported may be ascribed to errors in temperature measurement due to mud circulation. The reduced circulation in the smaller diameter hole through the high-pressure zone means that the mud measured by maximum thermometers will be closer to true formation temperature, that is, higher than in the larger hole where the circulation was faster.

Formation rupture. Most sedimentary rocks have low tensile strength; therefore the walls of the hole tend to rupture when the presure of fluids (water or drilling mud) in the borehole exceeds a certain critical value. When the rupture occurs, a crack is formed that is held open by the fluid pressure. Such fractures are usually vertical and extend radially from the borehole. If additional fluid is pumped in, the fractures propagate themselves to long distances (hundreds of meters) from the borehole. When the pressure in the fluids is reduced, the fractures tend to heal because the weight of the overburden causes the walls to come together.

The effect of fractures can be extremely beneficial or extremely harmful to oil and gas wells. When the producing formation has low permeability, it is advantageous to purposely fracture the walls of the borehole. This process is called *hydraulic fracturing*. It consists of rapidly pumping large volumes of fluid (water or oil) that carry propping material such as sand or small ceramic pellets into the well bore at pressures high enough to open the fractures. The grains of proppant keep the fractures open after pumping stops. Usually the fluid viscosity is increased by a thickening agent that helps keep the grains suspended and inhibits the entry of the fluid into the porous formation. Hydraulic fracturing enormously increases the effective surface of the well bore, increasing oil and gas production. This is especially important in tight (low-permeability) sandstones that, without fracturing, would be noncommercial.

The fractures not only extend radially from the well but also extend vertically above and below the permeable formation. Consequently, if there is a water-bearing bed close to the oil-or-gas-bearing unit, the fractures may reach it. Hydraulic fracturing is not successful in very unconsolidated formations that behave plastically.

Hydraulic fracturing has a bad effect during secondary and tertiary-recovery processes. When it opens a crack from the injection well to the producing well, most of the injected fluid goes straight through the crack, bypassing the oil in the sand. All sands have some beds that are more permeable and others less; they are usually separated by impermeable partings or shale beds. They then act as independent reservoirs. In waterflooding and tertiary recovery, it is important to complete the wells selectively; that is, in such a manner that fluids can be injected into the different subreservoirs separately. If the injection well is fractured, this becomes impossible.

Formation rupture is particularly disastrous in the case of a drilling well. If the weight of the column of mud in the hole is great enough to rupture the walls, the mud will run out into the fracture. This is called *lost circulation*. When it occurs, the level of mud in the hole drops and the pressure in the lower parts of the hole decreases. If there is oil or gas in any permeable formation, it may enter the hole rapidly, expelling the remaining mud and causing a blowout. When drilling abnormally pressured formations, it is necessary to increase the mud weight to overbalance the pressure in the formation fluids. If the mud weight is increased too much, however, lost circulation occurs and the well becomes even more likely to blow out.

It can be shown by the theory of elasticity that the circumferential tensile stress in the wall of a hole, σ_θ, is maximum at the wall and is there equal to the pressure in the hole, p_w (Figure 13–38). Rocks have a low tensile strength, so it would take only a small pressure inside the hole to cause the wall to fail in tension if it were not that the horizontal stress in the rock, σ_h, and the pressure of the pore fluids, p, also are resisting the pressure in the borehole.

Consequently, the pressure in the well bore necessary to cause rupture will be:

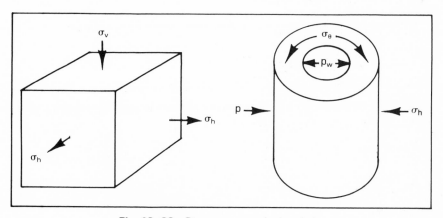

Fig. 13–38 Stresses around a borehole

$$p_w^* = \sigma_h + p + T \qquad (13\text{--}4)$$

where:

p_w^* = critical pressure in the well-bore fluid causing rupture, kPa or psi

σ_h = horizontal compressive stress, kPa or psi

p = pressure in the pore fluids, kPa or psi

T = strength of the rocks, kPa or psi

In a basin not subject to compressive tectonic horizontal stress, the horizontal stress will be the same in every direction and will be caused by the vertical stress. The vertical stress is the weight of the overburden less the buoyant effect of the water in the pores. Sedimentary rocks have a density of about 2.3, so the weight of a column of rock is 0.23 kgf per sq cm per meter (22.5 kPa per meter) less the weight of a column of water, which is 0.10 kgf per sq cm per meter (9.8 kPa/m). The vertical stress therefore is 0.13 kgf per sq cm per meter, which is 12.7 kPa per meter. In English units, the weight of the overburden is about 1.0 psi per foot, less the weight of a column of water, which is 0.43 psi per foot, or 0.56 psi per foot.

The horizontal stress (in a relaxed elastic solid) depends on the vertical stress as follows:

$$\sigma_h = \sigma_v \left(\frac{\mu}{1 - \mu} \right)$$

where μ is Poisson's ratio, which depends on the elastic properties of the rock. Poisson's ratio ranges between 0.15 and 0.4; a possible value would be 0.3. Consequently the horizontal stress will be about half the vertical stress. The pressure necessary to cause rupture becomes:

$$p_w^* = \tfrac{1}{2}\sigma_v + p + T \qquad (13\text{--}5)$$

where:

p_w^* = stress necessary to cause rupture

σ_v = vertical stress

p = pore pressure

T = tensile strength of the rock

σ_v and p increase with depth, while T increases only slightly. Expressing the stresses and pressures in terms of depth, we have:

$$\frac{p_w^*}{Z} = \frac{\tfrac{1}{2}\sigma_v}{Z} + \frac{p}{Z} + T \qquad (13\text{--}6)$$

where:

Z = depth.

Inserting the average values given above:

σ_v = 0.56 psi per foot, 12.6 kPa per meter

p = 0.43 psi per foot, 9.7 kPa per meter

The value p_w^* then becomes 0.71 psi per foot, 16 kPa per meter, plus the tensile strength of the rock.

This number is often taken as a common value for the fracture pressure. The vertical stress does not vary much, but the tensile strength ranges from a few psi to several hundred (a few kPa to maybe a thousand). The pore pressure also ranges between wide limits, from less than 0.2 psi per foot (4.5 kPa/m) to over 0.9 psi per foot (20 kPa per meter).

The effect of pore pressure is very great at both ends of the scale. At the lower end, when the pore pressure is low, the sand is very easily fractured. According to equation 13–6 (neglecting the tensile strength), if the pore pressure is 0.2 psi per foot (4.5 kPa/m), then the fracture pressure will be 0.48 psi per foot (10.8 kPa/m). This is only slightly more than the weight of a hole full of water. In the case of a depleted oil sand with low pressure, it is very important to keep the bottom-hole pressure low when starting a waterflood. After the pore pressure in the vicinity of the hole has been built up, it will be possible to raise the injection pressure.

At the other end of the scale, we have the case of abnormally pressured oil and gas sand. If the pore pressure is 0.8, which it often is, then the fracture pressure will be 1.08 psi per foot (24.5 kPa/m).

Figure 13–39 shows the relation between fracture gradient and pore pressure gradient noted when hydraulically fracturing some South Texas Vicksburg sands. At shallow depths (less than 1000 ft) the tensile strength becomes important.

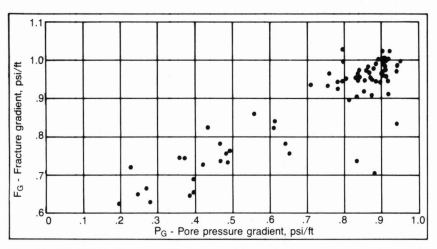

Fig. 13–39 Relation between fracture gradient and pore pressure gradient for a number of southern Texas Vicksburg sands *(after Salz)*

When the rocks are under compressive stress, as in a subduction zone, the horizontal compressive stresses are not equal and they may exceed the vertical stress. Hottman and Smith (1978) give data on the Gulf of Alaska. They compute that the compressive stress in a NW–SE direction is 1.2 psi per foot (31.7 kPa/m) and that in a SW–NE direction it is 1.2 psi per foot (27.2 kPa/m). The pore pressure was 0.85 psi per foot (19.2 kPa/m).

When drillpipe is run into the hole, it acts like a loose-fitting piston. If it is run in too fast it may increase the pressure at the bottom of the hole several hundred psi. At 5,000 ft, 500 psi is 0.1 psi per foot. Such increases in pressure may cause the walls of the hole to rupture. Large amounts of drilling mud are lost into the rupture, and the level of mud in the hole drops. When the pipe stops, there may be insufficient mud pressure at the bottom of the hole to withstand the pressure of the fluids in the formation and the well blows out, even if formation pressures are normal.

Fracture gradients are not easy to measure. In the case of water injection wells, it is possible to increase the pressure step-wise and notice the rate of increase of injection rate. When rupture occurs, the increase in rate for a given pressure suddenly increases, as shown in Figure 13–40. Such measurements are not possible on a drilling well.

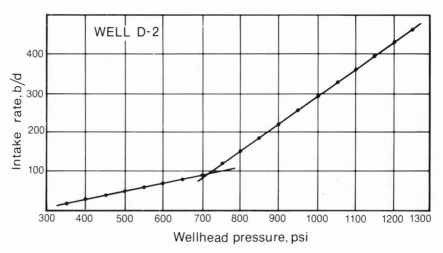

Fig. 13–40 Formation rupture test of water injection well 1,300 ft deep at Bradford, Pa. Water intake rate increases linearly with pressure up to 700 psi. At this pressure, rupture occurred and the well took water at a faster rate (*after Dickey and Andresen, API*)

It is possible to make a test in the open hole below the casing seat after setting the casing in cement. Some open hole is drilled, the drillpipe is pulled back into the casing, and the annulus is closed tightly around the drillpipe. Then mud is pumped slowly into the drillpipe and the increase in pressure is noted. At a certain critical pressure there will be a small leak of mud, followed by a more or less sudden drop in pressure when rupture occurs. Figure 13–41 is an example of such a test.

As the pore pressure in the fluids increases, the formation fracture

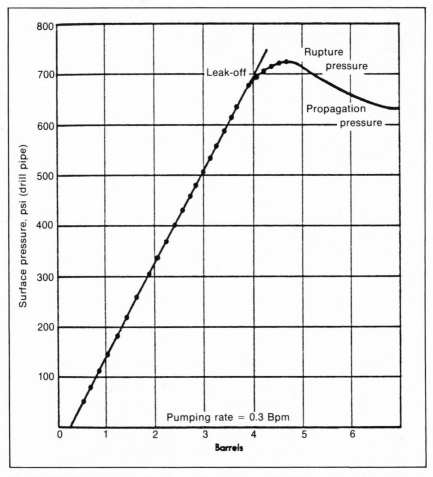

Fig. 13–41 Casing seat leak-off test. Casing seat at 3,000 ft; mud weight = 9.6 ppg (density = 1.15) *(after Moore)*

gradient also increases. However, the deeper the well and the higher the pressure, the less margin there is between the mud weight necessary to control the pressure and that which would cause a rupture. It is not uncommon to be in a situation where, if the mud is 16 lb per gal (density 1.92), the well will try to blow out and if it is 17 lb per gal (density 2.04) the walls will rupture and mud will be lost to the formation. If much mud is lost, the danger of a blowout is even greater.

In most cases, neither the pore pressure gradient nor the fracture gradient can be measured or even estimated with assurance. It is necessary to depend on experience with wells in the same area that have lost circulation on the one hand and kicked on the other. Figure 13–42 shows the estimated values for Roger Mills County, Oklahoma. Operators there plan their casing and mud program based on these estimates.

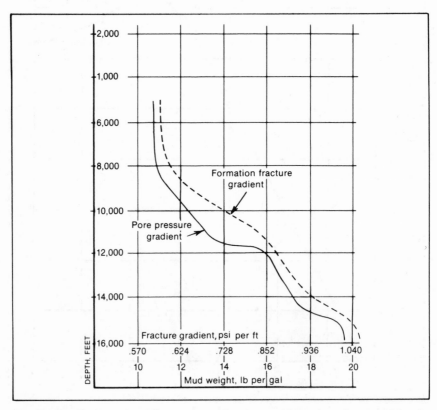

Fig. 13—42 Estimated pore pressure and fracture gradients based on experience in Roger Mills county, Oklahoma (after Magcobar)

Prevention of blowouts—handling kicks. A *kick* is an influx of fluid into the borehole. It may be water, oil, or gas which enters from a permeable formation while it is being drilled because the pressure in the hole is less than that in the formation. If the volume of formation fluid is small, as from a thin, tight gas sand, drilling may be continued safely in spite of frequent kicks. This is often done in central Oklahoma. If a large amount of fluid has entered that has formed a bubble, displacing a large amount of drilling mud, it must be circulated out of the hole or a blowout may occur. A blowout is an uncontrollable influx of fluid from the formation into the borehole. Formation fluids are all lighter than drilling mud. The more mud they displace, the lighter the column and the more rapidly they enter the hole.

When this happens, it is important to stop the influx of the fluids as quickly as possible by raising the pressure at the bottom of the hole. One way to do this is to add weighting material to the mud. However, if the pressure of the mud anywhere in the hole exceeds the pressure at which the rock will fracture, the mud will leave the borehole and enlarge the fracture. This will decrease the mud column in the hole, decreasing the pressure at the bottom, and making the situation even worse than before, as explained in the preceding pages.

If we close the annular space between the drillpipe and the casing at the surface with the blowout preventers, then no fluid will be able to leave the well bore at the surface. The pressure will build up until the pressure at the bottom of the hole is equal to the formation pressure, and then no additional formation fluid will enter from the formation (Figure 13–43).

We now must circulate the bubble of formation fluid out of the hole without increasing the pressure beyond that necessary to hold back the fluids. The static pressure at the top of the drillpipe is the excess pressure above the mud weight necessary to hold back the fluids. The pressure in the annulus at the wellhead will be more because of the bubble of formation fluid. We now start the pumps and notice the drillpipe surface pressure, which will be increased slightly by the friction of the mud going down the hole. We then hold the drillpipe pressure constant with the choke and circulate the bubble out without raising the pressure above that necessary to hold back the fluids. If the bubble is gas, it will expand on the way up the hole, causing the casing pressure measured at the surface to increase and mud to flow into the pits by heads.

When the bubble is out, weighting material may be added to the mud, but only enough to balance the formation pressure. Adding too much may cause loss of circulation higher up the hole. The additional

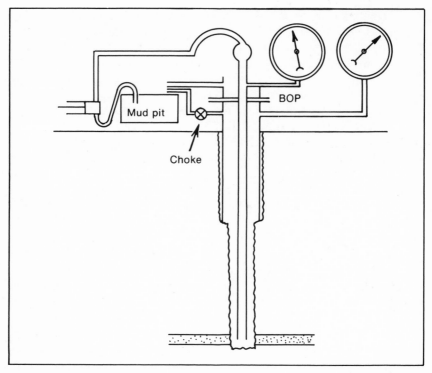

Fig. 13–43 Well showing blowout preventers, choke, and pressure gauges on drillpipe and casing

mud weight necessary to hold back the fluids can be calculated from the drillpipe pressure. Because water of 8.33 lb per gal exerts a pressure of 0.434 psi per ft, each additional lb per gal mud weight exerts an additional pressure of 0.052 psi per ft in depth. (Each additional 0.1 grams per cc exerts an additional pressure of 9.8 kPa/m.)

Shallow high-pressure zones. Abnormally high pressure can develop at any depth. Where the delta of a great river is prograding, coarser sediments are deposited on top of finer sediments, which had originally been deposited when the shoreline was farther back. It may result that the upper sediments are more permeable and therefore compact faster than the underlying finer sediments. When this happens, the finer sediments are undercompacted for their depth and their pore water is sustaining part of the pressure of the overburden. Off the Gulf Coast, there are many places where there is evidence of abnormal pressures. In one case a well blew out at 1,000 ft. It cratered and destroyed the platform. Later, surveys with high-frequency, high-resolution seismic equipment showed a high amplitude reflection (bright spot) which may have indicated free gas in the high-pressure horizon (Figure 12–44).

Economic value of abnormally pressured reservoirs. Some people (Timko and Fertl, 1971) have suggested that overpressured reservoirs are small and possibly not worth drilling, considering the extra cost of handling the high pressure. This may be valid on the Gulf Coast where high pressures occur in a geological facies of fine and lenticular sands. It is not true generally because many giant fields are overpressured, notably Ekofisk, Norwegian North Sea; and Altamont, Utah. These

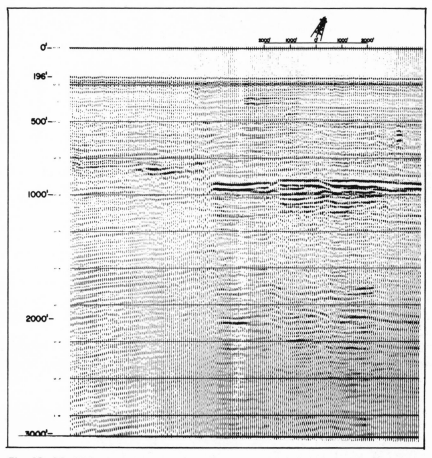

Fig. 13–44 High-resolution seismic profile over the site of a shallow blowout *(after Aquatronics)*

are in reservoir rocks of very low permeability, and the high pressure increases their oil production rate and therefore their economic value.

Geological significance of abnormal pressures. It was pointed out by Hubbert and Rubey in 1959 that when the pressure in the pore water approaches the weight of the overburden, the overlying strata are practically floating.

The weight of the overburden S is sustained by the stress in the skeleton of the solid grains σ and the pore pressure in the interstitial fluids (Figure 13–45)

$$S = σ + p \qquad (13–7)$$

As p increases, σ decreases and may become very small. That is, the solid skeleton is supporting very little weight, and the overlying strata are floating. They can slide under weak lateral forces, such as gravity sliding if the area is tectonically tilted.

Underwater slumping is very common. Contorted bedding may be due to a bed of sand 10 to 100 cm thick peeling off the front of a delta and sliding down the slope, crumpling on itself at the bottom. Most, if not all, low-angle thrust faults probably take place in a zone of abnormally high pressure.

High pore pressure facilitates landslides. Figure 13–46 shows the characteristic features of the kind of slump that causes problems with highways and other structures on mountainsides. Many of the typical features shown can be matched in the growth faults of the Gulf Coast.

The slumping may have taken place at the upper part of the continental slope. The rotation of the slump block gives rise to the back

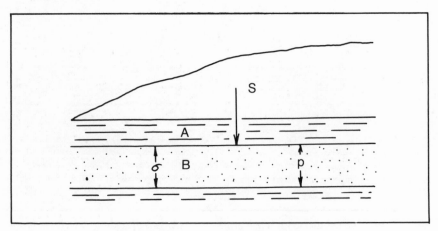

Fig. 13–45 Vertical stress in sediments

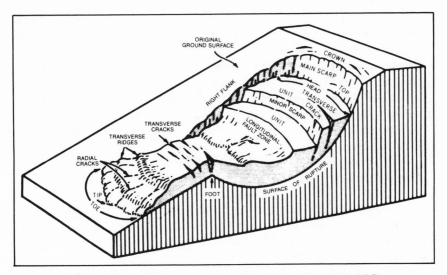

Fig. 13—46 Typical terrestrial landslide *(after Landslides, NRC)*

tilt, which causes the roll-over. Seismic cross sections clearly show the faults and dips of the stratified beds. They also show where the shale has become chaotic below the fault planes (Figure 13–47). Some of these featureless shale zones may be caused by diapirism deep below

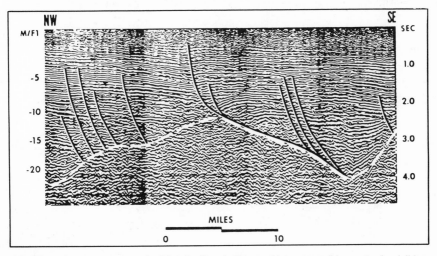

Fig. 13—47 Growth-type faulting in South Texas. Note resemblance to landslides *(after Bruce)*

the surface, while others may represent the toe zone of the slump block where the fault emerges at the surface, part way down the continental slope.

Shale in the chaotic zone is undercompacted and contains fluids at pressures almost equal to the weight of the overburden. An undercompacted bed is less dense than a normally compacted bed. It therefore is unstable and has a tendency to be forced upward.

At point A in Figure 13–48, the downward stress is $h_1\rho_1$ where h_1 is the thickness and ρ_1 the density of the upper bed. It is matched by the upward stress $p + \sigma$. At point B, the downward pressure is $h_2\rho_1$ and the upward pressure is:

$$(h_1\rho_1) - (h_1 - h_2)\rho_2$$

If h_1 is greater than h_2 and ρ_1 is greater than ρ_2, then at point B the upward pressure will exceed the downward pressure by $(h_1 - h_2) \times (\rho_1 - \rho_2)$. The unconsolidated muds will burst upward through bed 1, forming a shale diapir.

Such structures form continually a few km off the mouth of the Mississippi River. Low-density clays (Units I, II, and III) have pushed up through the denser bar sands (Figure 13–49).

Mud volcanoes are similar to the mud lumps but larger. In many respects they resemble salt domes. They are common in areas of recent rapid sedimentation, including the Caspian Sea area of Russia, northwest Colombia, eastern Venezuela, Trinidad, Burma, and the Copper River valley of Alaska. Shales with abnormally high pore pressures have been found to underlie all these areas (Figure 13–50).

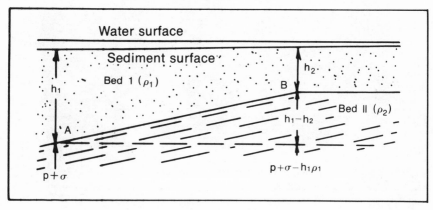

Fig. 13—48 Light, undercompacted shale underlying heavier normal material

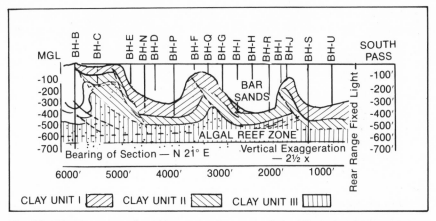

Fig. 13—49 Structure of mud lumps at the mouth of the Mississippi River and elsewhere on the continental shelf of the Gulf of Mexico *(after Morgan et al.)*

It was formerly believed that mud volcanoes were caused by escaping gas because gas sometimes bursts out in large eruptions. Others have abscribed them to tectonic (presumably horizontal) forces. However, it is more likely that they are simply shale diapirs formed when undercompacted shale intrudes overlying strata. The pore water of abnormally pressured shales often contains large amounts of dissolved methane. As the mud travels upward to the surface, the pressure de-

Fig. 13—50 Mud volcano at Bog Boga, Russia, ca. 1900

creases and the gas comes out of solution, forming bubbles that burst loudly and sometimes ignite.

Mud volcanoes are not, as was originally thought, a surface indication of oil at depth. They simply occur in areas of thick shales and sand sediments, which are also favorable for oil and gas.

As compaction and diagenesis proceeds, rocks become less permeable. At very great depths, all the shales and some of the sandstones must become impermeable to water. But compaction and recrystallization release water which has to get to the surface somehow.

For example, if there are 15,000 m of tertiary sediment on the Gulf Coast, temperatures reach 400° C and low-grade metamorphism is taking place. The water must burst it way upward. It is hot and contains silica, calcite, and other minerals in solution. As it migrates upward and cools, the minerals are precipitated, forming hydrothermal veins.

REFERENCES

Alberta Oil and Gas Conservation Board, 1965, Pressure-depth and temperature-depth relationships: Alberta Crude Oils Rept. OGCB 65–9, Calgary.

Allen, T. O., and A. P. Roberts, 1979, Production operations 2, Oil and Gas Consultants International Inc., Tulsa, pp. 141–170.

Barker, Colin, 1972, Aquathermal pressuring: role of temperature in development of abnormal pressure zones: AAPG Bull., v. 56, p. 2068–2071.

———, 1978 Personal communication.

Berry, Fred A. F., 1959, Hydrodynamics and geochemistry of the Jurassic and Cretaceous systems in the San Juan basin, northwestern New Mexico and southwestern Colorado: Stanford, Ca., Ph.D. dissertation, Stanford University.

Boatman, W. A., Jr., 1967, Measuring and using shale density to aid in drilling wells in high pressure areas: Jour. Petrol. Technol., Nov., p. 1423–1429.

Bolt, D. B., Jr., 1972, How to detect overpressure when drilling world wide: World Oil, May, 1972, p. 107–111.

Bond, D. C. and Keros Cartwright, 1970, Pressure observations and water densities in aquifers and their relation to problems in gas storage: Jour. Pet. Techn., Dec., p. 1492–1498.

Bradley, J. S., 1975, Abnormal formation pressure: AAPG Bull., v. 59, p. 957–973.

Bredehoeft, J. D., and Bennett, 1972, Potentiometric surface of Tensleep sandstone in Big Horn basin, Wyoming: U.S. Geol. Surv. open file.

Bruce, C. H., 1973, Pressured shale and related sediment deformation, mechanism for development of regional contemporaneous faults: AAPG Bull., v. 57, p. 878–886.

Campbell, John M., 1977 Discussion of tentative metric unit standards, JPT, December, pp. 1594–1610.

Carstens, Halfdan, and Henning Dypvik, 1981, Abnormal formation pressure and shale porosity, AAPG Bull., February, pp. 344–350.

Chiarelli, A., 1978, Hydrodynamic framework of Eastern Algerian Sahara—Influence on hydrocarbon occurrence. AAPG Bull., v. 62, p. 667–685.

Dickey, Parke A., and K. H. Andresen, 1950, The behavior of water-input wells, *in* P. D. Torrey, ed., Secondary recovery of oil in the U.S.: p. 317–341.

————, 1963, Effect of underground waters in localizing oil accumulations, *in* Economics of the petroleum industry: Dallas Southwestern Legal Foundation, p. 50–81.

Dickey, P. A., and W. C. Cox, 1977, Oil and gas in reservoirs with subnormal pressures: AAPG Bull., v. 61, no. 12, p. 2134–2142.

Dickinson, George, 1953, Geological aspects of abnormal reservoir pressures in Gulf Coast Louisiana: AAPG Bull., v. 37, no. 2, p. 410–432.

Dumont, A. E., and V. S. Purdy, 1976, Use of seismic data can cut arctic drilling costs: World Oil, January, p. 71.

Fertl, W. H., 1976, Abnormal formation pressures, Elsevier, New York, 382 p.

Fowler, W. A., Jr., 1970, Pressures, hydrocarbon accumulation, and salinities—Chocolate Bayou Field, Brazoria County, Texas: Jour. Pet. Techn., April, p. 411–423.

Geertsma, J., 1978, Some rock-mechanical aspects of oil and gas well completions, Proceedings, European offshore petroleum conference and exhibition, London, October 24–27, pp. 301–310.

Goebel, L. A., 1950, Cairo Field, Union County, Arkansas: AAPG Bull., v. 34, p. 1954–1980.

Griffin, D. G., and D. A. Bazer, 1969, A comparison of methods for calculating pore pressures and fracture gradients from shale density measurements using the computer, Jour. Petrol. Technol., v. 21, p. 1463–1474.

Hill, Gilman A., W. A. Colburn, and J. W. Knight, 1961, Reducing oil-finding costs by use of hydrodynamic evaluations, *in* Economics of petroleum exploration, development and property evaluation: International Oil and Gas Educ. Center, Prentice-Hall, Inc., p. 38–69.

Hottman, C.E., and J.H. Smith, 1979, Relationship among earth stresses, pore pressure, and drilling problems, offshore Gulf of Alaska, SPE 7501.

Hollenshead, C. T., and R. L. Pritchard, 1961, Geometry of producing Mesaverde sandstones, San Juan Basin, *in* J. A. Peterson and J. C. Osmond, eds., Geometry of sandstone bodies: AAPG, p. 98–118.

Hottman, C. E. and R. K. Johnson, 1965, Estimation of formation pressure from log-derived properties: Jour. Pet. Techn., June, 1965, p. 717–772.

Hubbert, M. K., and D. G. Willis, 1969, Mechanics of hydraulic fracturing: Soc. Petrol. Engs. Paper 210.

————, and W. W. Rubey, Role of fluid pressure in mechanics of overthrusting: Geol. Soc. America Bull., v. 70, p. 165–206.

Jordan, J. R., and O. J. Shirley, 1966, Application of drilling performance data to overpressure detection: Jour. Pet Techn., v. 18, no. 11, Nov., 1966, p. 1387–1394.

Low, Philip F., 1959, Viscosity of water in clay systems, *in* A. Swineford, ed., Proceedings of the 8th national conference on clays and clay minerals: Pergamon Press, p. 170–182.

Masroua, Luis F., 1973, Pattern of pressures in the Morrow sands of northwestern Oklahoma: Tulsa, Ok., M.S. Thesis, Univ. of Tulsa, 78 p.

Masters, J. A., 1978, Deep basin gas trap, West Canada. Oil and Gas Jour., Sept. 18, 1978.

Mathews, W. R., and J. Kelly, 1967, How to predict formation pressure and fracture gradient: Oil and Gas Jour., Feb. 20, 1967.

McLatchie, A. S., R. A. Hemstock, and J. W. Young, 1958, The effective compressibility of reservoir rock and its effect on permeability: Jour. Petrol. Technol., v. 10, no. 6, p. 49.

McNeal, R. P., 1965, Hydrodynamics of the Permian Basin, *in* Fluids in subsurface environments: Tulsa, Ok., Mem. 4, AAPG, p. 308–326.

Millikan, C. V., and C. V. Sidwell, 1931, Bottom-hole pressures in oil wells: Trans. AIME, v. 92, p. 194–205.

Moore, Preston L., 1974, Drilling practices manual: Petroleum Publishing Company, Tulsa, Ok., p. 448.

Morgan, J. P., J. M. Coleman, and S. M. Gagliano, 1968, Mudlumps; diapiric structures in Mississippi delta sediments, *in* Diapirism and diapirs: AAPG Mem. 8, p. 145–161.

National Research Council, Highway Research Board, 1958, Landslides in engineering practice, *in* E. B. Eckel, ed.: Washington, D.C., 232 p.

Pennebaker, E. S., Jr., 1968, Detection of abnormal pressure formations from seismic field records: API Pap. no. 926–13–C, Southern District Meeting, San Antonio, Tx., March 6–8, 17 p.

————, 1968, An engineering interpretation of seismic data: SPE Paper 2165.

Pollard, T. A., 1977, SI—the international system of units, JPT, December, pp. 1575–1594.

Reynolds, E., 1970, Predicting overpressured zones with seismic data: World Oil, Oct., p. 78–82.

Russell, W. L., 1972, Pressure-depth relations in Appalachian region: AAPG Bull., v. 56, no. 3, p. 528–536.

Sabins, F. F., Jr., 1963, Anatomy of stratigraphic trap, Bisti Field, New Mexico: AAPG Bull., v. 47, no. 2, p. 193–228.

Schultz, A. L., W. T. Bell and H. J. Urbanosky, 1974, Advancements in uncased hole wire-line formation-tester techniques, Soc. Petrol. Engrs.: SPE Paper 5035, 11 p.

Timko, D. J., and W. H. Fertl, 1971, Hydrocarbon accumulation and geopressure relationships and prediction of well economics from log calculated geopressures: Jour. Petrol. Technol., v. 21, p. 923–933.

————, 1972, Implications of formation pressure and temperatures in the search and drilling for hydrocarbons, 4th Canadian Well Logging Symposium, Calgary, Ala., May, Paper E, 10 p.

————, 1972, How downhole temperatures, pressures affect drilling: World Oil, October, p. 73.

Todd, T. W., 1963, Post-depositional history of Tensleep Sandstone (Pennsylvanian), Big Horn Basin, Wyoming: AAPG Bull., v. 47, no. 5, p. 599–616.

Wallace, W. E., Jr., 1965, Abnormal subsurface pressures measured from conductivity or resistivity logs: Log Analyst, v. 5, no. 4, March, p. 26–38.

Young, Allen, Philip F. Low, and A. S. McLatchie, 1964, Permeability studies of argillaceous rocks: Jour. Geophysical Research, v. 69, no. 20, p. 4237–4245.

14

Drillstem and Transient Testing

After a well has penetrated the sand, it is important to find out if it will produce gas or oil (or water) and at what rate. In consolidated formations it is possible to run a special tool downhole on the drillpipe without removing the mud. This is called a *drillstem test*. The well is allowed to produce for only a few hours. If the rocks are soft, it is necessary to run casing and perforate opposite the formation to be evaluated. Then, the well can be produced for several days. If, at the same time, the pressures at the bottom of the hole are recorded, it is called a *transient test*.

1. DRILLSTEM-TEST TOOLS

The drillstem test tool is designed to determine whether a formation will produce fluids by dropping the pressure in the hole so that the fluids will come out and then collecting a sample of the fluids.

Figure 14–1 shows diagrammatically the simplest form of drillstem tester. The device is attached to the drillpipe and run into the hole with the valve closed. The mud is unable to enter the pipe, so the inside of the pipe is empty at atmospheric pressure. A tailpipe of predetermined length hits the bottom of the hole. When this happens, a cone-shaped mandrel slides into a rubber ring, called a packer, forcing it to expand and engage the wall of the hole.

When the packer is seated, the valve is opened, either by dropping a weight which breaks a disk or by lifting or rotating the drillpipe. The mud below the packer is under the pressure of the column of mud to the surface, so it rushes into the drillpipe. If the formation below the packer is porous and permeable, the pore fluids will also enter the hole and follow the mud up the drillpipe. If the formations are permeable

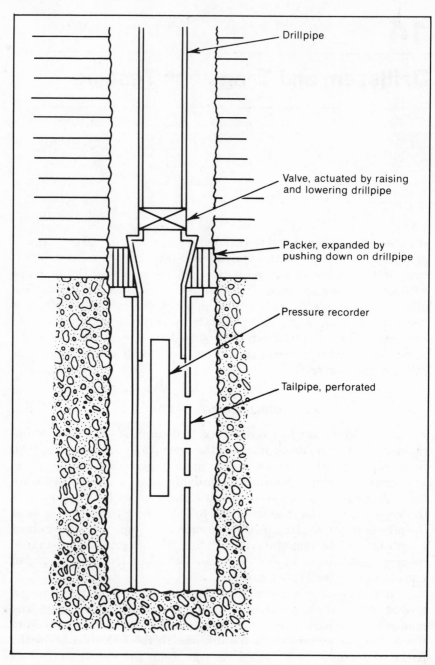

Fig. 14–1 Drillstem-test tool

and productive, the gas or oil may follow all the way up and flow at the surface.

After a short period, the valve is closed and the pressure below the packer is allowed to build up. It is measured by the recorder installed in the tool. It is customary to open and close the valve twice during the test period. Finally, the packer is retracted, which lets the column of drilling mud back on the formation, and the tool is pulled out of the hole.

If the formation produces fluid, it will blow the air out of the drillpipe at the surface. If the fluid is gas, it will come out at the surface. If it is oil or water, it will be found in the lower part of the drillpipe as it is pulled out, often mixed with mud. If the formation is impermeable, only a small amount of drilling mud will be found in the drillpipe above the tool. If the packer fails to seat and shut off the formation, only drilling mud will be recovered. Sometimes oil will flow out to the surface.

There are many variations in design of the tool. If the well is very deep, so that the mud pressure at the bottom of the hole is very high, it is customary to fill the drillpipe partly full of water so the pressure differential across the packer will not be so great. This is called a *water cushion*.

It is possible to have two packers so that only an interval is tested instead of the whole hole below the packer. These are called *straddle packers*. Sometimes the packer is expanded by pumping fluid into it, instead of by a mandrel. These are called *inflatable* or *balloon* packers. They are especially useful if the hole is large or irregular because they will expand to a larger diameter.

Sometimes a valve is provided to bypass the packer. When the test is finished, it is possible to pump mud down the annulus, which forces the formation fluid up the drillpipe to the surface. This is called *reversing out*. An additional pressure gauge can be placed outside the tailpipe so that any pressure drop across the perforations can be recognized.

If the formations are relatively unconsolidated, it is impossible to get the packer to hold. In this case, it is customary to run casing and cement it and then to gun perforate opposite the formations that are considered to be promising from electric log interpretations. The drillstem tester is then run with straddle packers on either side of the perforated intervals.

2. DRILLSTEM-TEST PRESSURE RECORDS

A recording pressure gauge was first inserted in the drillstem test tool in order to ascertain whether the packer seated and the valve opened properly. Over the last 20 years, it has become evident that with good pressure records it is possible to determine lots of important facts about the formation, especially its permeability in place, the pressure of the pore fluids, and whether it has been damaged by the drilling operation. In 1977, the Schlumberger Corporation introduced a tool that measures formation pressure in several different formations successively.

Figure 14–2 is a typical pressure record. The time reads from zero at the left, increasing to the right. The operation usually takes 2 to 4 hours, not counting the time running into and out of the hole.

Starting at the lower left corner, the pressure in the tailpipe increases (upward) with time (to the right) as the tool descends through the mud. Finally, at 1, the tailpipe is on bottom. The valve is then opened, and the pressure drops to a low value as the fluids start to enter the hole at 2. After a few minutes, the valve is closed and the pressure builds up to a value at 3, which should be close to the actual formation pressure. Then the tool is opened again, and flow from the formation resumes at 4. After a longer time—30 to 60 minutes—at 5 the valve is closed again. The pressure now builds up more slowly and usually does not reach the formation pressure at 6 because, during the long second flow period, a substantial amount of fluid has been pro-

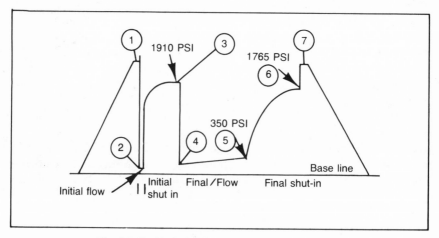

Fig. 14–2 Typical drillstem-test pressure chart

duced from the formation. The valve may then be opened for a third time, although usually there are only two flow periods—a short initial one called the preflow followed by a much longer final flow. If the final flow period is long enough to produce a considerable amount of fluid, and the final shut-in is long enough to give an accurate pressure build-up curve, then information can be obtained on permeability and well-bore damage, as will be shown later.

The packer is then retracted, which lets the weight of the mud back in the hole, at 7. As the tool comes out of the hole, the pressure decreases.

Figure 14–3 is an actual record of a test which produced gas, oil, and water. The time on this chart goes from left to right. The initial flow period (2) was 5 minutes; the initial shut-in period was 90 minutes; the initial shut-in pressure (3) was 859 psi; the final flow period was 60 minutes; the final shut-in pressure (7) was 834 psi (5,750 kPa).

Figure 14–4 is a record of a sand of very low permeability, although gas flowed at the surface one minute after opening the valve. The initial flow (2) was 15 minutes; the initial shut-in period was 60 minutes. The pressure built up very slowly and only attained 1,094 psi (7,542 kPa) when the final flow started. The valve was opened, and pressure during the final flow period (also of 60 minutes) was very low and did not build up. When the valve was finally closed, the pressure reached 1939 (937 kPa) psi after 120 minutes and it was still building.

Figure 14–5 is a test of a sand of very high permeability producing heavy crude oil. The initial shut-in period was 60 minutes, and the

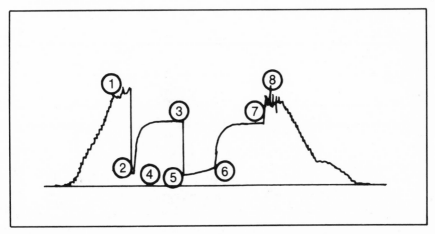

Fig. 14—3 Drillstem test of productive gas sand *(after Lynes United Services, Ltd.)*

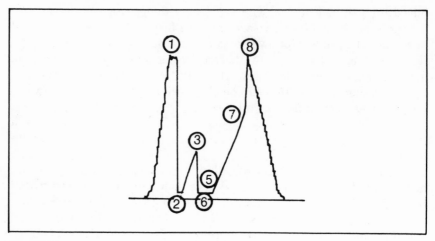

Fig. 14–4 Drillstem test of low-permeability sand *(after Lynes United Services Ltd.)*

pressure built up to 486 psi (234 kPa). The final flow period was 60 minutes, during which time the pressure increased from 151 to 310 psi. The final shut-in period was 90 minutes, and the pressure built up to 435 psi (207 kPa).

Figure 14–6 is an example of a gas well which produced 511 to 700 Mcf per day. The initial shut-in pressure was 563 psi (3,881 kPa), while the final shut-in pressure was 539 psi (3,715 kPa). In spite of a very long (240-minute) final shut-in period, the pressure did not build back to the initial; so the 120-minute flow period may have depleted the

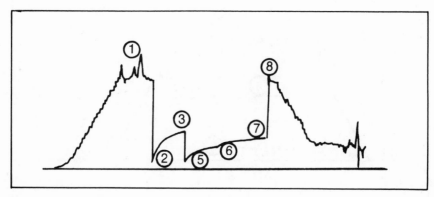

Fig. 14–5 Drillstem test of permeable sand with heavy oil *(after Lynes United Services Ltd.)*

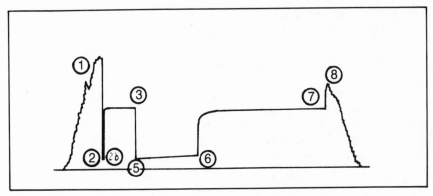

Fig. 14—6 Drillstem test showing depletion *(after Lynes United Services Ltd.)*

sand. It is suspected that the reservoir may be very small, and the production of a small amount of gas drew the pressure down. A quantitative interpretation would be necessary to be sure of this.

Figure 14–7 is a low-pressure, shallow gas sand where serious formation damage is suspected. The well produced 487 Mcf per day through the drillpipe. The flowing pressure was only 140 psi, decreas-

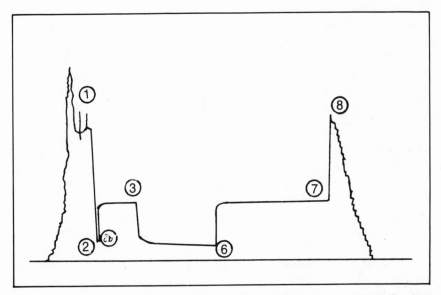

Fig. 14—7 Drillstem test showing formation damage *(after Lynes United Services Ltd.)*

ing to 95 psi at the end of the final flow period. When the valve was closed, the pressure built up very rapidly to 345 psi.

3. TRANSIENT WELL TESTING

If a considerable amount of gas or oil is produced from the discovery well, the fluid pressure in the pores in the vicinity of the hole is reduced. If the well is then shut in, the fluids will flow into the low-pressure zone until the static reservoir pressure is reached. The rate of pressure restoration in the well depends on the permeability of the rock; the pressure finally reached is the original pressure in the reservoir less any depletion that may have occurred. Permeability and pressure are two of the most important measurements needed to evaluate the reservoir. If the reservoir is so small that the amount of fluid produced during the test is an appreciable fraction of the total volume of fluid in the reservoir, the pressure will never be fully restored. However, this pressure drop may exist only in a subreservoir (channel or beach), and the reservoir as a whole may contain very much more fluid and not be depleted at all.

The time available for fluid production during a test depends on the situation at the well. If a drillstem-test tool is in an open, uncased hole, the production time is between 30 minutes and 3 or 4 hours. If the well is cased, the production period may range up to several days. The shut-in time should be at least equal to the production time, especially if the production time is short.

In the case of a drillstem test, it is customary to install an Amerada bottom-hole pressure gauge in the test tool. In the case of an extended test of a cased well, an Amerada gauge may be lowered on a wire line. However, more sensitive gauges may be lowered on electrical cables. They give much more accurate measurements.

A well test can be performed by producing fluid at a constant rate and noticing the decline in pressure (drawdown test). It is, however, often preferable to produce at a constant rate for a period of time and then shut the well in and read the pressure buildup. The changes in pressure with time will be as shown in Figure 14–8. The initial pressure is the reservoir pressure, P^*. The well is supposed to be produced at a constant rate q for a time t. However, if the production rate changes, an approximation of the average rate may be used, dividing the total volume in barrels or cubic meters by the time in hours. Or the effective production time may be calculated by dividing the cumulative production by the final rate. When the well is shut in, the pressure builds up. The time after shutting the well in is called Δt.

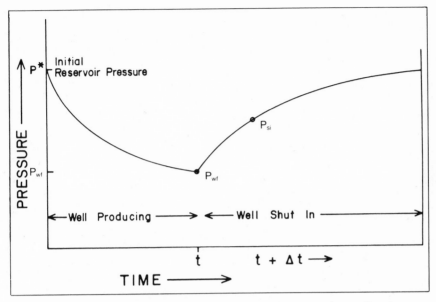

Fig. 14–8 Variation of pressure with time during a transient well test

Horner (1951) showed that for a single well in an infinite, homogeneous reservoir producing a fluid with a small and constant compressibility, the pressure after shutting the well in is given by the following formula:

$$P_{si} = P^* - 162.6 \frac{q\mu B}{kh} \log_{10} \frac{(t + \Delta t)}{\Delta t} \qquad (14\text{--}1)$$

where:

P_{si} = pressure at time Δt, psi
P^* = initial reservoir pressure, psi
q = production rate, stb
μ = viscosity, cp
B = formation volume factor
k = permeability, md
h = permeable formation thickness, ft
t = time well was flowing at production rate q prior to shutin, hr
Δt = time since shutin, hr

A typical Horner plot is shown in Figure 14–9. The pressure is read

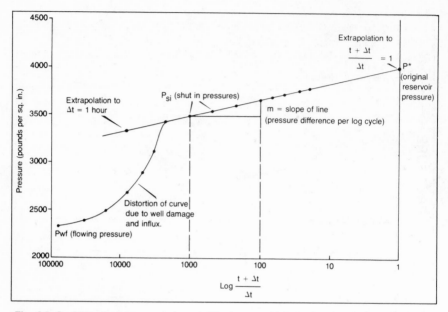

Fig. 14–9 Plot of pressure vs. $(t + \Delta t)/\Delta t$ (Horner plot) made during a transient well test

at intervals after shutting the well in and it is plotted against $(t + \Delta t)/\Delta t$ on a logarithmic scale. The pressure at the instant of shutin is P_{wf}. For a short while after shutin, the pressures are usually too low to fit the theoretical curve. This is because the permeability of the formation in the immediate vicinity of the hole may have been reduced by damage from the drilling mud. This is called *skin effect*. (The pressure drop across the skin can be estimated from the pressure during the flow period and the pressure shortly after closing the well in.) The pressures are also reduced by the flow of fluids into the well bore.

The slope of the straight line portion of the curve (m) is equal to the coefficient of the logarithm term in equation 14–1:

$$m = \frac{P^* - P_{si}}{\log \dfrac{t + \Delta t}{\Delta t}} = \frac{162.6 \, q\mu B}{kh}$$

or,

$$kh = \frac{162.6 \, q\mu B}{m} \qquad (14\text{--}2)$$

Consequently, the difference in pressure for one cycle of the $(t + \Delta t)/\Delta t$ value gives a value from which kh can be calculated. If the thickness of the permeable formation is known from the logs, then the average permeability can be determined. This value will represent the effective permeability to the oil or gas back in the formation. It is, therefore, a better value than can be obtained from cores.

The well-bore damage can be estimated from the pressure during the flowing period, P_{wf}, and the pressure shortly after closing the well in. If the straight-line portion of the curve is extrapolated back to the pressure one hour after closure, the skin factor (s) will be:

$$s = 1.151 \frac{P_{1hr} - P_{wf}}{m} - \log \frac{(k)}{(\phi \mu c r_w{}^2)} + 3.23 \qquad (14\text{--}3)$$

where:

ϕ = porosity
c = compressibility of oil, psi^{-1}
r_w = radius of well, ft

It is very important to estimate the skin effect on a discovery well. It has often happened that the discovery well had a very poor initial production because of well-bore damage. Later, offset production wells were completed more quickly, suffered less damage, and had much higher initial production.

When wells are stimulated, as by facturing, the skin effect is negative.

If the reservoir is of limited extent, the straight-line portion of the curve will change slope at large values of Δt (small values of $(t + \Delta t)/\Delta t$). When extrapolated to 1, the pressure value will be lower than P* and will be the average pressure in the reservoir as a result of the drawdown from production. Many papers describe elaborate calculations in an effort to determine the size and shape of the reservoir from drawdown and buildup tests.

One of the principal reasons for making an extended well test is to decide whether the reservoir is big enough to warrant another well. In order to get any reliable idea of the size of the reservoir, it is necessary to produce an appreciable fraction of the petroleum in the reservoir. If the reservoir is large, the sort of limited test possible on the first well will never achieve this. However, even if no depletion is seen, it is possible to calculate the minimum reservoir volume proven by the test.

If one starts out by assuming the minimum volume required for the field is economic, the amount of oil that must be produced on the test may be calculated.

With increasing application of improved recovery methods, it is becoming obvious that most large sandstone reservoirs consist of a series of subreservoirs more or less separated from each other by hydraulic barriers. These are usually long and narrow because they were deposited as river channels or beaches. They may be side by side, like shingles, or stacked on top of each other.

The several treatises on well testing listed in the references all mention reservoir heterogeneity, but none of them addresses the problem of how to interpret the pressure transient test of a well that penetrated a subreservoir of small extent that formed a part of a large reservoir.

The permeability values determined by transient well tests will also be affected by reservoir heterogeneity. The transient test gives a value for kh, which is thickness multiplied by permeability. To get permeability, kh is divided by h (sand thickness), which is determined by electric logs. Unfortunately, logs do not distinguish between permeable and impermeable sands. If much of the sand body is too impermeable to contribute substantially to the flow, then h may be grossly overestimated. In that case, k will be underestimated.

The same calculations of permeability and damage ratio may be made on successful drillstem tests. The production and buildup are much shorter, so the estimates are much less accurate. Particular caution should be exercised if the test suggests a small reservoir.

Recently, many papers have appeared in the literature that discuss methods of mathematically interpreting pressure buildup tests. Only a few of these are listed in the references.

REFERENCES

Carslaw, H. S., and J. C. Jaeger, 1959, Conduction of heat in solids: Oxford, England, Clarendon Press.

Earlougher, Robert C., Jr., 1977, Advances in well-test analysis, Society of Petroleum Engineers, 264 pp.

Horner, D. R., 1951, Pressure build-up in wells: Proc. 3rd World Petrol. Congress, Sect. 2, The Hague. Reprinted in SPE Reprint Ser. No. 9, 1967, p. 25–43.

Johnston Testers, Inc., 1972, Review of basic formation evaluation: Johnston, P. O. Box 36369, Houston, Texas, 39 p.

Lynes, 1978. Drill-stem test analysis and interpretation: Lynes United Services Ltd., Calgary, Alberta, Canada, 63 p.

Mathews, C. S., and D. G. Russell, 1967, Pressure buildup and flow tests in wells: SPE of AIME Monograph, No. 1, Dallas, 167 p.

Miller, C. C., A. B. Dyes, and C. A. Hutchinson, Jr., 1950, The estimation of permeability and reservoir pressure from bottom-hole pressure build-up characteristics: AIME Trans. v. 189, p. 91–104. Reprinted in SPE Reprint Series No. 9, 1967.

Theis, C. V., 1935, The relationship between the lowering of piezometric surface and rate and duration of discharge of wells using groundwater storage. Amer. Geophy. Union Trans., v. 2, p. 519.

15

Fluid Behavior in Reservoirs

1. CAPILLARY PROPERTIES OF ROCK, OIL, AND GAS

Capillary properties. Water in sands is governed by capillary forces if the sands also contain oil or gas.

Forces between water molecules are balanced in the middle of the fluid but are unbalanced on the surface, Figure 15–1. This gives rise to a tension in the interface, called surface tension. Surface tension is measured by the force necessary to pull a wire out of the water (Figure 15–2). If the water completely wets the surface, the contact angle is 0°. If it does not wet the surface at all, the contact angle is 180°. Often, the surface is intermediate in wettability and there is a contact angle θ (Figure 15–3). A good deal depends on which fluid the surface was recently exposed to and the composition of the oil.

Capillary pressure, P_c, is the difference in pressure between the oil

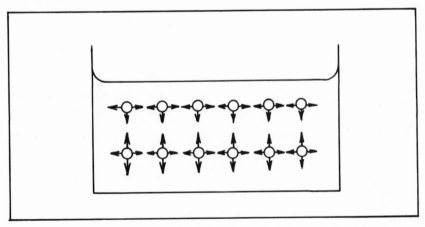

Fig. 15–1   Balanced and unbalanced forces between molecules

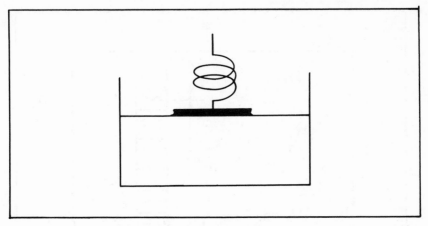

Fig. 15–2 Measuring surface tension

(or gas) and the water. It increases with increasing surface tension and with decreasing pore width.

If a small-diameter glass tube is placed in a beaker containing water and oil (Figure 15–4), the water level in the tube will rise above the level of the oil-water contact because the glass is preferentially wet by water. The water will rise until the pull of the tension in the glass-water interface is equal to the weight of the column of water. The upward pull is the circumference of the tube $2\pi r$ times $\sigma \cos \theta$, where σ is the surface tension and θ is the contact angle.

The weight of the column of water is the area of the tube πr^2, times the height of the column h_2, times the density of the water ρ_w, less the density of the oil ρ_o, times the force of gravity g or $\pi r^2 g h_2 (\rho_w - \rho_o)$. Therefore

$$2\pi r\sigma \cos \theta = \pi r^2 g h_2 (\rho_w - \rho_o)$$

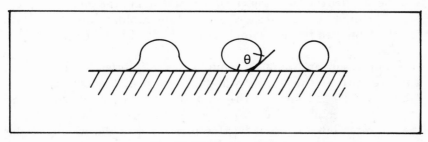

Fig. 15–3 Contact angle

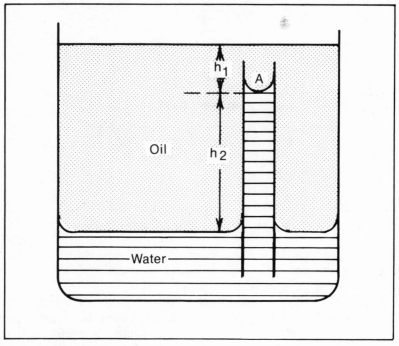

Fig. 15–4 Water will rise in the tube until the upward pull in the meniscus balances the weight of the water in the tube

Solving for h_2:

$$h_2 = \frac{2\sigma \cos \theta}{rg\,(\rho_w - \rho_o)} \qquad (15-1)$$

The pressure in the oil at the oil-water interface is higher than in the water across the interface. This is because the interface is curved, and the smaller the radius of curvature the higher the pressure difference. The difference in pressure is called the capillary pressure, P_c. It may be computed in the example of Figure 15–4 as follows:

The pressure in the oil at point A at the interface will be $h_1\rho_o g$. The pressure at the flat oil-water interface at the bottom of the beaker will be the same in the oil and in the water. It will be the weight of the column of oil, $(h_1 + h_2)\,\rho_o g$. The pressure in the tube at the oil-water interface will be this pressure less the weight of the column of water, $h_2\rho_w g$. Consequently:

$$P_c = h_1\rho_o g - (h_1\rho_o g + h_2\rho_o g - h_2\rho_w g)$$
$$P_c = h_2 g(\rho_w - \rho_o) \tag{15–2}$$

Substituting the value of h_2 from equation 15–1,

$$P_c = \frac{2\gamma \cos \theta}{r} \tag{15–3}$$

This pressure difference will be zero at the oil-water contact and will increase with the height above the oil-water contact according to equation 15–2 (Figure 15–5). On the average, water weighs about 0.5 psi per foot, and oil weighs about 0.4, so the capillary pressure will be about 0.1 psi per foot (2.2 kPa/m) of height above the oil-water contact.

It is usually assumed that a natural oil sand was originally saturated with water and that the oil, by some unknown route, entered the reservoir. Quartz and most minerals are preferentially wet by water, so the grains are coated with water. The amount of oil that entered the reservoir would be no more than could displace the water at the available capillary pressure difference.

As the water enters the sand, it pushes the oil out of the way until the pressure difference between the oil and the water stops further movement, as at A (Figure 15–6). If more oil enters the reservoir so that the column of oil below increases in height, the difference in pressure between the oil and water will increase, and the oil will move farther back in the pore until the new smaller radius of curvature of

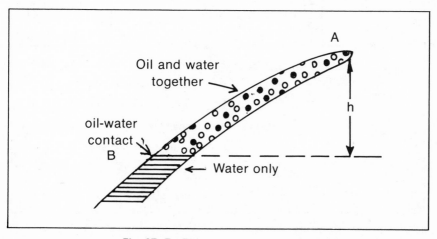

Fig. 15–5 Diagram of an oil reservoir

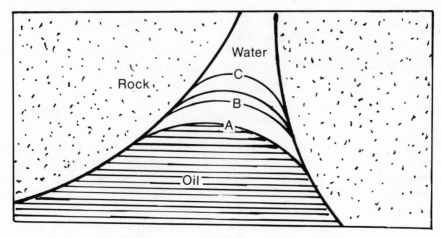

Fig. 15—6 Curvature of oil-water contact in sands

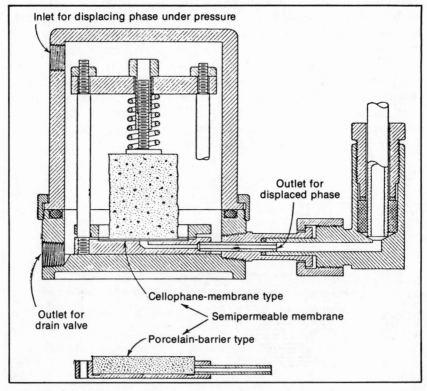

Fig. 15—7 Capillary pressure device *(after Russell and Dickey, courtesy John Wiley and Son)*

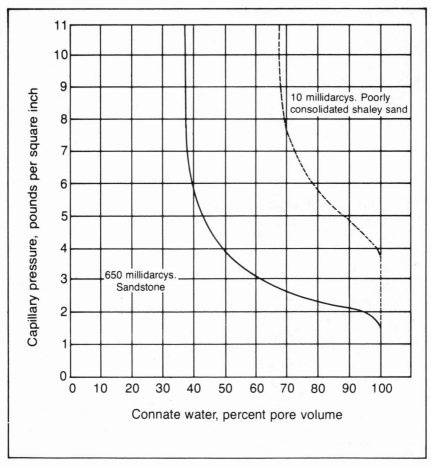

Fig. 15—8 Curve of water saturation vs. capillary pressure *(after Russell and Dickey, courtesy John Wiley and Son)*

the interface stops further entry. For this reason, oil cannot enter pores smaller than a certain width.

Capillary pressure in sandstone is measured by saturating a clean sample of rock with water and placing it in a closed pressure vessel on a plate permeable to water but not to oil (Figure 15–7). Oil (or gas) is admitted to the vessel and the pressure is increased stepwise. The water which saturates the core is able to seep out into a fine graduated tube.

At first no oil enters the pores, but when the oil pressure is enough to force the oil into the larger pores some water is expelled. Each time the pressure is increased stepwise, a little more oil enters, as shown in Figure 15–8. Finally, when the 850 md sand contains 38% water and 62% oil, no more oil will enter the pores no matter how high the pressure is raised. This amount of water is called the *irreducible water saturation*. In the case of the shaly 10-millidarcy sand, the water between the clay platelets could not be displaced, and the irreducible water saturation was 68%. This water was immobile, and the sand would have produced clean oil.

2. RELATIVE PERMEABILITY

The concept of permeability as a property of the rock, defined by the constant in the Darcy equation, was presented in Chapter 3. Often oil, gas, and water are all found together in the same set of pores. These are three fluids which are mutually immiscible. The concept of permeability now becomes somewhat more complicated.

Imagine a piece of sandstone with a permeability of 100 md. If it contains only gas and no oil and we measure its permeability to gas, we get a value which is the same as its absolute permeability, that is, 100 md. Now let us introduce some oil into the sandstone, filling, say, 50% of the pores. Obviously, the oil will impede the passage of gas through the rock, and its measured permeability will be somewhat less. Now introduce more oil into the rock, so that it is 100% saturated with oil. The permeability of this rock to gas will now be zero because it will be impossible to force any gas through it without displacing some of the oil. However, now that the rock is 100% saturated with oil, it will have 100% of its specific permeability, that is, 100 md permeability to oil. The permeability to gas drops to zero as the percent of liquid saturation in the rock increases, while the permeability to liquid increases as the liquid saturation increases (Figure 15–9). It is important to note that the two curves do not cross at 50%, but rather at 73% liquid saturation. At this point the permeability to gas and to liquid is the same and only about 10% of the specific permeability for either liquid alone. Liquid oil will wet the rock surface while gas does not. The small amounts of oil up to 50 or 60% liquid saturation are attracted into the fine pores, leaving the large pores available for the gas to flow through. Consequently, when the pores are 40% filled with liquid, the permeability to gas is decreased only 10%. When the liquid fills 75% of the pores, it seriously impedes the flow of gas without being able to flow itself very well. At this point, the permeability to either gas or liquid is only about 10% of the specific permeability. The fact that the liquid

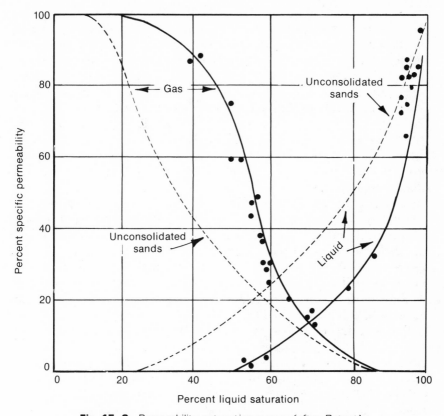

Fig. 15–9 Permeability-saturation curves *(after Botset)*

wets the rock and the gas does not makes gas a very ineffective driving fluid. When 50 to 60% of the pore volume is still full of liquid, gas goes freely through the pores without removing any more oil.

3. PRIMARY PRODUCTION MECHANISMS

Dissolved gas drive. When the average reservoir is first penetrated by the drill, the available pore space contains about 30% water, mostly in the smaller interstices and on the grain surfaces, and 70% oil. Dissolved in the oil is a varying amount of gas. Frequently there is more than enough gas in the reservoir to go into solution in the oil, and in this case a gas cap will be present above the oil.

When the oil well is completed and production begins, a low-pressure area develops in the vicinity of the well bore. This area rapidly expands and, after a certain amount of oil has been withdrawn, the pressure in the fluids becomes less than the saturation pressure of the dissolved gas. When this occurs, the gas comes out of solution in the pores of the sand. This pressure is called the *bubble point.* At first, the small disconnected bubbles expand, pushing the oil toward the well. Later, however, the bubbles coalesce and the gas starts to move as a separate flowing phase, also toward the well. When this happens, the oil-producing rate is substantially reduced. The relative permeability of the sand to gas increases rapidly, while that to oil decreases. Consequently, the gas production of the well increases and the oil production decreases. The production of fluids has been largely brought about by the pressure gradient toward the well maintained by the expansion of the gas originally dissolved in the oil. When this gas is able to travel freely to the well bore, driving only small quantities of oil with it, the efficiency of the recovery mechanism falls off very rapidly. Furthermore, as the gas itself is exhausted, its flow eventually also decreases. This mechanism of depletion of oil pools is known as *dissolved gas drive.*

Figure 15–10 shows the production history of a portion of the Rodessa field in Texas. The field was drilled mostly during 1937, and by the early part of 1938 approximately 100 wells were producing. The maximum production came during the later part of the drilling program, and it amounted to more than 10,000 bbl (1,400 tons) per day in 1938. Only two years later, in early 1940, oil production had declined to about 2,000 bbl (300 tons) per day. This decline in production was brought about by a tremendous decline in the reservoir pressure, which fell from more than 2,500 psi (175 kg per sq cm) when the field was new to only about 200 psi (14 kg per sq cm) in 1940. Notice the behavior of the gas-oil ratio, which increased from about 500 cu ft (99 cu m per ton) per barrel (probably close to the amount of gas originally in solution in the oil) to almost 8,000 cu ft per barrel (1,520 cu m per ton) during the period of rapid pressure decline. Eventually, when the gas itself became exhausted, the gas-oil ratio fell to about 2,000 cu ft per barrel (400 cu m per ton). When the gas is gone, as much as 70 to 80% of the original oil in place may still be in the reservoir.

Water drive. Many oil fields are surrounded by water, usually downdip, at a lower elevation than the oil. When the size of the *aquifer*—the permeable reservoir rock containing the water—is relatively small, the effect of the surrounding water, usually called *edge water*, is negligible. If the reservoir rock is filled with water for a long distance away from the oil so that there is a very large volume of water in the aquifer, it will tend to move into the oil reservoir when the pressure

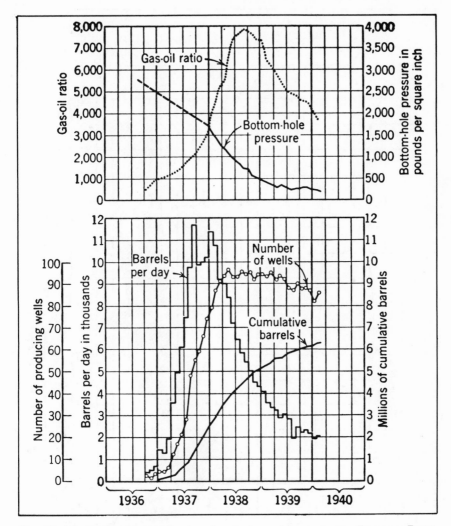

Fig. 15–10 Production history of a dissolved gas-drive reservoir, Rodessa, Texas (after Mullane)

declines. This is because the water expands slightly as the pressure is taken off of it, and it is able to displace the oil, occupying the pores previously occupied by oil. When the water is able to encroach into the field in this manner, quite a different reservoir performance mechanism occurs.

Figure 15–11 shows the production history of the North Searight

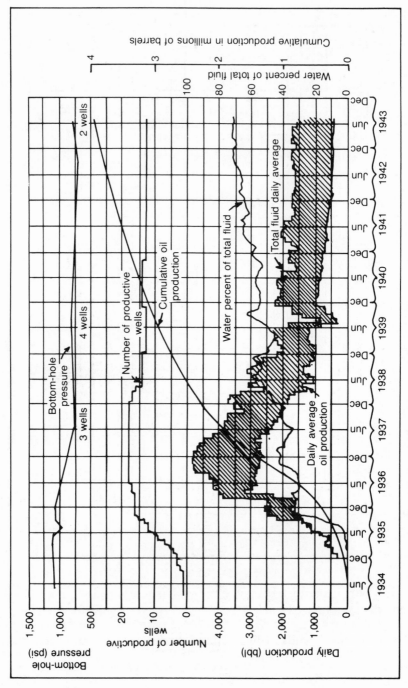

Fig. 15–11 Production history of north Searight field *(after Mullane)*

field in Oklahoma. The producing rate increased rapidly during the drilling just as in the preceding example. Shortly after the oil production hits its maximum, water started to appear and was produced in increasing quantities, at first somewhat less than that of the oil, but later the water production vastly exceeded the oil production. In spite of this large production of fluid (both water and oil), the bottom-hole pressure in the reservoir dropped only slightly, from a little over 1,000 psi to about 760 psi (70 to 53 kg per sq cm), and this pressure was maintained for many years.

This reservoir performance is called *water drive*, and it is much more effective than gas drive. Between 10 and 60% of the pore volume remains filled with oil at the time the wells produce too little oil to be worth pumping any more.

Gravity drainage (and gas-cap drive).   If the reservoir has a very high permeability and a comparatively steep dip, substantial amounts of oil flow down through the rock by gravity alone. This mechanism is called gravity drainage, and in pools with favorable geology it is the most efficient of all. It results in the formation of an artificial gas cap above the oil. If all of the gas produced with the oil is injected into the top of the structure, it maintains the pressure in the gas cap and therefore the whole reservoir. This keeps most of the dissolved gas in the oil. The oil viscosity remains low, and the differential pressure between the reservoir and the wells remains high so that production rates are maintained better. Figure 15–12 shows the production history of a gas-cap drive field.

Compaction drive.   When the pressure in the pore fluids in the reservoir drops, there is an increase in the pressure on the solids because the pore fluid pressure no longer helps sustain the weight of the overburden. This results in a small reduction in the volume of the pores to about 7×10^{-6} of the pore volume per psi (10^{-4} pv per kg per sq cm). A drop in reservoir pressure of 1,000 psi (70 kg per sq cm) would result in a decrease in the volume of the reservoir of 0.7%. It therefore contributes slightly to the recovery.

Over certain oil fields in Tertiary, relatively unconsolidated sediments, there has been a subsidence of the surface. It has amounted to over 20 ft at Wilmington, California, and Lagunillas, Venezuela. In this case, the subsidence has resulted from rearrangement of the grains. If the sands compact, the reservoir volume is decreased. If, as seems more likely, it is the interbedded shales which compact, then their pore water is injected into the adjacent oil sands. These effects amount to a substantial volume of oil being forced out of the sand toward the producing wells.

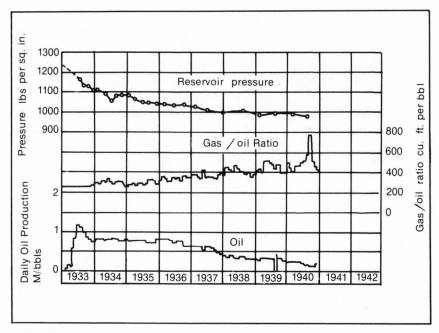

Fig. 15–12 Production history of gas-cap drive field *(after Clark, courtesy SPE)*

Primary recovery efficiency. An elaborate study was conducted by a committee of the American Petroleum Institute in an effort to determine the average efficiencies of the different reservoir producing mechanisms. Using fields for which good data was available, the conclusions were as shown in Table 15–I.

Water drive, on the average, recovers 600 bbl per acre-ft (800 cu m per h-m), while solution gas drive recovers about 150 bbl per a-f (200 cu m per h-m). Gravity drainage, which is infrequent, is the most efficient.

The maximum recovery efficiency in each case was about 3 times the minimum. The effects of several rock and fluid properties were investigated statistically. It was found that the important variables, in order of their importance, were original oil in place, mobility (that is permeability divided by viscosity), water saturation, and the ratio of initial pressure to final pressure.

In the case of both water- and gas-drive reservoirs, the more original oil in place, the greater the efficiency. The greater the mobility, the greater the recovery efficiency. This means that the more permeable the reservoir or the less viscous the oil, the greater is the recovery in

TABLE 15–I

RECOVERY EFFICIENCY OF SEVERAL PRIMARY PRODUCTION MECHANISMS
(FROM API STATISTICAL STUDY OF RECOVERY EFFICIENCY,
J. J. ARPS, CHAIRMAN, 1967)

	Sandstones			Carbonates		
Water drive	Minimum	Median	Maximum	Minimum	Median	Maximum
Recovery, bbl per acre-foot	155	571	1,641	6	172	1,422
(cu m per h-m)	(199)	(735)	(2,113)	(8)	(221)	(1,831)
Recovery efficiency, percent*	28	51	87	6	44	80
Solution gas drive						
Recovery, bbl per a-f	47	154	534	20	88	187
(cu m per h-m)	(60)	(198)	(688)	(26)	(113)	(241)
Recovery efficiency*	9	21	46	15	18	21
Gas cap drive	All rock types combined					
Recovery, bbl per a-f	68	289	864			
(cu m per h-m)	(88)	(372)	(1,113)			
Recovery efficiency*	16	32	67			
Gravity drainage						
Recovery bbl per a-f	250	696	1,124			
(cu m per h-m)	(322)	(896)	(1,448)			
Recovery efficiency*	16	57	63			

*Percent stock tank oil originally in place.

percent of original oil in place. The water saturation had the opposite effect on water- and gas-drive reservoirs. In the case of water-drive reservoirs, the higher the water saturation, the poorer was the recovery; in the case of gas-drive reservoirs, the higher the water saturation, the better the recovery. This is because the presence of the water in the pores helps the gas push on the oil instead of blow past it. If the ratio of initial to abandonment pressure is high, less oil is recovered from water-drive fields while more oil is recovered from gas-drive fields.

In estimating the recoverable oil, the recovery efficiency term in Table 15–I should be modifield by these other factors, if they are known.

A study by a special committee of the American Petroleum Institute (Craze and Buckley, 1946), using data from some of the same reservoirs, reached the conclusion that well spacing was not a factor in recovery efficiency; that is, drilling more wells into a reservoir did not recover more oil—it simply recovered it faster.

This conclusion gave rise to an enormous amount of controversy. The geological and engineering factors were often confused by economic interests. In general, the major oil companies accepted these findings and argued for wide spacing; one well to each 40, 80, or even 160 acres (16, 32, or 64 hectares). This required fewer wells and reduced the development costs. Smaller companies, on the other hand, generally needed larger cash flows, which required closer spacing even though the development cost was greater. The well spacing in California generally was less than that in the Midcontinent, apparently because the huge thickness of pay sand, sometimes more than 1,000 ft (300 m), was sufficient to pay the cost of the extra wells.

Craze and Buckley started with the basic assumption that the reservoir rock in most reservoirs is homogeneous and isotropic. If that were true, then one properly placed well could deplete the whole reservoir. More wells simply would result in producing the oil faster. With more experience in secondary recovery, it is now recognized that reservoirs are never homogeneous and isotropic. If the subreservoirs cover less than 40 acres, as they often do, then a 40-acre spacing will not recover all the oil. In recent years there has been a large amount of infill drilling; that is, drilling new wells in an old and depleted field mainly in connection with secondary recovery. This has resulted in substantial increases in recovery from the pool. Some of the increase is due to the extra wells, which drain previously untapped portions of the reservoir. We may conclude that Craze and Buckley were wrong, and that closer spacing does increase the recovery, especially in the case of secondary recovery. For a good sweep, it is necessary to have at least one injection and one producing well in each little beach or channel that makes up the main reservoir.

4. ESTIMATION OF OIL RESERVES

It is the joint responsibility of development geologists and engineers to estimate the recoverable reserves of an oil or gas field. In the case of remote fields in hostile environments, it is necessary to decide if the reserves are sufficient to warrant development. In the case of a new discovery in an oil-producing region, the oil reserves must be estimated in order to raise the money for the development of the field.

Estimation of reservoir volume. The first step in calculating oil or gas reserves is to estimate the volume of the reservoir. To do this, contours are drawn on the thickness of oil-saturated sand. The area within each contour line is measured and expressed in acres or hectares. Each area is then multiplied by the contour interval to give the volume in acre-feet or hectare-meters. One acre-foot is the same as 7,758 42-gallon barrels and one hectare-meter is 10,000 cubic meters. If the total volume of sand is multiplied by the porosity, it will give the volume of the pores; if the product is then multiplied by the oil saturation, it will give the volume of oil in place in the reservoir. Since the oil saturation is usually calculated from the electric logs, some companies multiply the total sand volume by $\phi(1 - S_w)$, where ϕ is the porosity and S_w is the water saturation.

Inasmuch as there is shrinkage of the oil resulting from the escape of gas as it is produced, it is usual to divide the total oil in place by the formation volume factor. This converts *reservoir barrels* to *stock-tank barrels*. The result of the calculation gives *stock-tank oil initially in place* (sometimes denoted as STOIIP).

A major source of error in estimating reservoir volume is the tendency to include sands of low permeability. Every sand body consists of a series of strata of different permeabilities, ranging from some maximum down to zero. The wireline logs do not determine permeability, so normally it is impossible to estimate the thickness of sand with sufficient permeability to produce commercial amounts of oil. This tendency to include noncommercial sands when estimating reservoir volume may have given rise to serious underestimations of recovery factors. Maybe our reservoirs have produced more efficiently than we thought, and maybe we have been leaving less oil behind.

In the case of carbonate reservoirs with intergranular porosity, it is possible to estimate reservoir volume in the same way as estimating sands. On the other hand, in the case of fractured and vuggy reservoirs, it is almost impossible to estimate the porosity. Therefore, it is difficult to come up with a reliable value for oil in place.

Another source of error is not evaluating correctly the volume of the wedge of sand as it dips gently into the edge water. On most oil-field maps, the oil-water contact is drawn as a single line where the oil-water contact is at the top of the sand. Downdip from this line, there will be no oil, updip from the line, the oil-saturated part of the sand gradually thickens. To assume that the entire thickness of sand is saturated updip from the oil-water contact will result in an over-estimation of the volume.

Oil-water contacts are by no means always horizontal planes; they may dip scores or even hundreds of meters per kilometer. As the field develops, efforts should be made to determine the tilt of the oil-water contact.

Another method of estimating the total oil in place is called the *material-balance equation*. It is assumed that the total cumulative production at the surface will be equal to the expansion of the fluids in the reservoir as a result of the pressure drop. As the pressure drops, gas is released from solution in the oil, causing a shrinkage of the oil. The gas is highly compressible, so that as its pressure drops its volume expands.

One form of the material balance equation is as follows:

$$N_p [B_o + (R_p - R_s)B_g] = N [(B_o - B_{oi}) + (R_{si} - R_s) B_g] \quad (15\text{--}4)$$

$$\underbrace{\qquad\qquad}_{\substack{\text{volume of produced oil}\\ \text{and gas}}} \qquad \underbrace{\qquad\qquad}_{\substack{\text{expansion of the oil plus}\\ \text{originally dissolved gas}}}$$

where:

N_p = cumulative oil production
B_o = oil formation volume factor at the current pressure
R_p = cumulative gas-oil ratio
R_s = solution gas-oil ratio at the current pressure
B_g = gas formation volume factor; this is the volume of gas at the surface, atmospheric pressure and 60°F, divided by the volume of the same amount of gas at the reservoir pressure and temperature.
N = oil originally in place, reservoir barrels, at bubble point.
B_{oi} = oil formation volume factor at the original pressure
R_{si} = original solution gas-oil ratio

The reduction of pressure in the reservoir results in a small elastic expansion of the water and the rock minerals. The reservoir itself is compressed slightly because the decreased fluid pressure supports less of the weight of the overburden. This results in a reduction of the reservoir volume of about 10×10^{-6} pore volume per pore volume psi (1.4×10^{-6} pv/pv/kPa). If the pressure drops 1,000 psi, the reservoir volume will contract 10×10^{-3}, or 1 percent. Considering that the reservoir volume cannot be estimated to this accuracy, it seems hardly worthwhile to take the compressibility into account.

Much more important is the effect of the encroachment of water from the aquifer downdip. The reduction in pressure in the reservoir causes a reduction in pressure in the aquifer. This reduction in pressure causes the aquifer to compact, which forces the water into the reservoir. The amount of compaction is about the same as that of the reservoir. In the case of a thousand-pound (7,000 kPa) pressure drop, one percent of the aquifer volume would be forced into the reservoir. In the case of permeable aquifers covering thousands of square kilometers, this amounts to a large volume of water. However, it takes time for the pressure drop to reach all parts of the aquifer, especially if it is large. Usually it is difficult to determine the size and reservoir characteristics of the aquifer because reservoir data are not ordinarily obtained for wells that produce only water. In the case of many pools, especially those in small stratigraphic traps or multiples of stratigraphic traps, the aquifer is so small there is no advance at all. In the case of many fields, the water advances a few hundred meters and then stops. When it stops advancing, the field begins to produce by the inefficient dissolved-gas drive mechanism. Water can be injected into the aquifer to keep it coming, and this is often successful. However, downdip water injection is often unsuccessful because of inhomogeneities in the reservoir or an asphalt layer at the oil-water contact.

Another very valuable method of estimating oil reserves consists of plotting the daily or monthly production of an oil field or oil-producing property. If the production rate is unrestricted—that is, the wells are allowed to produce to their maximum capacity—the rate will decline. The decline rate is logarithmic; it follows the equation $q - A = K (t - B)^{-n}$ where q is production rate; A, K, B, and n are constants; t is time.

If the production rate vs. time is plotted on log-log paper, the plot will form a straight line that can be extrapolated to the *economic limit*, that is, the production rate q at which it is no longer profitable to operate the wells.

Often the production decline curve is plotted on semilog paper, which will be a straight line for a constant percentage decline per year. The production rate is plotted on the up-and-down logarithmic scale and the time on the horizontal arithmetic scale. This method is somewhat less realistic but more convenient, as the time scale is linear.

The decline curve method (Fig. 15–13) of prediction requires the condition that no changes in operating method will take place. Where production is partly restricted for technical or economic reasons, production decline curves cannot be used. However, they provide very

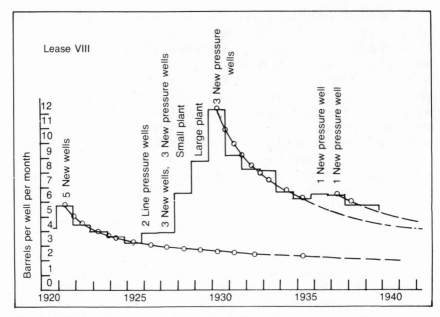

Fig. 15–13 Production decline curve of an oil-producing property subjected to gas injection

valuable and often significant information. A change in the position or slope of the curve indicates a change in the recovery mechanism. If the production curve of a dissolved gas-drive reservoir reaches a low value and then flattens, this indicates water encroachment, either from edge water or possibly water from abandoned, improperly plugged wells. The production decline curves can be used to evaluate quantitatively the result of anything done to the operation, including secondary recovery.

Decline curves can also be used in the case of a secondary recovery operation.

5. ESTIMATION OF GAS RESERVES

It is usually easier to estimate gas reserves than to estimate oil reserves. It is only necessary to measure the pressure in the reservoir, produce a known amount of gas, and then measure the pressure again. Boyle's law says that, at a given temperature, the product of pressure times volume of an ideal gas is constant (PV = k).

If we produce a certain amount of gas ΔV, the reservoir pressure will drop correspondingly:

$$V - \Delta V = (P - \Delta P)k$$

The remaining volume of gas in the reservoir can then be calculated. The more complete formula for an ideal gas is:

$$pV = nRT \tag{15-5}$$

where:

p = pressure, psi or kPa
V = volume, cu ft or cu m
T = absolute temperature, °R. or °K.
n = number of pound-moles or kilomoles (one pound-mole is the molecular weight of the gas expressed in pounds)
R = universal gas constant, which for English units is 10.732 psi, cu ft/lb-mole, °R, and for metric units is 8.314 × 10^7 cu m Pa/gram-mole, °K.

This formula is applicable only for pressures near atmospheric. At higher pressures the molecules are packed closer together and interact with each other. It is necessary to introduce a factor Z that depends on the composition of the gas and that varies with both temperature and pressure. The useful formula is

$$pV = Z\,n\,R\,T$$

or

$$\left(\frac{P}{Z}\right) V = n\,R\,T \tag{15-6}$$

One way to determine Z is to place a sample of gas in a cylinder and compress it with a piston, changing the pressure and maintaining the temperature constant at the reservoir temperature. If the chemical composition of the gas is known, it is possible to ascertain the Z-factor from tables or curves.

If the gas field is strictly volumetric (there is no water influx) then:

$$\frac{P_{sc}\,G_p}{T_{si}} = \frac{P_i\,V_i}{Z_i\,T} - \frac{P_f\,V_i}{Z_f\,T} \tag{15-7}$$

where:

P_{sc} = absolute pressure at standard conditions, 14.7
G_p = cumulative gas produced, cu ft or cu m
T_{sc} = temperature at standard conditions, °R or °K.
P_i = absolute initial pressure in reservoir, psi or kPa
V_i = absolute initial pore volume of gas [$\phi\,(1 - S_w) \times$ reservoir bulk volume]

Z_i = gas deviation factor at initial reservoir pressure
T = temperature in reservoir, °R or °K
P_f = absolute final pressure, psi or kPa
Z_f = gas deviation factor at final pressure

For any given reservoir, P_{sc}, T_{sc}, P_i, V_i, Z_i, and T do not change during production. Equation 15–7 can therefore be written

$$\frac{P_f}{Z_f} = mG_p + \frac{P_i}{Z_i} \qquad (15\text{--}8)$$

The value m is a constant. This is the equation of a straight line. We can, therefore, plot cumulative production against pressure divided by Z. The total gas in place can be determined graphically by extrapolating the curve to zero pressure to get the total gas initially in place or to any abandonment pressure to get the total recoverable gas (Figure 15–14).

6. WATERFLOODING

The recovery efficiency of pools producing under dissolved gas drive is notably low—the statistical study just quoted shows that it averaged less than 25 percent of the original oil in place. Water drive fields, on the other hand, averaged about 50 percent. It has therefore been possible to obtain a second crop of oil by injecting water through special wells. The water then pushes the oil to the producing wells.

In Pennsylvania in the 1860s, water from the ground water occasionally entered the sand through wells whose casing had become defective. This water travelled to neighboring properties, causing oil wells to produce water instead of oil. Laws were passed by the Pennsylvania legislature prohibiting operators from permitting water to enter oil wells. It was noticed that the water was often preceded by a great increase in oil production. In 1880, the first petroleum geologist and engineer, John F. Carll, pointed out that deliberate water injection might result in increased oil recovery because "all the oil cannot be withdrawn from the reservoir without the admission of something to take its place."

The Bradford pool is a large stratigraphic trap in northern Pennsylvania that has produced about 500 million bbl (80 million tons) of oil. The sand was deep for those days—about 2,000 ft (600 m)—so the wells were widely spaced initially. It has a low permeability, about 10 md. Primary recovery during the 1870s was relatively poor. Beginning about 1900, certain operators would rip the casing near the surface in

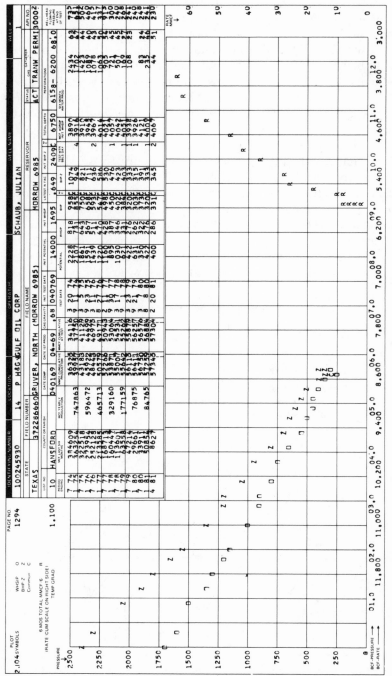

Fig. 15-14 Production record of a well in the north Gruver Morrow Sand gas field. Pressure and rate decline with cumulative production *(courtesy Dwight's Reports)*

one well and permit the ground water access to the sand. This increased the production of the surrounding wells. The practice was legalized in 1921, and it became common to drill a line of wells, run tubing to bottom, and inject water through the tubing. A parallel line of producing wells was drilled a few hundred feet away. As the water reached these wells and they watered out, a new line of producing wells was drilled, and the wells in the first line were converted to intakes.

The next step was to drill alternating injection and producing wells, in what is called the 5-spot pattern (Figure 15–15). Water was pumped down the tubing under pressure. This intensive flood pattern was very successful and has generally become the standard method.

Most reservoirs have a well-defined oil-water contact. In many fields, a line of wells has been drilled to inject water into the aquifer just below it. This is called *peripheral* water injection. It often gives rise to an even updip movement of the oil-water contact. The most successful of these operations is in East Texas, where the pressure has

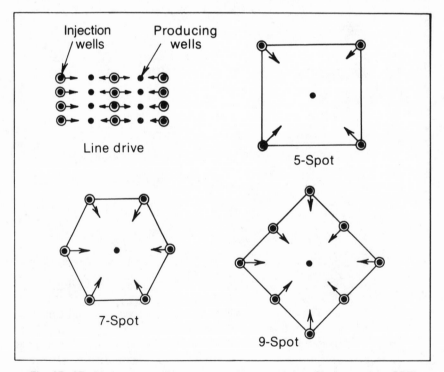

Fig. 15–15 Various secondary-recovery patterns *(after Clark, courtesy SPE)*

been maintained and a remarkably high recovery has been obtained (Salathiel, 1973). In many other fields, however, peripheral floods have not been successful. Usually, the advance of the water is too slow to sustain good rates of oil production. Often the heterogeneities in the sand cause the water to move preferentially in one direction, and an even updip advance of the oil-water contact does not occur.

In spite of its great success at Bradford, waterflooding was slow to be adopted in the rest of the country. It was introduced in Oklahoma in the 1930s and in Illinois in the 1940s. It was not until the 1950s that it became widespread in Texas, first in North and then in West Texas.

Early attempts at flooding in many western fields were unsuccessful. There were three common reasons for failure, and these remain important today.

(1) Operators did not want to spend much money on an experiment, so they usually took the poorest well in the field and pumped water down it. After several years, nothing happened; so the operator said "waterflooding does not work in this field." Sometimes the well was poorly located, but more often it was in poor mechanical condition.

(2) The field had already flooded. In the Venago fields of Pennsylvania, drilled in the 1860s and '70s, and in many of the shallow fields of Illinois and Oklahoma developed in the early 1900s, there were many old wells which had never been properly plugged. These admitted ground water that entered the sand in a slow, irregular manner. Production was sustained by the slow flood, but there was no sudden increase in oil production followed by large amounts of water. The wells simply produced a few barrels of oil daily with a gradually increasing water-oil ratio. Many companies embarked on ambitious projects of redrilling only to find that the new wells produced large amounts of water and very little oil.

(3) The geology was not properly taken into account. Most sands are very heterogeneous with wide variations in permeability. The water simply rushed through the most permeable sands. Some carbonates had open fractures and others big vugs; so the water travelled through the openings, leaving the oil behind.

The waterflooding process. The waterflooding process is shown schematically in Figure 15–16. The map view above shows a unit cell of 5-spot pattern with water injection wells at the corners and the producing well at the center. Before the flood starts, the sand contains about 25% of pore space filled with interstitial water, 50% oil, and 25% gas, as shown in the bottom section. As the flood advances, it pushes the oil ahead, increasing the oil saturation and forming an *oil seal* or *bank*. This happens because at first the relative permeability of the

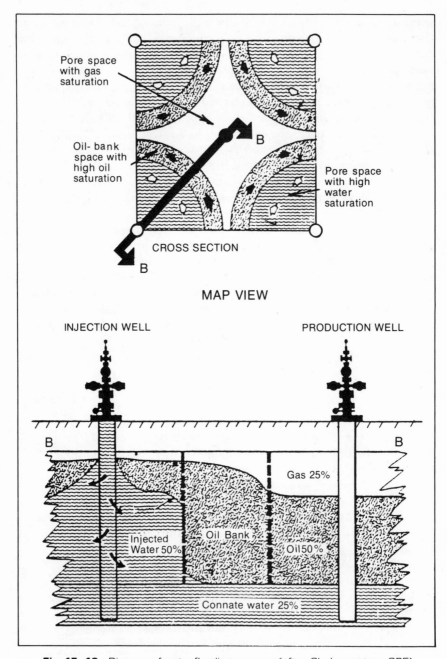

Fig. 15–16 Diagram of waterflooding process *(after Clark, courtesy SPE)*

sand to oil is greater than its relative permeability to water, so oil constitutes most of the flowing stream. Behind the bank, the water saturation increases to 60 or 70% of pore volume. At this saturation, the permeability to oil drops rapidly to near zero and very little oil flows. Some oil, usually about 25 to 30% of pore volume, remains behind as small droplets or is attached to oil-wet surfaces. A small amount of gas is also trapped.

Neglecting gravity effects and the capillary pressure gradient, the basic formula for water displacing oil in a porous medium is

$$f_w = \cfrac{1}{1 + \cfrac{\mu_w}{\mu_o}\cfrac{k_{ro}}{k_{rw}}} \qquad (15\text{--}9)$$

where:

f_w = fraction of water in the flowing stream
μ_w = viscosity of water
μ_o = viscosity of oil
k_{ro} = relative permeability of the sand to oil
k_{rw} = relative permeability of the sand to water

This formula says that the fraction of water in the flowing stream depends on the ratio of the viscosities of water and oil and on the ratio of the relative permeability of the sand to water and to oil.

Figure 15–17 is an example of a water-oil relative permeability relationship. At a water saturation of about 25% and an oil saturation of 75%, the relative permeability to oil is 1.0—that is, only oil will flow. As the water saturation increases, the permeability to oil decreases. At about 75% water, 25% oil, the permeability to oil becomes zero and only water will flow. These numbers apply to a rock whose interior surfaces are mainly water wet. They would be somewhat different if the rock were mainly oil wet.

This results in the relationship shown by Figure 15–18, which shows that when the water saturation begins to exceed 25% the fraction of water in the flowing stream increases rapidly until it is all water at a water saturation of about 75%.

These and similar, more complex formulas can be used to calculate the behavior of the advancing flood if the parameters can be estimated. The viscosities of oil and water are easily measured. The relative permeabilities are more difficult to determine because they depend on the state of wettability of the rock, which is usually poorly known. The original oil saturation is also seldom known. Theoretically, there

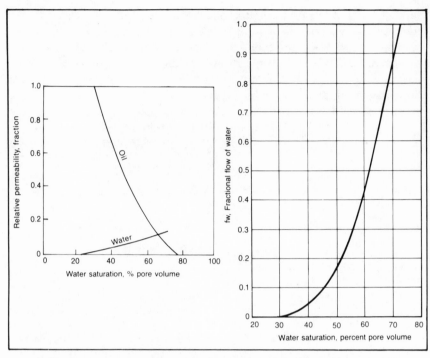

Fig 15–17 Water-oil relative permeability curve, water-wet rock (*after Craig, courtesy SPE*)

Fig. 15–18. Fraction of water in flowing stream, water-wet rock (*after Craig, courtesy SPE*)

should be a rather sudden increase in water saturation right at the flood front, followed by a slower increase (Figure 15–19).

Sweep efficiency. Unfortunately, the formulas are not realistic because the flood advance is not smooth and even as is assumed in the theory. Even in the laboratory using uniform porous media, discrete fingers of water push ahead of the oil. If the viscosity of the oil is much higher than that of the water (an *adverse mobility ratio*), the situation becomes unstable. Figure 15–20 is a laboratory model of water displacing oil with a mobility ratio of 80 to 1. The pressure drop per unit distance is lower in the water finger than in the oil bank. Consequently, the pressure at the end of the finger becomes higher than that in the surrounding oil. This causes the finger to push ahead even faster, so that fingering is a self-aggravating phenomenon. After the fingers break through to the producing well, they widen laterally. Eventually, most of the oil is recovered at the cost of a high water-oil ratio.

In a natural oil reservoir, the situation is worse. All oil sands have

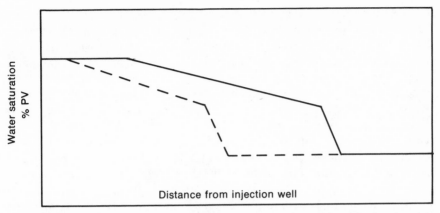

Fig. 15—19 Diagram showing change in water saturation as flood front advances (after Craig, courtesy SPE)

a wide range of permeabilities. The water first goes straight through to the producing well by the easiest path. If flooding is continued, additional oil will be recovered from the less permeable areas but with the production of a great deal of water which is cycled uselessly through the permeable paths where it removes little oil. Much oil is also trapped against zones of pinchout.

Reservoir heterogeneities. All sandstones are layered; that is, they

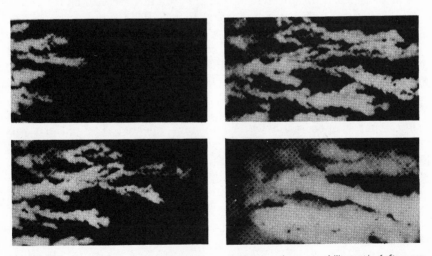

Fig. 15—20 Viscous fingering of water into oil with an adverse mobility ratio (after van Meurs and van der Poel, courtesy SPE)

consist of multiple beds with similar grain size and other physical properties. These layers may be from a few inches to many feet thick. They are usually separated by layers of impermeable shale that range in thickness from paper-thin to many feet. In any one layer, the permeabilities range from some maximum value down to zero. The different layers have different average permeabilities. Consequently, when water is injected, it is accepted by the different layers in proportion to their permeability. Sometimes there is vertical communication between layers, but usually there is none (or at least very little). Consequently, there develops a series of waterfloods, one in each layer, that are practically independent of each other. First the most permeable layer waters out, then the next, and so on.

It sometimes happens that one layer is enormously more permeable than the rest. In this situation, there is a rapid breakthrough of water, followed by a long period of oil production with high water-oil ratios.

If the different layers have different lateral dimensions as they would, say, in a point bar or a series of shingled beaches, the flood will progress rapidly in one direction and slowly or not at all in another. In such cases, the injection well pattern should be altered to conform to the average direction of the preferential permeability.

In some cases, the rocks are broken by a series of vertical fractures or *joints* that cut individual sandstone beds. They are smooth, straight, and parallel, spaced a few inches to a few feet apart. They have a constant direction that may vary only a few degrees over a large area. Their origin is quite unknown, and they do not seem to be related to local structural features. Sometimes sharp folds may have several sets of fractures that are clearly caused by the folding, but normally the regional joint pattern goes right over the folds as if they were not there.

Most joints appear to be closed at depth, even though they are opened by weathering at the surface. In some fields, notably Spraberry, Texas, the cores showed open joints. Burbank, Oklahoma, has a set of joints that opened when water was first injected. The pattern was changed so the lines of water injection wells are parallel to the joint pattern. At Bradford, Pennsylvania, the pattern is chosen similarly, so the fractures opened by hydraulic fracturing run between injection wells and not from injection to producing wells.

Carbonates have many different kinds of porosity. In the case of textures consisting mostly of calcite or dolomite grains of sand size, they behave like sands. There have been many successful waterfloods in this type of rock, e.g., the San Andres of West Texas. On the other hand, if the porosity is partly or largely in the form of open vugs and channels, the injected water will go right through the largest channels, leaving behind the oil in the smaller channels and interstitial pores. Most waterfloods in this type of rock have been unsuccessful.

If the field had an initial gas cap, the original oil saturation would have been very low in the zones occupied by gas. If water is injected into these zones, the porosity will quickly fill with water. When this happens, the relative permeability to water will be high, and the water will move preferentially through the gas cap. It will quickly break through to the producing wells, bringing no oil with it. The flood will be a failure, even if much oil remains in the lower part of the sand.

In many fields, the bottom layers of the sand originally contained mostly or entirely water. Their relative permeability to water was therefore much higher than that of the overlying oil sands. In such cases, the injected water travels preferentially through the water-saturated sands, resulting in a failure or a flood of low profitability.

Determination of reservoir heterogeneity. Much of this book has been devoted to explaining how the geologist determines the physical properties of the reservoir. He assembles information from cuttings, cores, and wireline logs. These are plotted first in the form of cross sections so the different layers can be identified and correlated and their vertical and lateral limits defined. They are then plotted on a series of maps, one or several for each of the vertical subdivisions.

Often, not enough geological information is obtained during the original drilling. In this case, it is customary to drill and core several new wells. Often, a *pilot flood* is attempted. It is preferable to drill all of the wells in the pilot area and obtain all information possible, including complete cores. When the behavior of the pilot is placed into its geological framework, plans can be made for the full-scale waterflood and economic forecasts can be made.

Secondary recovery usually requires a shorter spacing than primary so that the injection and producing wells will both be in the same subzone of the reservoir. The infill wells will often provide a source of the detailed geological information that is necessary for development.

It is always desirable to make selective well completions. In the Bradford field, it was customary to shoot the less permeable layers with nitroglycerine, leaving the more permeable layers unstimulated. Hydraulically fractured wells are more difficult to complete selectively. It is sometimes possible to make multiple completions—that is, subdivide the reservoir vertically into a few zones, separate them in the well bore by packers, and run a separate tubing string to each zone. It is expensive, but sometimes justifiable, to drill separate wells, completing each one in certain zones.

The best and probably the only way to cope with reservoir hetero-geneity is to use repeated patterns. If a producing well is surrounded by injection wells, as in the 5-spot or 9-spot patterns of Figure 15–15, the water will crowd the oil toward the producing well. Even if there are permeable channels, the oil is still forced on.

The temptation is to drill as few new wells as possible because wells are the largest item in secondary-development expense. Usually, wa-ter injection starts as a peripheral or line drive, but these projects often fail because part of the water simply goes downdip into the aquifer and part traverses preferential channels, watering them out and leaving much of the reservoir unswept. Even when pattern floods are started, they are commonly reverse 7-spot or reverse 9-spot; that is, there are more producing than injection wells. This is asking for trouble. In most cases it is better to have an equal number of injection and producing wells.

Gas injection. It was found in about 1900 that the injection of gas into certain wells increased the oil production of the others. A detailed study of the results of air injection in Pennsylvania (Dickey and Bossler, 1940) showed that additional recoveries of 10 to 100 bbl per acre-foot (13 to 130 cu m per h-m) could be obtained with intense gas drives. The recovery increased as well spacing decreased from 6 acres to 2 acres per well (2.4 to 0.5 h).

The fields in Pennsylvania were discovered in the 1860s. When injection was begun in the 1930s, oil production had declined to a small fraction of a barrel per day and very little gas was available. Conse-quently the operators injected air. The recovery process was essential-ly the same as gas, but certain mechanical problems developed, including oxidation of the equipment and the formation of explosive mixtures.

Reinjection of gas produced with the oil was widely practiced in the U.S. and even more in foreign operations between 1930 and 1950. It tended to maintain the reservoir pressure. Large amounts of gasoline were produced in vapor form along with the gas, which was condensed out and sold. Some additional oil was recovered as a result of gas injection.

7. ENHANCED RECOVERY (TERTIARY RECOVERY)

In the U.S. it has been customary to produce an oil field by the primary mechanism until it was no longer profitable and then drill new wells for water injection. Such redevelopment results in *secondary-recovery* oil. When gas is injected either into the gas cap or elsewhere or water is injected peripherally when the field is still in its early stages, the operation is called *pressure maintenance*.

After a field has been exhausted by waterflooding, it becomes a candidate for *tertiary recovery*. The newer methods are not necessarily applied as a third development of the field, so it has become customary to call them *enhanced-recovery methods*.

Even waterfloods leave a large amount of oil behind—probably around 30% of the pore volume or 50% of the original oil in place. Adding the fields where water has not or cannot be applied, Geffen (1975) has estimated that 67% of the original oil in place still remains underground. The total oil discovered up to 1974 in the U.S. is estimated as 435 billion barrels, of which 103 billion have been produced and 40 billion remain to be recovered. Of the 292 billion barrels still left in the ground, it is estimated that 20 to 60 billion barrels may be recovered by other methods still to be perfected. In spite of very elaborate laboratory and field tests, none of the newer methods (except steam) can be considered commercially successful as of 1980. However, with a greatly increased price of oil, some of them may become so.

Solvent drive. The residual oil is left in the reservoir mainly because interfacial tension makes it form discontinuous globules which get trapped in the pores. The relative permeability to oil drops to virtually zero while there is still a large amount of oil (30 to 50 percent of the pore volume) left in the rock. If the driving fluid were completely miscible with the oil, there would be no interfacial tension and almost perfect piston-like displacement would occur.

Light hydrocarbons—propane to hexane (LPG)—are completely miscible with the oil. They can be injected as gas or liquid. In either case, they dissolve in the oil and form a bank of light hydrocarbons that pushes the oil toward the producing well.

Another method is to inject methane enriched with ethane through hexane. If the pressure is sufficiently high, the C_2–C_6 hydrocarbons will dissolve in the oil, and this is supposed to form a bank that will miscibly sweep the reservoir. A third method is to inject dry gas (mostly methane) at still higher pressures. If the oil contains considerable C_2–C_6 hydrocarbons, they will transfer to the gas and form a bank that is miscible at very high pressures (more than 4,000 psi or 28,000 kPa).

Light hydrocarbons (LPG) would be too expensive to inject into the reservoir until all gas-filled pore spaces were filled. Consequently, it is customary to inject a slug of LPG and follow it with dry gas. It is hoped that the LPG will form an expanding ring that will push the oil ahead of the gas.

Solvent drives have not been successful mainly because the LPG tends to enter and flow through the more permeable beds. Because it has much less viscosity than the oil it is displacing, this *bypassing* is a self-aggravating process. The faster it goes in the more permeable beds, the less resistance to flow they have and the more the bypassing.

The same *adverse mobility ratio* exists between the gas and the LPG banks. These methods looked very promising in the laboratory where the LPG was confined in pipes. In field tests, the LPG generally appears in one or more producing wells when only a small fraction of the residual oil has moved.

Bypassing can be mitigated somewhat by injecting LPG and water alternately. The water enters the more permeable streaks and makes them less permeable to LPG. Another method is to perforate selectively and separate different layers of sand with packers, injecting gas into each zone at a different pressure.

In the case of gravity drainage from steeply dipping beds or from a carbonate reef with great vertical permeability, LPG looks promising. Gravity segregation will help form a more stable bank between the gas and the oil.

Carbon-dioxide drive. Carbon dioxide has the curious property of dissolving in crude oil. The crude then separates into two fractions: light and heavy. If the pressure is high enough and there is sufficient CO_2, it will dissolve in the oil and there will be a single liquid. With high pressures and rich mixtures of CO_2 and oil, there will be two liquid phases. If the pressure is low, there will be a gas phase and a liquid phase. (Figure 15–21). If CO_2 is injected at sufficient pressure, it will dissolve in the oil and form a bank consisting of light hydrocarbons and CO_2. The CO_2 bank is then driven by water. Good results have been obtained by injecting alternating slugs of CO_2 and water (Figure 15–22).

If the CO_2 is injected as a gas in too small a quantity, or at too low a pressure, an immiscible gas drive will result. Although any gas is rather ineffective as a recovery medium because it is nonwetting and because of its low viscosity, the presence of the water remaining after a waterflood increases its efficiency. The water fills pores that would otherwise be full of gas and forces the gas to push on the residual oil. Carbon dioxide goes into solution in the residual oil and causes it to swell, also improving the immiscible gas-drive process.

There have been a large number of small CO_2-injection projects, most of them of the low-pressure immiscible type. One large-scale proj-

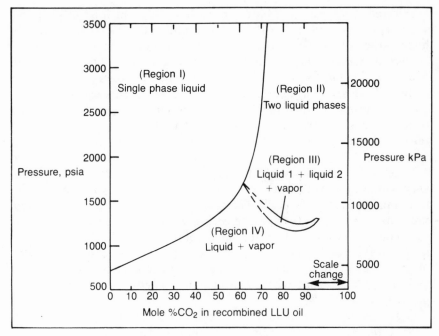

Fig. 15–21 Phase equilibria for mixtures of CO_2 and Levelland reservoir oil

ect is in the Kelly-Snider field of Texas, where Chevron operates the Sacroc unit. The large amounts of gas required are brought in by a special pipeline from North Texas and New Mexico where there are some CO_2 gas fields. It is expected that the incremental recovery will be about 6% of the original oil in place.

Another large CO_2 project is in the Levelland field. Carbon dioxide

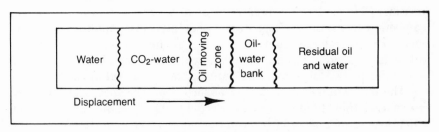

Fig. 15–22 CO_2 flooding process

is injected at pressures greater than the minimum miscibility pressure, which is 1200 psi (8100 kPa). Light and intermediate-weight hydrocarbons transfer to the gas phase, forming a miscible bank.

Surfactants. It is possible to add a detergent to the flood water that will lower the interfacial tension at the interface between the oil and the water. There are many surface-active agents, called *surfactants*, that have this property. In sufficient concentration, they can lower the interfacial tension to almost zero. When this happens, the droplets of residual oil can be deformed easily so they can move with the injected water.

The main difficulty with this method is that surfactants, by their very nature (the word is a contraction of "surface-active agents"), collect at interfacial surfaces. The area of the surfaces between the oil and the water and the water and the rock is enormous. Consequently, the surfactant is immediately adsorbed out of the water. Even if surfactant is continually added behind as the flood progresses, the bank of depleted water in front widens. There is no way to keep the active chemical at the flood front where the displacing action should be taking place.

Caustic soda and sodium hydroxide have also been added to the flood water. Besides reducing surface tension, they appear to undergo reactions with certain oils that promote emulsification.

Modest improvements in recovery of oil by surfactants have been frequently reported. Surfactants offer little or no promise of recovering all or most of the residual oil.

Micellar polymers. An emulsion of water in oil as a driving medium was first suggested by Marathon Oil Co. and has since been developed by other companies. The emulsion is miscible with the oil, so it gives a piston-like displacement. Its viscosity is equal to or greater than that of the oil, so the mobility ratio is favorable. The emulsion is stabilized by high concentrations of a surfactant, which tends to surround the droplets of water in structures called *micelles*.

The emulsion is then pushed by a bank of water whose viscosity is increased by thickening agents. Most of these are types of chemical compounds consisting of aggregates of smaller compounds called *polymers*. The surfactant maintains a low surface tension between the micellar emulsion and the thickened water so that it is also displaced very efficiently. The thickened water is then followed by plain water.

The surfactants are not only adsorbed on the rock and oil-water interfaces; they tend to be destroyed by the metallic ions, such as calcium and magnesium, both in the pore water and attached to the clay minerals. It is therefore customary to precede the emulsion with a chemical solution called a *preflush*.

There have been a number of field tests, some of which are promising. A post-test evaluation well showed that a micellar-polymer project at Sloss, Nebraska, reduced the residual oil after waterflood from 30 to 8% of pore space. Some lower permeability zones showed a residual oil saturation of 16%. In this case the rock degraded the polymer so the good mobility ratio of the thickened water was lost.

Results at a Wilmington, California, pilot project have been favorable. In a waterflooded reservoir the produced water-oil ratio decreased from 100 to 10, and the oil production rate increased from 50 to 200 b/d as the oil was mobilized. Promising results have been obtained in Germany, but other operations have been discouraging. The chemicals needed are pretty expensive. And like all secondary and tertiary recovery methods, reservoir heterogeneities cause bypassing. Many of the tests did not have repeated patterns, that is, similar or equal numbers of injection to producing wells.

Steam injection. Heavy oils with a gravity between 10 and 20° API (density 1.0 to 0.93) are highly viscous and contain very little gas. Consequently, the original dissolved-gas drive was very ineffective, recovering only 10 or 20 percent of the oil in place. The injection of steam into heavy oil sands has given excellent results, especially in California and Venezuela. The bypassing due to sand irregularities is less fatal in the case of steam because the warming effect should extend on either side of the channel. However, as always, geological factors have a profound effect on steam injection (Lennon, 1976).

The most successful method of steam injection is the steam soak or "huff-and-puff" method. The story is told (Giusti, 1974) that it was discovered by Compania Shell de Venezuela in their Mene Grande field. An experimental steam flood was started, but the sands were both shallow and soft; so fissures opened and the steam appeared at the surface of the ground. It was decided that the project was a failure, and the wells were put back on the pump. The wells into which steam had been injected immediately started to produce more oil than they ever had.

In Venezuela, it is customary to build large steam-generating stations and to lay insulated steam lines to the wells to be injected. Steam is injected for several weeks, and the production cycle may be a year or more. In California, a portable boiler may be brought to the well and injects steam for a few days. In both cases, the second and third shots of steam produce less oil than the first.

It is also possible to inject steam continuously from permanent injection wells to producing wells. In this case, the steam condenses. Going out from the injection well, there is first a steam zone, then a bank of hot water, and finally a bank of cold water.

Steam injection is the only enhanced-recovery process that has been commercially successful on a large scale. Van Poollen stated (1979) that more than 400,000 b/d were being produced by steam injection. This process is extremely important because it is the only practical way to produce very heavy oil (10 to 20° API). There are huge fields of heavy oil in Venezuela and Alberta, and large-scale steam injection plants are being planned.

The sand must be fairly thick (over 10 ft, 3 m). If it is too thin, the heat losses in the top and bottom wells will be too great. The spacing must be close, at least for continuous injection, although it is not critical for stimulation. The loss of heat down the tubing is substantial and limits the depth to which steam injection is practical. However, down-hole steam generators have been suggested.

In-situ combustion.   It is possible to burn the oil in the reservoir by injecting large quantities of air. Sometimes the oil is ignited by heating the bottom of the injection well, but usually the oxidation of the oil heats it enough so that it ignites spontaneously. A diagram of the process is shown in Figure 15–22. The air passes through the burned region until it reaches the burning front. The hot products of combustion are driven ahead. They vaporize the water and hydrocarbons, leaving only the nonvolatile material called *coke*. This coke burns and provides the heat. Farther out, the vapors condense, forming banks of water and light hydrocarbons with (hopefully) a bank of oil ahead.

Behind the burning front, the rock is very hot; this heat gradually dissipates up and down into the adjacent formations. It is helpful to inject water intermittently, which vaporizes and carries the heat forward to the combustion zone.

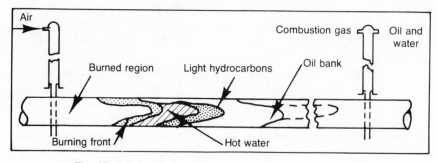

Fig. 15–22 In-situ combustion process *(after van Poollen)*

As of 1972, more than 100 field tests of burning oil in place had been mentioned in the literature. Farouq Ali (1972) analyzed 24 in detail and summarized the results. Most are small, experimental projects, and there are still no really large-scale projects.

The process requires driving air through the sand, keeping a burning zone expanding from the injection well. The biggest problem and expense is injecting enough air to keep the fire going.

The temperature in the combustion zone ranges from 650° to 1,200° F (500 to 800° C). It moves from about 1/10 to 1 ft (3 to 30 cm) per day. The fuel required to keep the fire going has ranged from 0.8 to 2.5 lb per cu ft of rock (13 to 40 kg per cu m) of rock. The air requirements range up to 15 million cu ft per acre-ft (500,000 cu m per h-m) of rock.

Large increases in oil production usually occur, and the average oil recovery has been about 50 percent of the oil in place. The air-oil ratio has ranged from 3,000 to 20,000 cu ft per bbl (600 to 3,500 cu m per T). The produced water-oil ratio was between 2 and 5. Sometimes a water-in-oil emulsion was produced. The API gravity of the oil has usually been increased slightly and its viscosity decreased.

The gas produced is largely nitrogen with some carbon dioxide. Oxygen content is usually very low until the burning zone passes the producing well.

The sweep efficiency has usually been rather good laterally, although there is a tendency for the burning zone to move through the upper part of the sand. Heat losses to the rock above and below are small because heat flow by conduction is extremely slow. But when the burning zone reaches the producing well, the heat and products of combustion cause severe corrosion of equipment and caving of the hole walls.

The many failures have disclosed the need for adequate appraisal of the geology before trying a fire flood. Most important is the amount of oil in place, which should exceed 700 bbl per acre-ft (900 cu m per h-m). The oil should not be too light to lay down a layer of coke to provide the fuel. The next most important requirement is that the sand have adequate permeability to admit the large amount of air required. It should be deep enough that injection pressures can be high enough to force the air into the sand. Great heterogeneities in the sand body have permitted severe bypassing.

Geological limitations on enhanced-recovery methods. There are certain limitations to enhanced-recovery methods, resulting from the geology of the reservoir. These have not always been taken into account; if they had been, many disastrous failures could have been avoided.

The greatest cause of trouble in either secondary or enhanced-recovery methods is heterogeneity in the reservoir. Most reservoirs consist of a complex of channels and bars with huge differences in permeability, both vertically and horizontally. In recent years, most attempts at enhanced recovery have been preceded by detailed geological studies of cores and well logs. These are interpreted in the light of depositional environment, which enables predicting the geometry of the subreservoirs. The communication from one well to the next has been investigated with pulse tests. As the project progresses, it is customary to inject tracers so the preferred directions of flow can be identified.

Because of these heterogeneities, the well spacing for enhanced recovery needs to be short. Most methods require the formation of an oil bank. However, in any but an ideally homogeneous reservoir (which does not exist in the real world), the farther the fluids go, the greater is the opportunity for fingering and bypassing that destroy the bank. In the case of chemical (surfactant) flooding, the farther the fluids go, the more the chemicals are degraded or adsorbed by the reservoir rock. Well spacing will have to be between 2 and 20 acres.

The spacing also implies a depth limitation. Even if the price of oil reaches $50 to $100 per barrel, it may be uneconomic to drill to more than 10,000 ft on a 20-acre spacing.

Enough experience has now been obtained that it is possible to define the types of reservoirs which are suitable for each of the different enhanced-recovery methods. Table 15–II is a screening guide showing the limits on oil gravity and viscosity, reservoir porosity and permeability, and oil saturation.

The amount of residual oil remaining after primary or secondary recovery is obviously one of the most important parameters. It is also one of the most difficult to ascertain. Conventional cores are flushed and their fluid content is changed from that in the reservoir. Electrical and other wire-line logs give only approximate values for oil saturation. Attempts are being made to use a pressure core barrel that holds the fluids in the core during its trip to the surface. But this is not the whole answer because flushing ahead of the bit still occurs.

It is very likely that the residual oil remaining after primary production and waterflooding is different physically and chemically from the oil that was produced. The geochemical process of maturation produces methane and light hydrocarbons. These cause precipitation of asphaltenes. The maturation process also results in the formation of solid, high molecular-weight bitumens. Thus, a substantial fraction of the residual oil is not oil at all but some solid hydrocarbon. It should

TABLE 15–II

LIMITATION OF ENHANCED-RECOVERY METHODS

Process	μ_O cp	°API	ϕ%	k md	S_O %P.V.
Steam Inj. (contin.)	200 to 1000	10-20	≥30	>1000	>50
Steam Inj. (cyclic)	1000 to 4000	10-40	≥30	—	>50
Fire Flood (forward)	<1000	10-30	≥20	>300	>50
Fire Flood (COFCAW)	<1000	10-30	≥20	—	>50
Polymer Flood	<20	20-40	—	≥50	>25
Surfactant Flood	≤20	≥25	>20	≥50	>25
CO_2 Flood	<10	30-45	—	>10	>25

μ = viscosity   ° API = gravity
ϕ = porosity S_o = saturation
k = permeability

not be counted on to move like the original oil in primary and secondary recovery.

8. RESERVOIR BEHAVIOR IN CARBONATES

When the porosity in carbonate (limestone or dolomite) is altogether intergranular, the behavior of oil, gas, and water is essentially the same as in sandstones. The concepts, mechanisms, and formulas used for sandstones all apply.

On the other hand, many carbonates have interstitial porosity and porosity in fractures or vugs. The behavior of fluids in the two types of porosity is naturally very different. The permeability of fractures is very great, and it increases as the square of the channel width. A fracture only 0.1 mm wide will have a permeability of 833 darcies, whereas the permeability of the limestone matrix may be 0.001 darcies (1 md) or even less.

Fractured reservoirs therefore characteristically have a very high productive rate, especially at first. The huge fields of Iraq (Kirkuk) and Iran (Gach Saran and others) have production rates of more than 20,000 bbl (3,000 cu m) per day per well. The fractures are interconnected, and pressure changes are transmitted quickly throughout the reservoir for many miles.

In these fields there is considerable interstitial porosity and most of the oil is in the matrix, with the fractures serving mainly as permeable channels to the wells. Other less-prolific examples are the chalk fields of Ekofisk (Norway), Kirk (Texas), and Pine Island (Louisiana), where the porosity is 20 percent or more but the permeability of the matrix is very low.

In such reservoirs the fluid behavior, the total oil in place, and the recoverable oil are almost impossible to predict. Some people have attempted to treat the system mathematically as a pile of blocks, each with its own porosity and permeability, separated by crevices with infinite permeability but negligible volume (Figure 15–23).

In the case of dissolved gas mechanism, as the pressure drops, the gas will force oil in the interstitial porosity in the matrix blocks out into the crevices, through which it will flow to the well bore. Recovery efficiency will be low, as in the case of dissolved gas drives. When water enters the crevices, it will tend to block the production of oil from the matrix blocks.

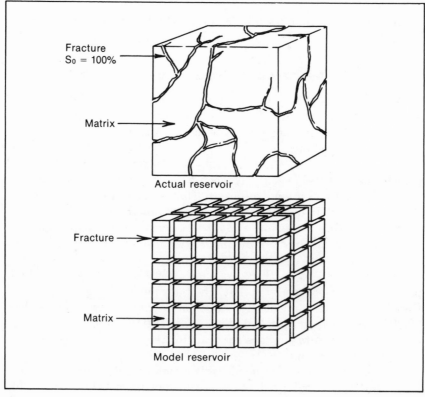

Fig. 15–23 Fractured reservoir with oil contained both in the fractures and matrix and a representative model used in calculations *(after Warren and Root)*

On the other hand, if the internal surfaces of the interstitial pores are preferentially wet by water, then water from the crevices will be drawn into the pores by capillary forces, and some of the remaining oil will be expelled. This process is called *imbibition*. If the water-oil contact in the crevices rises vertically at a slow rate, the level of the water drawn into the matrix will remain above it and the oil expelled will feed upward into the oil column. Even if the rate of rise of the water-oil contact exceeds the critical rate and covers the block, some oil will still be expelled and enter the water in the crevices as droplets. These may rise to the water-oil contact and join the oil. Imbibition appears to have been important in secondary recovery by waterflooding in some fractured reservoirs, both limestone and sandstone. On the other hand, most efforts at waterflooding fractured reservoirs have failed.

In some cases, the secondary development was poorly planned, possibly because there was insufficient information on the reservoir. It is important to take into account—
- The amount of interstitial porosity
- The shape and size of the pores
- Their oil saturation
- The wettability of their surfaces
- The spacing of the fractures
- Their width
- Their orientation
- The percent of total rock volume occupied by fractures.

To get this information, cores are indispensible, and much of this information can best be determined immediately as the core is withdrawn from the hole.

9. RESERVOIR SIMULATION

A widely used approach to predicting the performance of an oil reservoir and the estimation of the reserves is *mathematical modeling* or *reservoir simulation*. It has the advantage that heterogeneities can be noted. The reservoir is subdivided into small cells which may be one-, two-, or three-dimensional. The dimensions and location of each cell are specified. A value for the rock properties' porosity and permeability is assigned to each cell. The initial oil, gas, and water saturation and pressure are specified. Then the changes in pressure and fluid volumes as the field is produced are calculated on a computer using appropriate formulas. The actual production history, as far as it goes, is then compared with the computed history. One or another parameter may be changed to make the computed correspond with the actual production. When the computed history agrees with the actual history, the future may be predicted.

For reservoir simulation to be useful, the parameters must be reasonably well known. If one has a wrong value, the others will be adjusted to compensating wrong values in order to get a history match.

Most of the information required for reservoir simulation is supplied by the geologist (Harris, 1975). He can be most reliable if he knows how each parameter affects the calculations, and applies quantitatively reliability estimates. The geologist normally provides information on:

- Overall shape of reservoir
- Subdivisions of reservoirs by faults and shale layers
- Size and shape of aquifer
- Porosity and its lateral and vertical variations
- Permeability and its variations

The engineer normally provides information on:

- Gas, oil, and water saturation
- Pressure and pressure variations
- Drill-stem and other production tests
- Past production history.

REFERENCES

Al Naquib, F. M., R. M. Al-Debouni, T. A. Al-Urhagim, and D. M. Morris, 1971, Water drive performance of the fractured Kirkuk field of northern Iraq: Soc. Petrol. Eng. SPE 3437.

Aquilera, Roberto, 1980, Naturally fractured reservoirs, PennWell Pub. Co., Tulsa, 703 p.

American Petroleum Institute, 1942, Standard procedure for determination of permeability of porous media: Code 27, 2nd edition: Div. Prod., Dallas, Texas, April, 21 p.

Arps, J. J., and others, 1967, A statical study of recovery efficiency: Amer. Petrol. Inst. Bull. D14, 33 p.

Botset, H. G., 1940, Flow of gas-liquid mixtures through consolidated sands: Amer. Inst. Min. Engrs., Trans., v. 136, p. 91–105.

Boyce, B. M., 1972, Ekofisk phase I production looks good: Oil and Gas Jour., June 12, p. 87.

Bruce, W. A., and H. J. Welge, 1947, Restored state method for determination of oil-in-place and connate water: Amer. Petrol. Inst., Drill. and Prod. Practice, p. 166–174.

Carll, J. F., 1880, The geology of the oil regions: Second Geol. Surv. of Pennsylvania, Rept. III, p. 251.

Clark, Norman J., 1960, Elements of petroleum reservoirs: Soc. Petrol. Engrs., Dallas, 250 p.

Craig, F. F., Jr., 1971, The reservoir engineering aspects of water flooding. Soc. Pet. Engrs., Dallas, 134 p.

Craft, D. C., and M. F. Hawkins, 1959, Applied petroleum reservoir engineering: Englewood Cliffs, N. J., Prentice-Hall, 437 p.

Craze, R. C., and S. E. Buckley, 1945, A factual analysis of the effect of well spacing on recovery: Amer. Petrol. Inst., Drill. and Production Practice, p. 144.

Dake, L. P., 1978, Fundamentals of reservoir engineering, Elsevier, Amsterdam, 443 p.

Dickey, P. A., and R. B. Bossler, 1944, Role of connate water in secondary recovery of oil: AIME Trans., v. 155, p. 175.

Dickey, P. A., and R. B. Bossler, 1950, Oil recovery by air and gas repressuring in Pennsylvania, *in* Paul D. Torrey, ed., Secondary recovery of oil in the U.S., 2nd edition: Amer. Petrol. Inst. Div. of Production, Dallas, p. 444–462.

Doscher, T. M., and F. A. Wise, 1976, Enhanced crude oil recovery potential—an estimate: Jour. Petrol. Techn., May, p. 575–585.

Elkins, L. F., and A. M. Skov, 1960, Determination of fracture orientation from pressure interference: AIME Petrol. Trans. TP 8137, v. 219, p. 301–304.

Ervin, P. S., 1957, A successful water flood of a fractured reservoir: Oil and Gas Jour., June, p. 122–125.

Farouq Ali, S. M., 1972, A current appraisal of in-situ combustion field tests: Jour. Petrol. Tech., April, p. 477–485.

Frick, Thomas C., Editor, 1962, Petroleum production handbook, Vol. II, Reservoir engineering: Soc. Petrol. Engrs., Dallas, 49 chapters.

Geffen, Ted M., 1975, Here's what's needed to get tertiary recovery going: World Oil, March.

Giusti, Luis E., 1974, CSV makes steam soak work in Venezuela field: Oil and Gas Jour., Nov. 4.

Groves, D. L., and B. F. Abernathy, 1968, Early analysis of fractured reservoirs compared to later performance: Soc. of Petrol. Engrs. SPE Paper 2259, 14 p.

Hester, C. T., J. W. Walker, and G. H. Sawyer, 1965, Oil recovery by imbibition water flooding in the Austin and Buda Formations: Jour. Petrol. Techn., August, p. 919–925.

Hubbert, M. K., 1940, The theory of ground water motion: Jour of Geology, v. 48, p. 785–944.

Interstate Oil Compact Commission, 1974, Secondary and tertiary oil recovery processes. Oklahoma City, 187 p.

Kelton, F. C., 1953, Effect of quick-freezing on saturation of oil well cores: AIME Trans., v. 198, p. 21–22.

Lennon, R. B., 1976, Geological factors in steam-soak projects on the west side of the San Joaquin basin: Jour. Petrol. Tech., July, p. 741–748.

Mattax, C. C., and J. R. Kyte, 1962, Imbibition oil recovery from fractured water-drive reservoir: Soc. Petrol. Engrs. Jour., June, p. 177–184.

Mattax, C. C., R. M. McKinley, and A. T. Clothier, 1975, Core analysis of unconsolidated and friable sands: Jour. Petrol. Tech., December, p. 1423–1432.

McCarthy, Danny W., et al., 1981, Tertiary oil recovery in economics in Louisiana, SPE-DOE Symposium on Enhanced Oil Recovery, Tulsa, pp. 597–612.

Mullane, J. J., 1944, Reservoir performance: Amer. Petrol. Inst., Drill. and Production Practice, p. 53–65.

Muskat, Morris, 1937, The flow of homogeneous fluids in porous media: New York, McGraw Hill Book Company, 763 p.

Ribout, M., J. Ezbrayat, and W. Machtalere, 1977, High accuracy BHP data improve reservoir analysis. World Oil, August 1, 1977, p. 51–64.

Rose, W. D., and W. A. Bruce, 1949, Evaluation of capillary character in petroleum reservoir rock: Amer. Inst. Min. Engrs., Jour. Petrol. Technol., May.

Russell, C. D., and P. A. Dickey, 1950, Porosity, permeability, and capillary properties of petroleum reservoirs, *in* Parker D. Trask, ed., Applied sedimentation: New York, John Wiley, p. 579–615.

Salathiel, R. A., 1973, Oil recovery by film drainage in mixed wettability rocks. Jour. Petrol. Techn. v. 25, p. 1216–1224.

Scheidegger, A. E., 1957, The physics of flow through porous media: New York, Macmillan.

Schilthuis, R. J., 1938, Connate water in oil and gas sands: Trans. Amer. Inst. Min. Engrs. v. 127, p. 199–214.

Staggs, H. M., and E. F. Herbeck, Reservoir simulation models—an engineering overview: Jour. Petrol. Techn., December 1971, p. 1428–1436.

van Meurs, P. and C. van der Poel, 1958, A theoretical description of water-drive processes involving viscous fingering, Trans. AIME v. 213, p. 103–112.

Warren, J. E., and P. J. Root, 1963, The behavior of naturally fractured reservoirs, Trans. AIME, v. 228, p. 245–255.

Woods, Dalton J., 1963, Louisiana's Sabine drilling action to expand: Oil and Gas Jour., December 9, p. 124–130.

Yellig, William F., 1981, Carbon dioxide displacement of a West Texas reservoir oil, SPE-DOE Symposium of Enhanced Oil Recovery, Tulsa, pp. 197–212.

16

Application of Reservoir Geology to Waterflooding and Enhanced-Recovery Operations

1. METHODS OF SUBDIVIDING RESERVOIRS INTO UNITS OF SIMILAR PROPERTIES

When gas, water, or other fluids are injected into a reservoir to displace the oil, they do not spread out uniformly from the injection wells, but enter the most permeable strata and follow permeable zones directionally. In many cases, flowmeter tests have shown that most of the fluid enters or leaves the wellbore at a few vertically restricted intervals (Boesi, 1977). Normally, neither sandstone nor carbonate reservoirs are homogeneous; on the contrary, they usually consist of multiple units of different shapes and permeabilities. Very often the subdivisions are more or less isolated from each other by impermeable rocks. It is obvious that these heterogeneities must be taken into account in planning development and in designing the enhanced-recovery operation.

Subdivision vertically into time-rock units. The first step in defining the component sand bodies within a reservoir is to line up the electric logs of all the wells and subdivide the reservoir into small time-rock units. Thick, complex delta deposits are cyclic. Each cycle is usually terminated by a transgressive marine shale. These transgressive shale beds can be correlated because they are more continuous than the sands that come and go laterally. They are approximately parallel to each other.

Other marker beds which might be used when they are present are coals, indicating marsh deposits. In some deltaic sequences, there are thin limestones that can be correlated long distances. In carbonates, kicks on the wire-line logs can be used to correlate even when their lithologic significance is unknown.

These transgressive shales, or some other markers, divide the reservoir vertically into time-equivalent subdivisions. It has become customary to call them *packages* of sand.

Figure 16–1 shows electric and gamma-ray logs of three wells 600 m

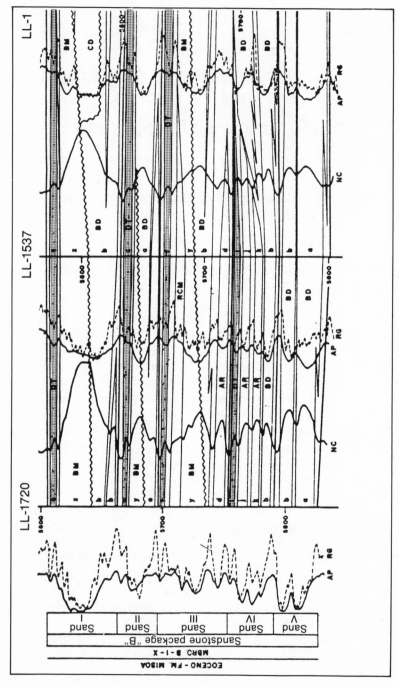

Fig. 16—1 Electric logs of three wells in the Eocene Misoa of the Bolívar coastal fields, Venezuela, showing correlation horizons and transgressive deposits (DT) *(after Zamora)*

apart in the B–1 member of the Eocene Misoa of Lake Maracaibo. A pilot secondary recovery project was planned for this area (Zamora, 1977). The lithology consists of grey shales and sands that come and go rapidly. For a long time, the Misoa in this area seemed to defy correlation, but it is now interpreted as the delta of a large river system. On the left of each well are the SP (AP) and gamma-ray (RG) curves, and on the right is the short normal resistivity log (NC). The horizontal lines are the horizons used to correlate. DT stands for transgressive marine deposits that separate the 230-ft interval into five packages of sand.

2. IDENTIFICATION OF DEPOSITIONAL ENVIRONMENTS

The next step is to identify the depositional character of each of the sand bodies to predict their size, shape, and direction. Figure 16–2 shows how this was done for well LL–1290 which was cored. It is located 10 km from the pilot in sediments of similar age and depositional environment.

Starting at the bottom between 4,390 and 4,397 ft, there is a shale characterized by woody material and animal burrows. It is interpreted to be interdistributary bay deposits. Over this, the sand body from 4,373 to 4,390 is shown by the SP, GR, and resistivity curves to be cleaner and coarser in the upper part. It contains woody material, convolute (disturbed) bedding and small crossbeds. Interpreted to be a distributary mouth bar, it is overlain by another interdistributary bay deposit from 4,370–73. The sand above from 4,363–4,370 contains worm burrows and chunks of shale. It is interpreted as a crevasse splay. Above it is another interdistributary shale and another crevasse splay sand.

Above this, from 4,328 to 4,348, is a well-developed sand. The base is abrupt and appears to be erosional. The lower part is the coarsest and most permeable. It contains chunks of shale, woody fragments, small crossbeds, and wavy horizontal bedding. It is clearly a channel, either a distributary channel fill or a point bar deposit.

Continuing upward, it is overlain by another interdistributary shale and a crevasse splay sand. Above this, from 4,287–4,309, is another channel.

On the right, the values of porosity and permeability are tabulated. The permeabilities are mostly below 50 md. They are best in the lower parts of the channel deposits and the middle of the crevasse splays. In this package there are seven permeable sand bodies, each separated from the others by impermeable shale.

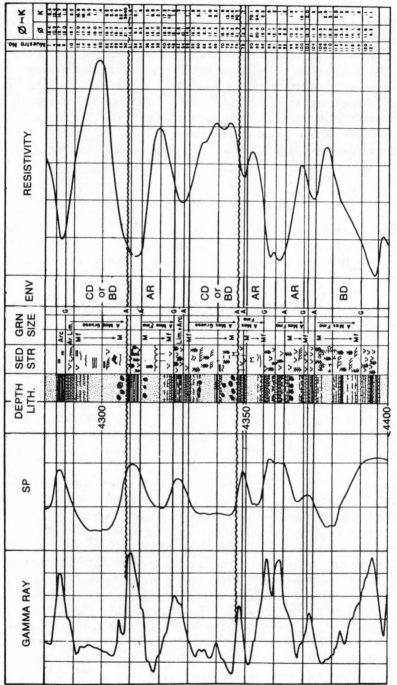

Fig. 16–2 Environmental interpretation of B-1-X sands of Misoa formation *(after Zamora)*

The characteristic electric log patterns in the detailed cross section, Figure 16–1, are then compared with those exhibited by the well which was cored, Figure 16–2. Distributary bar sands are recognized in the lower part of the package in sands IV and V. Laterally adjacent in the same time zones are interdistributary shales and crevasse splays. Overlying these, sands I, II, and III consist of point bar and distributary channel deposits.

This study was for the purpose of planning a small pilot waterflood involving a dozen or so wells. The results were presented in the form of a series of fence diagrams of which Figure 16–3 is an example.

Sometimes, instead of drawing a fence diagram, it is better to cut out pieces of the electric logs of all the wells corresponding to one sand unit. These are pasted on a map at their proper locations. It is now possible to line up directional trends of logs with similar characteristics. Figure 16–4 is a log shape map of a single sand in the Temblor II Zone in the Coalinga, California, field.

3. MAPPING INDIVIDUAL SAND UNITS

Several different parameters have been used to show on maps the distribution of sand bodies with similar reservoir properties. Most thick sand bodies first must be subdivided into small intervals or packages of time-equivalent deposits, and these must be further subdivided into individual sand units. One or more maps are then drawn on each unit, showing where the permeable sand bodies lie. The simplest map simply shows net sand thickness.

The most important parameter in determining the behavior of the fluids in a reservoir is permeability. Sands or carbonates of less than a certain permeability will not contribute, at least economically, to the oil production. The cut-off permeability differs from one reservoir to another, depending on the average permeability of the reservoir. Sometimes it is taken as one-tenth the maximum permeability.

Unfortunately, permeability cannot be determined directly from any wire-line log. It is necessary to take cores, compare them with wire-line logs, and plot permeability against some other measurable property to get an empirical relation. Frequently, the resistivity curve gives an indication of permeability. The coarser beds of sand contain less clay. Because they are more permeable, they contain more oil, or they may be more easily invaded by drilling mud filtrate of high resistivity.

An example of how resistivity was used to map permeable sand bodies is an elaborate study of a large reservoir in the Bolivar Coastal

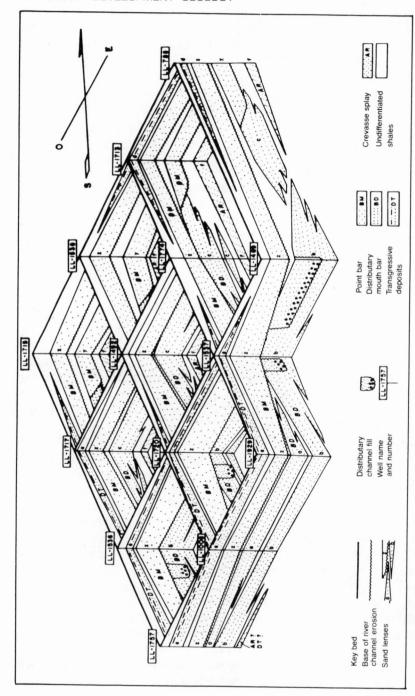

Fig. 16—3 Fence diagram showing lateral development of sand 1, package B, Misoa sandstone, Venezuela *(after Zamora)*

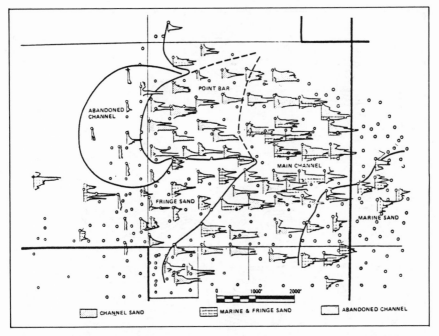

Fig. 16—4 Mx sand log shape map, section 29, Coalinga field, California *(after Lennon, courtesy JPT, copyright SPE)*

field, Venezuela (Jam, 1977). The LL–3 is a Miocene reservoir 20 km from north to south and 10 km from east to west. It contained more than 6,000 million bbl of oil in place and had produced 872 million to 1977. It dips to the south and is bounded by faults on the east and west. There are more than 600 producing wells. A program of edge water injection at the south end of the pool was started, but it gave very poor results. The water did not advance uniformly. Some wells watered out quickly, while other were not affected. A detailed geological study of the reservoir was undertaken.

Because of the large number of wells, it was advisable to use computer methods. Most of the SP and resistivity logs were digitized, and these data were combined with the master tape file of the well logs. Cross sections were then made using a Calcomp plotter. Nine shale units could be correlated across the field, so the producing formation was divided into packages labelled by letters A to F.

The computer was then programmed to count the thickness of sand in each interval, using the resistivity curve to give three categories: 7 to 12, 12 to 20, and over 20 ohmmeters. Maps were then generated and

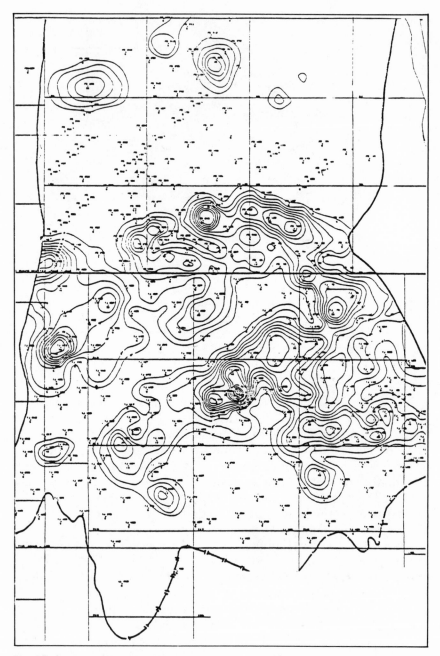

Fig. 16—5 Computer-generated map showing thickness of sand in the A-01 unit with resistivity greater than 20 ohm-m *(after Jam)*

plotted, one for each package. The most permeable sands were in packages A–10, A–01, and E–01. Figure 16–5 is the computer-generated map showing the thickness of sand with resistivity greater than 20 ohmmeters in the A–01 package. Two well-defined channels striking SW-NE are clearly shown. Most of the injection water had followed these channels. In other packages other channels appeared, mostly with a N-S orientation (Figure 16–6).

Another commonly used parameter to express sand quality is $\phi(1 - S_w)$ where ϕ = porosity and S_w is water saturation. Porosity can be determined by the density log, and S_w can be determined by the resistivity or induction logs. The resistivity values should be corrected for shaliness using the Waxman-Smits formula. In the case of the Laguna sands of the Bolivar Coastal field (De Andrea and Soria, 1977), there was a good correlation between permeability and $\phi(1 - S_w)$, Figure 16–7. A relation between resistivity from the 16-in. short normal electric log and $\phi(1 - S_w)$ was also determined, Figure 16–8.

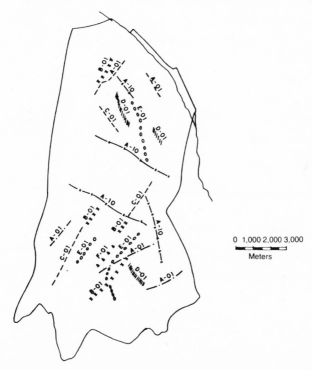

0 1,000 2,000 3,000

Meters

Fig. 16–6 Map of lower Lagunillas sand in LL-3 reservoir, showing principal sand-body orientations *(after Jam)*

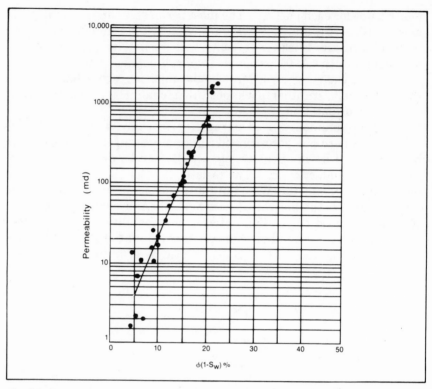

Fig. 16—7 Relation between permeability and $\phi(1 - S_w)$ *(after De Andrea and Soria)*

These relations apply to these particular sands only, and different sands might have different values. However, it may be that these curves could be used to determine sand permeability quantitatively, although roughly, in other fields. Van Veen (1977) shows almost identical relations between permeability, resistivity, and $\phi (1 - S_w)$ for the Eocene sands of the Bolivar Coast, although they are petrographically quite different from the Miocene LL–3 sands.

In the case of the southwest Bachaquero waterflood, it was decided to use a permeability of 350 md as the cutoff point between good and poor sand. This corresponds to a value of $\phi (1 - S_w)$ of 0.17 (Figure 16–7) or 14 ohms (Figure 16–8). When plotted on a map, $\phi (1 - S_w)$ shows two N-S trending permeable zones, which were interpreted to be offshore bars (Figure 16–9). Figure 16–10 shows that the water advanced into Sand 4 in two long tongues whose locations correspond to the location of the bars of permeable sand.

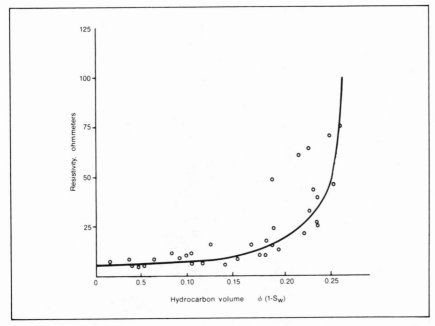

Fig. 16—8 Relation between resistivity as determined by 16-in. normal and $\phi(1 - S_w)$ (after De Andrea and Soria)

Van Veen used a cutoff of $\phi (1 - S_w)$ of 0.16 between good and poor which corresponded to a permeability of 300 md. His paper shows a water advance along the axis of the channel, Figure 16–11.

The Rangely Tensleep Sand reservoir in Colorado was discovered in 1933 but was not drilled up until 1943–1950 (Larson, 1974). It is the largest field in the Rocky Mountains, having produced 375,000,000 bbl by 1965. It produces from a section of 600 ft of interbedded sands, silts, and shales of the Pennsylvanian Weber formation. The field is on a large anticline, but there are many individual sand bodies of different types, separated from each other laterally and vertically by imperme-able beds (Figure 16–12). Channels, bars, dunes, and blanket sands are present. There is greater lateral permeability in the western part of the field where dunes and blanket sands are more common. Obviously, peripheral water injection would not be effective. Even the original 40-acre spacing was too wide to permit communication and a good sweep between injection and producing wells. Much oil will be trapped in pinchouts and feather edges.

Porosity, permeability, and net sand thickness maps were drawn

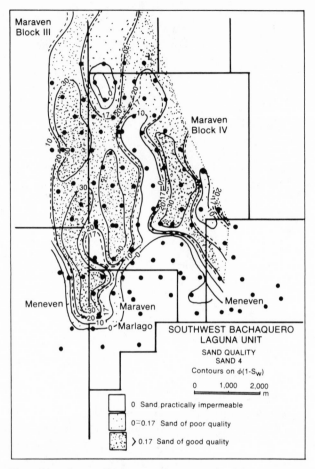

Fig. 16–9 Map of southwest Bachaquero Laguna unit with contours on $\phi(1 - S_w)$ showing sand quality *(after De Andrea and Soria)*

for many of the individual stratigraphic zones. Based on these studies, more than 150 infill wells were drilled on a 20-acre spacing, and water injection is proceeding on a modified 5-spot pattern.

Pressure transient tests are useful in determining continuity of permeability (Harris, 1975). The difficulty with these is that usually more than a single sand unit is open in the wells to be tested. One sand may communicate in one direction and another in a different direction. The repeat formation tester has been useful in distinguishing sand units at Rangely.

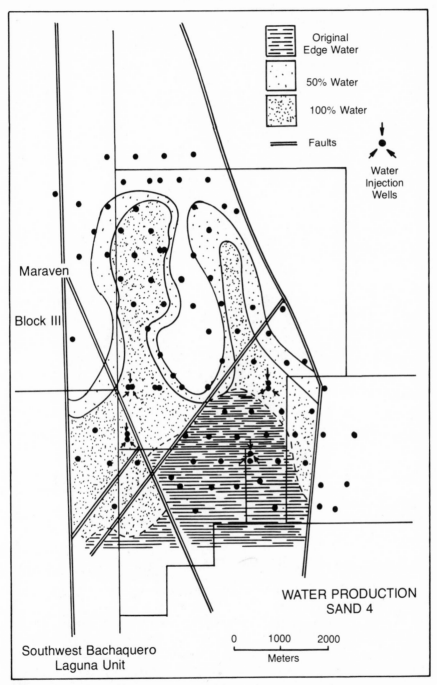

Fig. 16–10 Map of southwest Bachaquero Laguna unit shows percent water production *(after De Andrea and Soria)*

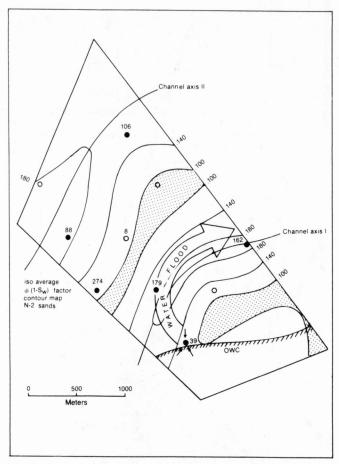

Fig. 16–11 Map of portion of VLA 8 block with contours on $\phi(1 - S_w)$ indicating sand quality. Water injected in well 39 followed channel axis *(after van Veen, courtesy SPE)*

4. BREAKDOWN OF CARBONATE RESERVOIRS

Subdividing a carbonate reservoir is a good deal more difficult than subdividing a sandstone reservoir. Carbonate permeability does not change with clay content like sand permeability; thus, permeability barriers are more difficult to recognize on wire-line logs. Several enhanced recovery projects were undertaken with the supposition that the carbonate reservoir rock was quite homogeneous with good vertical permeability. Water was injected downdip or gas into a gas cap with the idea that vertical movement of the gas-oil or water-oil contact

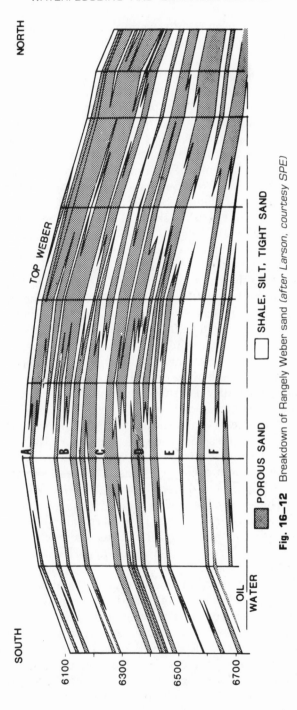

Fig. 16–12 Breakdown of Rangely Weber sand *(after Larson, courtesy SPE)*

would occur. When poor results were obtained, it was noticed that impermeable beds were present which acted as barriers to vertical flow.

Another difficulty in interpreting carbonates is that porosity can be increased or decreased by diagenetic changes. That is, pores may be enlarged by leaching or filled by precipitation of secondary calcite or other minerals, especially anhydrite.

On the other hand, carbonate textures are easy to recognize in cores. Depositional environments can be interpreted, and the reservoir can be subdivided into units of similar porosity and permeability.

West Seminole. The West Seminole field in Gaines County, Texas, produced oil from the dolomitic San Andres limestone of Permian age (Barrett et al., 1977). It is in the shape of a large dome with a gascap in the upper part and water below. The original oil in place is estimated as 172 million barrels and the gas in the gascap was 137 billion cu ft. The field was discovered in 1948 and developed by 54 wells on a 40-acre spacing. By 1962, the reservoir pressure had declined from 2,020 to 1,600 psi with the production of only 6 percent of the oil in place. Injection of gas into the gascap was started, but it did little to halt the rate of pressure decline. In 1969, a peripheral water injection was begun, but it also had little effect.

During 1974–76, a program of infill drilling and coring was conducted. A total of 28 wells were drilled and 6,000 ft of core was taken. Porosity and permeability measurements were made, and the core was slabbed and studied geologically.

The San Andres reservoir turned out to contain three major depositional cycles of transgression and regression (Figure 16–13). The lower two cycles were incomplete because they were truncated by unconformities. The lower parts of each cycle consist of clean skeletal fusilinid carbonates overlain by nonskeletal carbonates with oolites and hardened pellets. These beds are porous and permeable and can be traced across the field. They are overlain by impermeable micritic dolomite with thin beds of shale and anhydrite. Thus, there were three permeable zones, separated from each other by impermeable beds. Each of these zones also contained thinner impermeable beds.

At first it was feared that injection of water above the oil-water contact might move some of the oil into the gas cap, where it would be unrecoverable (Figure 16–14). The geological study was supplemented with a program of pressure measurements, transient tests, and interference tests. In general, the well tests confirmed the geological observations that there were strong barriers to vertical flow. The mathematical simulation, however, suggested that these barriers were not completely impermeable. The decision was to inject water in a

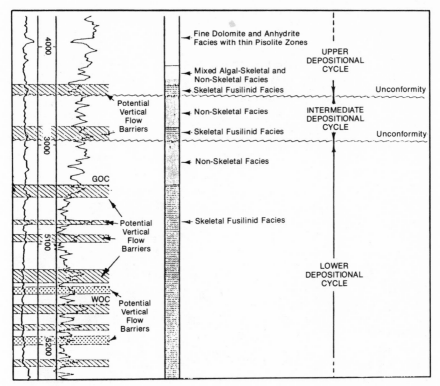

Fig. 16–13 Carbonate rock types in San Andres reservoir, West Seminole field, Texas *(after Barrett et al., courtesy SPE)*

pattern flood in the upper part of the structure, and to continue infill drilling. Some of the peripheral wells were recompleted in the oil zone above the oil-water contact, and turned out to be good producers.

Golden Spike. The Golden Spike pool in Alberta is in a very tall Devonian reef discovered in 1949. It covers only 1,385 acres but has a vertical thickness of 480 ft. Originally it contained 319 million bbl of oil with no gas cap or water leg. Initial pressure decline was rapid, so gas was injected at the crest of the reservoir which formed a secondary gas cap. Because the reservoir appeared to be quite homogenous, it seemed like an excellent place to apply miscible solvent extraction. Hydrocarbon solvent (LPG) was obtained by stripping the reservoir crude. Studies indicated that 95 percent of the original oil in place could be obtained by placing a bank of solvent between the gas cap and the oil zone, pushing it downward by injecting gas, and producing the oil from the bottom of the reservoir.

By 1972, production had declined and additional wells were drilled.

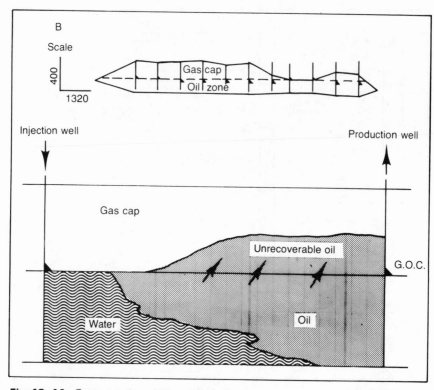

Fig. 16–14 Cross section of West Seminole showing gas cap. If the limestone were homogeneous, there would be danger of driving oil into the gas cap (*after Barrett et al.*)

Information from these wells indicated that the miscible flood was not performing as expected. By 1973, it was obvious that the upper surface of the oil was no longer horizontal. It was discovered that a horizontal permeability barrier about 250 ft below the top covered a large part of the area of the reservoir. The gas had underrun part of the barrier, leaving oil and solvent behind above it (Figure 16–15). By 1974, the gas had completely underrun the main barrier. Some of the oil previously left behind had drained down, but some additional oil was caught on top of a smaller horizontal barrier. By this time, all the injected solvent had dissipated into the gas cap, and it was no longer helping the recovery. Only about 60 percent of the volume of oil in the reservoir above the gas-oil contact had been recovered instead of the 95 percent that had been predicted.

Using the additional data obtained from the infill drilling, a detailed geological study of the reef was undertaken (Figure 16–16). The

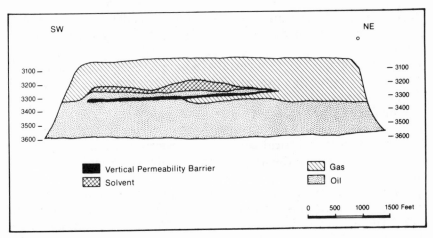

Fig. 16—15 Fluid distribution in Golden Spike reef in 1973 *(after Reitzel and Callow, courtesy SPE)*

lower 200 ft of the reef consisted of deep-water lime mud on which grew shallower water patch reefs. Porosity averaged 7 percent.

The next 125 ft was a reef rim composed of massive stromatoporoids (a type of calcareous alga) with a porosity which averaged 7 percent.

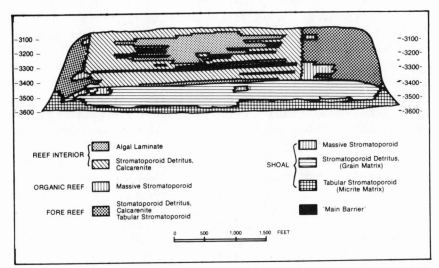

Fig. 16—16 Environmental facies of the Golden Spike D3 A pool *(after Reitzel and Callow, courtesy SPE)*

Outside this reef structure was a rim of fore-reef detritus averaging 3 percent porosity and an inner-reef detritus averaging 15 percent porosity.

A period of uplift and subaerial exposure formed a layer of 10 to 15 ft of lime mud over the interior but not over the fore reef. This formed the main barrier.

The top 250 ft consisted of sediments deposited in shallower water. There was less organic reef and fore-reef development and much algal laminate with low vertical permeability because of micrite beds and green shales. Capping the whole reef was a highly porous lime sand, 25 to 50 ft thick.

With the additional geological information, new simulation models were programmed that eventually explained the fluid distribution in 1975. The picture was supported by the well logs and well behavior. The poor performance of the miscible solvent was the result of the permeability barriers and the fact that it dispersed into the gas cap. However, it did effectively remove the oil above the main barrier before it dispersed. Small amounts of oil are still lodged above the permeability barrier, but at least some of it will drain down and be recovered. The ultimate recovery will still be very good—67 percent of the oil in place by the year 2020.

It was concluded that this sort of geological data ought to be obtained early in the life of the pool and the progress of any enhanced recovery scheme should be closely monitored. The large number of infill wells provided an abundance of data. One can't help wondering, however, whether the careful environmental reconstruction could have been made with data from the original wells and at least the possibility of unhomogeneities in the reservoir could have been predicted.

Elk Basin.   Elk Basin, Montana, Madison zone, is an example of an oil field that was originally developed without an understanding of the geology; the pressure maintenance and secondary recovery operations did not work out as expected (Wayhan and McCaleb, 1967).

The field is a large anticline, dipping about 25° on the west flank and 45° on the east. It produces from the Tensleep sandstone above and the Madison limestone below, each having about one-half billion barrels of oil in place. It was developed slowly between 1948 and 1960 because of small demand for the black, heavy crude oil.

The early wells were completed open hole through the entire 920-foot section which was believed to be essentially homogeneous, porous dolomite. There was a strong water drive, and a reservoir simulation study indicated that a producing rate of 25,000 bbl of fluid per day could be produced without dropping the pressure below the bubble point (1,140 psi at −700 ft datum, 7,860 kPa at −213 m).

It turned out that there were four different reservoirs in the Madison, separated by impermeable beds (Figure 16–17). Additional wells completed in single zones between 1958 and 1961 showed that the four zones were quite different lithologically and had quite different reservoir performance. The A zone had been exposed to erosion during post-Madison time. Portions had been leached by ground water while other portions had been solidly cemented by calcite and anhydrite. As a result, there was practically no lateral communication from one portion to the other. The A zone had been producing by a typical dissolved gas-drive mechanism. The pressure had dropped to 700 psi (9,830 kPa) by 1963, and gas-oil ratios were increasing. There was little effect from the water injection wells. The peripheral wells on the east flank had built up the pressure in some portions of the A zone to 1,000 psi (6,900 kPa) above virgin pressure, showing that water was being pumped into portions of the reservoir where there was no outlet. The peripheral water wells on the west flank had accomplished nothing because the B

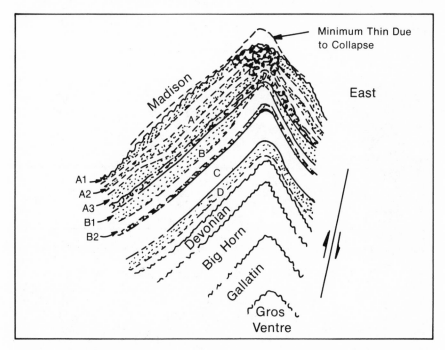

Fig. 16—17 Cross section of Elk Basin field, showing four different zones in Madison reservoir (after Wayhan and McCaleb, courtesy SPE)

and D zones already had a good natural water drive. The injected water was not entering the A zone because of its poor lateral permeability.

Burbank, Oklahoma. The Burbank field in Osage County, Oklahoma, is a large stratigraphic trap discovered in 1920. By 1950 it had produced 221,000,000 bbl (35×10^6 cu m) from the Pennsylvanian Burbank sand at a depth of about 3,000 ft (1,000 m). It had operated under dissolved gas drive, and it was estimated that it had produced only 25 percent of the original oil in place. A pilot flood was started in 1950, consisting of nine injection wells and four producing wells on a 5-spot pattern. Water was injected into the pilot flood at pressures equal to or less than hydrostatic (less than 0.433 psi per ft or 9.8 kPa/m). Production increased from 37 to 1,000 b/d by January 1951. Elated by their success, the operators began work immediately on a thousand-acre extension.

However, right from the beginning, one well (127–6) produced with a high water-oil ratio. Uranine dye was injected into the nearest injection well, and it was detected in wells to the northeast and southwest in a few days, indicating that open fractures existed. As development proceeded, the water production problems grew worse. In spite of a good water supply, the intake rate declined as a result of plugging. Pressures were increased gradually to 600 psi (4,140 kPa), and at 300

Fig. 16–18 Photograph of cores from Burbank, Oklahoma, showing fractures (after Hagen)

psi (2,120 kPa) surface pressure the rate of injection increased sharply. It was supposed that the plugging problem was solved, but actually what happened was that the fractures were opened (Figure 16–18).

Cores of the Burbank sand showed vertical fractures, probably joints. There were several cases of lost circulation in the shale immediately above the sand. Wells in the Stanley Stringer, a separate reservoir at the same horizon but a mile to the east, started to produce large amounts of water. The water also travelled westward to the Little Chief pool. The water never travelled north or south.

Evidence from the behavior of the wells and the tracer tests suggested that the fractures were oriented east-west and belonged to a joint system. The joints are very well developed in the hard limestones and sandstones at their surface outcrop, and they are visible both on the ground and from the air, as seen in Figure 16–19. The average direction of several thousand was measured and is very constant. Their average azimuth is 70° (N 70° E) with only a few less than 65° or more than 75°.

The field development of the waterflood utilized the old wells, which had been drilled on a square grid, NS and EW lines, with one well to every 10 acres (4 h). Every other well was recompleted as an

Fig. 16–19 Photograph of joints visible on the surface at Burbank field, Oklahoma (after Hagen)

injection well. This resulted in a 20-acre (8-h) 5-spot with the lines of injection and producing wells oriented NE-SW. Thus each producing well had an intake well directly east and another directly west (Figure 16–20). When the east-west fracture system was recognized, the pattern was changed. New intake wells were drilled on east-west lines. This made a staggered line drive, with the water moving north and south. Presumably it filled the fractures between the intake wells and then pushed the oil north and south at right angles to the fractures. Much better results were obtained, but most of the wells continued to produce at a high water-oil ratio. It is not clear whether the joints were originally closed and opened when the pressures were raised, or whether they were open at depth. It seems fairly certain that good oil recoveries at lower oil-water ratios would have been obtained if the injection pressures had been kept low at the beginning.

It has been known for many years (Dickey and Andresen, 1949) that there is a critical pressure which opens fractures in the reservoir rock, permitting large volumes of water to circulate uselessly without moving any oil. In spite of this, there are still many projects in which water is being injected at higher than fracture pressure. Methods of estimating the fracture pressure are discussed in Chapter 13. It was shown there that the fracture pressure depends on the pore pressure. The fracture pressure in a waterflood will increase as the reservoir pressure is restored.

If the injection rate is unduly high, it is a good indication that the water is going through fractures. The injection rate per foot depends on the permeability of the rock, which ranges from 0.1 to several

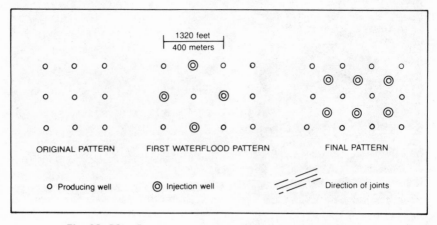

Fig. 16–20. Successive patterns used at Burbank, Oklahoma

thousand md. It also varies with the amount of fill-up of the reservoir and the amount of sand-face plugging by impurities in the water. Some water floods have been successful in low permeability sand at less than 10 bbl/d per ft. (0.5 m^3/m). If the injection rate exceeds 100 b/d/ft (15m^3/m), fractures should be suspected and the behavior of the well should be carefully studied.

Bullion-Clintonville. The Bullion-Clintonville field in Pennsylvania was developed between 1890 and 1910. A single pay sand at the Venango second sand horizon is 10 to 16 ft thick with a permeability between 100 and 400 md. In 1943, most of the original wells were still in operation, each pumping a fraction of a barrel of oil and a small but undetermined amount of water.

In that year, a 25-well pilot waterflood was started. The delayed drilling method was used; that is, the intake wells were completed first, and water was injected several months before the producing wells were drilled. The idea of this was to push the oil in the more permeable sand layers into a bank at the producing well location, where it had no place to go. It would stay there until the oil bank in the less-permeable beds had a chance to catch up. When the producing wells were drilled in 1944, they produced 30 b/d (4.7 m^3/d) water, which was exactly what the intake wells were taking, and less than 1 b/d (0.2 m^3/d) oil.

The first indication of trouble came after the intake wells were completed but before water was injected. Water was found standing in the tubing about 200 ft (60 m) above the sand in most of the wells. In those adjacent to a creek, there was 400 ft (120 m) of water. Apparently, water had been entering the sand from old, unplugged wells near the creek, probably for many years. It had been moving very slowly, for some of the old wells were still pumping. However, it had obviously filled the pores of the sand completely because, when injection started, the producing wells immediately started to produce the same amount of fluid that the intake wells were taking without any period of fill-up. The accidental entry of water had partly restored the original bottom-hole pressure.

Similar waterflood failures resulting from entry of water into the sand from old wells have occurred in Oklahoma. They might have been prevented if the operators had (1) noted the water-oil ratio of the old wells and (2) measured the bottom-hole pressures in the producing sand.

Conclusion. The foregoing chapter describes several waterflooding projects which failed because the geology of the reservoir was unknown and therefore not taken into account in planning the project. Very many other failures have occurred which were never described in the published literature. Neither the engineers nor the executives involved want to take credit for a failure.

On a more positive note, it is now becoming more generally recognized that the geology of a petroleum reservoir should be worked out in detail during the primary development and taken into account in planning secondary and enhanced recovery.

REFERENCES

Barrett, D. D., K. J. Harpole, and M. W. Zaaza, 1977, Reservoir data pays off: West Seminole Unit, Gaines County, Texas: SPE Paper 6738, 12 p.

Boesi, Tito, 1977, Construccion de un modelo geologico conceptual del yacimiento VLC-363: V Congreso Geologico Venezolano, Memorias, Vol. IV, Caracas, 1977, p. 1429–1444.

DeAndrea, Rafael, and Jose Soria, 1977, Revision petrofisica y sedimentologica del yacimiento de Laguna, Lago de Maracaibo: V Congreso Geologico Venezolano, Memorias, Vol. IV, Caracas, 1977, p. 1505–1528.

Dowling, Paul L., Jr., 1970, Application of carbonate environmental concepts to secondary recovery projects: SPE Paper 1987, 16 p.

Hagen, Kurt, 1972, Mapping of surface joints on air photos can help understand waterflood performance at North Burbank Unit, Osage and Kay Counties Oklahoma: M.S. Thesis, University of Tulsa, 85 p.

Harpole, Kenneth J., 1980, Improved reservoir characterization—a key to future reservoir management for the West Seminole San Andres unit, SPE paper 8274, Jour. Petrol. Tech., November, pp. 2009–2019.

Harris, D. G., 1975, Role of geology in reservoir simulation studies: Jour. Petrol. Techn., May 1975, p. 625–632.

Hewitt, C. H., 1966, How geology can help engineer your reservoirs: Oil and Gas Journal, Nov. 14, 1966, 8 p.

Jam, L., Pedro, 1977, Aplicacion de tecnicas de computacion al estudio geologico de un yacimiento petrolifero: V Congreso Geologico Venezolano, Memorias, Vol. IV, Caracas, 1977, p. 1393–1404.

Kamal, M. M., 1977, Use of pressure transients to describe reservoir heterogeneity: SPE Paper 6885, 12 p.

Key, Carlos E., 1977, La formacion Oficina en el campo Jobo: V Congreso Geologico Venezolano, Memorias, Vol. IV, Caracas, 1977, p. 1599–1616.

Larson, Thomas, C., 1974, Geological considerations of the Weber Sand reservoir, Rangely field, Colorado: Soc. Petrol. Engrs., SPE Paper 5023, 8 p.

Lennon, R. B., 1976, Geological factors in steam-soak projects on the west side of the San Joaquin Basin: Jour. Petrol. Techn., July 1976, p. 741–748.

Reitzel, G. A., and G. O. Callow, 1977, Pool description and performance analysis leads to understanding of Golden Spike's miscible flood: Jour. Petrol. Tech., July 1977, p. 867–872.

Richardson, Joseph G., Donald G. Harris, Robert H. Rossen, and Gus Van Hee, 1977, Synergy in reservoir studies: SPE Paper 6700, 13 p.

Sneider, R. M., F. H. Richardson, D. D. Paynter, R. E. Eddy, and I. A. Wyant, 1977, Predicting reservoir rock geometry and continuity in Pennsylvanian reservoirs, Elk City Field, Oklahoma: Jour. Petrol. Techn., July 1977, p. 851–866.

van Poollen, H. K., and Associates, 1980, Fundamentals of enhanced oil recovery, PennWell, Tulsa, 155 p.

van Veen, F. R., 1977, Prediction of permeability trends for water injection in a channel-type reservoir, Lake Maracaibo, Venezuela: Soc. Petrol. Engrs. SPE Paper 6703, 8 p.

Wayhan, D. A., and J. A. McCaleb, 1968, Elk Basin Madison heterogeneity—its influence on performance: Soc. Petrol. Engrs., SPE Paper 2214, 10 p.

Zamora, Lucas G., 1977, Uso de perfiles en la identificacion de ambientes sedimentarios del Eoceno del Lago de Maracaibo: V Congreso Geologico Venezolano, Memorias, Vol. IV, Caracas, 1977, p. 1359–1376.

17

Evaluation of an Oil Discovery:
How to Locate the Second,
Third, and Fourth Wells

In offshore areas, it is extremely expensive to build a platform and develop an oil field. Consequently, a promising show of hydrocarbons in a wildcat well cannot be considered a discovery until some idea of the size of the field and its productivity is obtained. Extensive well tests and several additional wells are required. It is customary to drill offshore wells with drillships solely for the information without any intention of completing them as producers.

In onshore areas, the situation is different. A discovery well can be completed as a producer. Each successful stepout well can be added to the field. There is less necessity for an early evaluation. However, the technology of evaluation is similar, even though the economics and philosophy are different.

In many onshore areas in remote and hostile environments, the situation is comparable to that in remote offshore areas. Due to lack of an infrastructure, that is, highways, pipelines, and other transport facilities, it becomes necessary to determine whether a discovery is commercial before proceeding with its development.

Each discovery must be evaluated on the basis of its particular geological environment. Few typical anticlinal structures are found these days. Pools are bounded by faults and stratigraphic pinch-outs more often than rollover into an oil-water contact. Consequently, it is usually necessary to drill several wells, some of them dry holes, before the structure is fully appraised.

1. REINTERPRETATION OF SEISMIC DATA

It will be assumed that good seismic data was used in choosing the location of the discovery well. Sometimes the first well is located on a structural high. This is determined by drawing structure contours on the depth below sea level of a well-defined reflecting horizon that can be traced over a considerable area. Sometimes the first well is drilled on a suggestion of a truncation or pinch-out appearing on a seismic section. Before drilling the well, the depth of the reflecting horizons is known only approximately. Their lithology and geological identification are not known at all. Seismic *events* are not all reflections from geological formations with contrasting lithology. Indeed, many events are composites of interfering waves and show no obvious correspondence with interfaces in the earth.

In recent years, huge efforts have been made to utilize seismic information to delineate stratigraphic traps in advance of drilling. AAPG Memoir 26 contains many articles on seismic modelling to help in the interpretation of the seismic record. Seismic modelling consists of representing subsurface geological configurations like sand lenses, pinch-outs, and reefs geometrically and mathematically and then predicting the response of reflected acoustic waves to that configuration. The result is a picture showing what a seismic section ought to look like from the assumed geological configuration.

Memoir 26 is written primarily from the point of view of exploration. However, modelling techniques are really more adapted to development geology in offshore areas such as the North Sea. After the discovery well is drilled, it is customary to return to the area and record detailed seismic lines over the structure. The cost of shooting a few extra kilometers of marine seismic profiles in order to detail the prospect is small in comparison with the enormous cost of setting a platform and drilling development wells. Interpretation of the seismic data may provide a picture of the size and shape of the reservoir.

Recent advances in the digital recording and mathematical filtering of seismic data have resulted in enormously improved records. It is possible to use information on frequency, phase, and amplitude of the reflected waves, which formerly was filtered out and disregarded.

Figure 17–1 is a seismic record especially processed to show phase. At about the center of the picture at a reflection time of 2.6 seconds, there is a marked downdip pinchout. Farther left, the interval above the prominent reflection widens, and there is a suggestion of progradation to the left near the left edge of the picture between 2.7 and 3.0 seconds. Deeper, in the center, there is a suggestion of an onlap pattern. These are all clearly shown and are fascinating stratigraphic patterns, but they would be difficult or impossible to interpret without information from wells near the profile to identify the events.

What we need from a well is average velocity to the reflecting

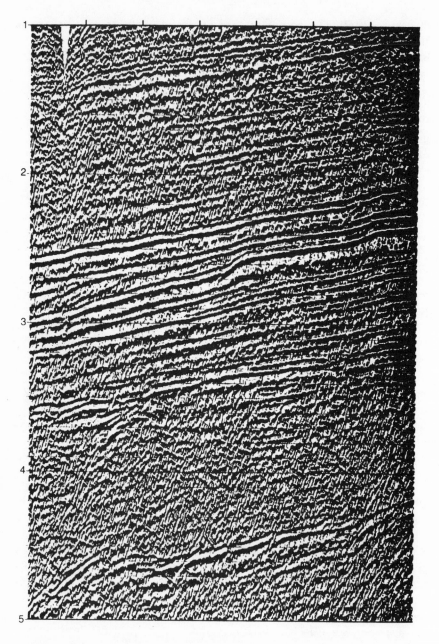

Fig. 17—1 Seismic phase display from East Texas. The downdip pinch-out in the center at 2.6 sec is the edge of the Edwards reef. To the left, the prograding character of the Woodbine sand is suggested *(after Taner and Sheriff)*

horizons for converting reflection times to depths and the interval velocity and density of each geological unit for preparing a synthetic seismogram, showing the expected character and amplitude of the reflections beneath the well location. The first step in reinterpreting the seismic data is to identify the reflecting horizons or "events" that are seen on the seismic section (two-way travel time instead of feet). To convert reflection time to depths, the velocity of the seismic wave must be known. This can be determined from a velocity survey in which a geophone is lowered into the hole while explosive charges are shot at the surface. Average velocity to various depths can then be computed.

A much more detailed record of the velocity within each individual geological unit (interval velocity) is obtained from a sonic log. Acoustic velocity in rocks ranges very widely from as little as 2,000 feet per second in unconsolidated shale through 10,000 to 15,000 feet per second in sandstones to 20,000 feet per second or more in limestones. The velocity of a seismic wave in sandstones is generally higher than in the adjacent shales, so the amount of sand in the section can be estimated. However, in some places, notably Nigeria, there is little or no contrast in velocity between sandstones and shales. The sonic log records travel time in microseconds per foot, which is the reciprocal of the velocity. If the sonic log is recorded digitally, the total travel time can be calculated by integrating the interval travel times. The total travel time to any depth can be multiplied by 2 to convert to the appropriate reflection time for that depth, and the resulting integrated log shows sonic velocity vs reflection time.

The sonic velocity log, together with a density log, can be used to generate a log of acoustic impedance (ρV, where ρ = density and V = sonic velocity). From the acoustic impedance log, a log of reflection coefficients (Figure 17–2) can be derived using the following equation:

$$R = \frac{\rho_2 V_2 - \rho_1 V_1}{\rho_2 V_2 + \rho_1 V_1} \qquad (17\text{–}1)$$

where:

R = reflection coefficient
ρ_1 = density of first unit
ρ_2 = density of second unit
V_1 = sonic velocity in first unit
V_2 = sonic velocity in second unit

If a density log is not available, density can be neglected, as it

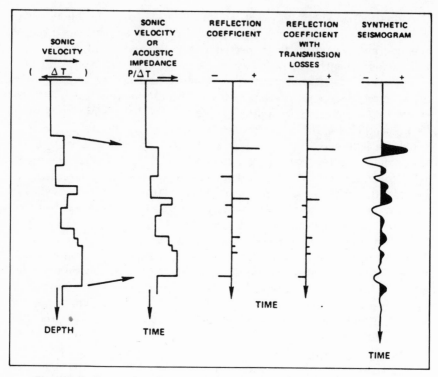

Fig. 17–2 Conversion of sonic log to synthetic seismogram (after Stone)

usually changes less than velocity. In this case, equation 17–1 simplifies to:

$$R = \frac{V_2 - V_1}{V_2 + V_1} \qquad (17-2)$$

The reflection coefficient log is convolved with a selected seismic wavelet to produce a synthetic seismogram (Figure 17–2).

It should now be possible to superimpose a correctly scaled synthetic seismogram on the seismic section and have it match fairly well. Sometimes the synthetic seismogram is shown repeated several times to make it look more like the seismic section (Figure 17–3). All the reflections on the synthetic seismogram were derived from the sonic log, and their depths were therefore determined exactly. It is now possible to compare the seismic section with the sonic and the other wire-line logs and the bit cuttings and decide which geological units are represented by the events on the seismic section.

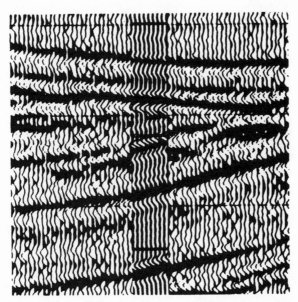

Fig. 17–3 Seismic section with synthetic seismogram superimposed at the center (*after Stone*)

The sonic and other wire-line logging devices resolve vertical changes in lithology to as little as one or two feet. However, they give no information whatever on lateral changes. A discovery well may be drilled only a few hundred meters from a pinch-out in one direction or from a much thicker and more productive formation in another direction. The seismic section, on the other hand, shows all kinds of lateral changes in reflection character. If we could interpret them, we could tell something about lateral changes in rock character.

The seismic section gives very poor vertical resolution because the wavelets are complex and are usually longer than 20 milliseconds (50 Hz). At 10,000 feet per second velocity, this represents a vertical thickness of 200 feet. Thus it appears impossible to identify thin beds.

Suppose we have a wavelet of 50 Hz frequency traversing a bed of 20,000 ft/sec velocity (Figure 17–4). If the bed has a thickness of more than one wavelength (200 feet), there will be two distinct reflections: a down-kick at the top of the bed and an up-kick at the bottom. Where the bed is less than one wavelength thick, the wavelets overlap. At a thickness of half a wavelength, the top and bottom can still be distinguished. At less than half a wavelength, there is only one wavelet with increased amplitude. Bed thickness now cannot be determined by the

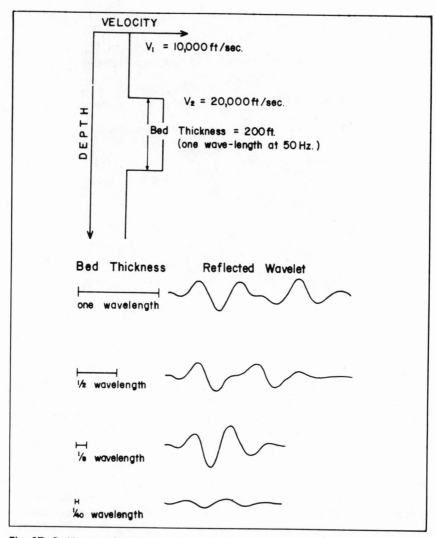

Fig. 17—4 How a reflected wave form changes as the interval between the top and bottom of a bed decreases (*after Widess*)

distance between two reflections but by the amplitude of a single reflection. The thickness for maximum amplitude is called the *tuning thickness,* and in ordinary Gulf Coast conditions it is about 60 feet (Ausburn et al., 1978). The bed can still be recognized down to a thickness of ⅛ wavelength. If it is possible to identify the producing formation with a reflection, it may be possible to predict its lateral changes from the change in character of the reflection. As it pinches out, it will go from two events to one and finally to none.

The resolution of a seismic section is limited vertically and laterally. We normally draw a diagram of a seismic section as if the acoustic wave travelled vertically down and was reflected vertically back to the surface. This is oversimplified. If the reflecting bed is dipping, the reflection will come—not from directly below the shot point—but from the updip side.

It is possible that more than one reflection may hit the geophone at the same instant from points ahead, behind, or to one side of the line. Correction for reflections from dipping beds is called *migration.*

The development of 3-D seismic recording techniques makes it possible to distinguish between reflections from on-line and off-line sources and locate the reflection points correctly by 3-D migration. It is also possible to construct seismic sections in any desired direction across a prospect. An interesting new technique is to show the reflection events as a series of horizontal slices at different reflection times (Figure 17–5). From these, detailed contour maps can be constructed (Brown, 1979).

The lateral resolution of a seismic section is also limited because an advancing wave front has approximately a spherical shape. The energy arriving as much as ¼ wavelength behind the front will interfere and cancel the earlier reflection. Consequently, the reflection will come not from a point but from an area defined by the radius of the spherical wave front, that is, the depth and the wavelength (Figure 17–6). This area is called the Fresnel zone (Figure 17–7). If a sand lens is one Fresnel zone wide, say about 1,000 feet, the seismic section will show it as somewhat wider. If the reflecting body is only ¼ of a Fresnel zone wide, the reflection will be reduced in amplitude and have the parabolic shape of a diffraction pattern.

Ausburn et al. (1978) give an example from two wells in the North Sea (Figure 17–8). One (Well B) had 112 m of sand, of which the top 30 m was gas filled. The other (Well A), 3,100 m distant, showed only a few sand stringers and was a dry hole. Two geological interpretations were possible. If the sand pinches out according to Model A, the field will be much smaller than if it disappears by a change to shale according to Model B. Using the velocity and density data shown in the figures, reflectivity values were calculated. These were then convolved with the basic wavelet, and two synthetic models were made based on the two differing interpretations (Figure 17–9). Ausburn et al. do not show the seismic section between the two wells, but they state that it looked more like Model B.

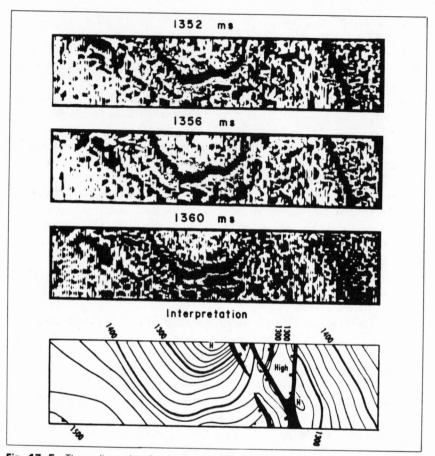

Fig. 17—5 Three-dimensional seismic data displayed in three successive horizontal sections taken at 4-ms intervals with an interpretation *(after Brown)*

A local thickening of a channel sandstone body that compacts less than the laterally equivalent shales will cause draping of the overlying

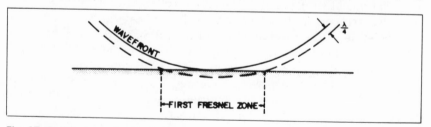

Fig. 17—6 Diagram of spherical front of acoustic wave showing Fresnel zone *(after Meckel and Nath)*

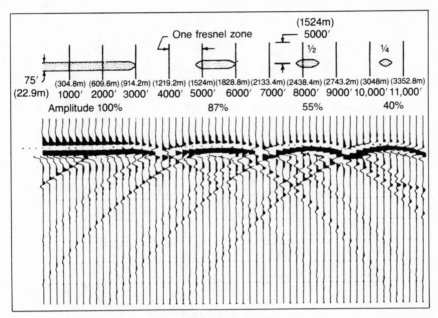

Fig. 17–7 Seismic model showing lateral resolution of reflections from sand bodies of differing width *(from Neidell and Poggiagliolmi)*

formations. On a seismic section, this will look like a local thickening of the interval between two reflections. The draping is much less sharp than the channel, so the method has been successful only where the geology is well known and the reflections are very good (Lyons and Dobrin, 1972).

Figure 17–10 is a seismic model of a sandstone-shale sequence that thickens and thins by the superposition of irregular channel sands. Wells A, B, C, and D were used to construct the model. Much significant data is shown by the seismic model. If only one of the four wells had been drilled, a seismic section could be used to extend the geological data laterally. If no well had been drilled, the significance of the seismic section would be difficult to evaluate.

Carbonate reefs are upstanding bodies that often have more than 100 feet of vertical relief. These show up very well in seismic sections. Figure 17–11 is an example of a reef in the Zama area of Alberta (Evans, 1972). A very marked local thickening of the internal between two horizons above and below the reef is evident. Sometimes the wave forms that represent the actual reef lithology can be recognized. Excellent success in finding reefs with the seismograph has been obtained in Alberta and Michigan.

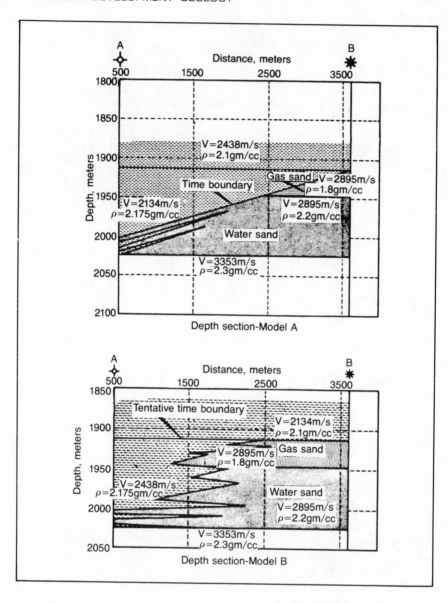

Fig. 17–8 Two interpretations of the lateral pinchout of a 112-m sand penetrated in wells A and B *(after Ausburn et al.)*

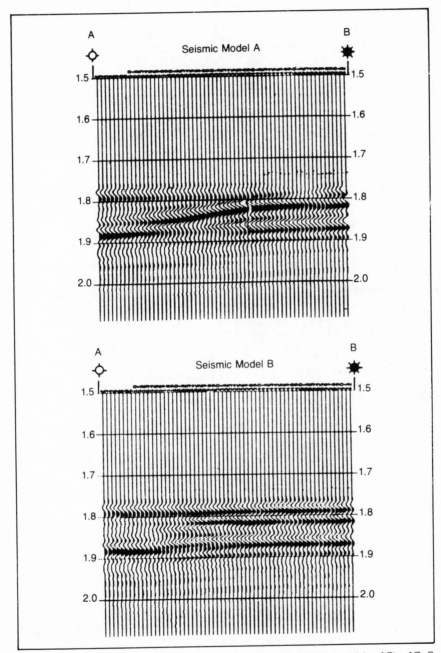

Fig. 17–9 Synthetic seismic sections based on the two differing models of Fig. 17–8 (after Ausburn et al.)

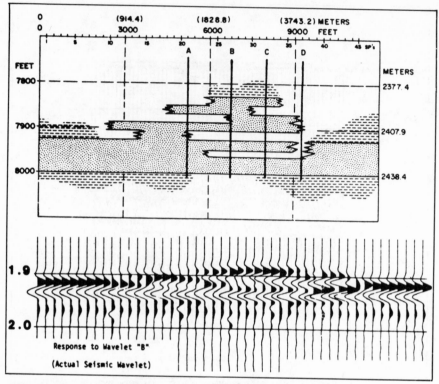

Fig. 17–10 Seismic model of the response to a sandstone body that thickens upward by the superposition of channel sands interfingering with shale *(after Neidell and Poggiagliolmi)*

2. WHERE TO DRILL THE SECOND WELL

The decision on where to drill the second well obviously is almost as important as the decision on where to drill the first one. If the confirmation test turns out dry, it will certainly be discouraging—perhaps unnecessarily so. For this reason it would seem better to move the second location in the direction of better structural and stratigraphic conditions.

If additional closely spaced seismic lines are shot, there will be a much more detailed map of the structure. If the log data from the first well helps in the processing of the seismic data, the map will be more accurate as well as more detailed. The second well should be drilled at a higher structural location.

The stratigraphic data are a good deal more difficult to interpret. The petrophysical data from the cores, samples, and wire-line logs will

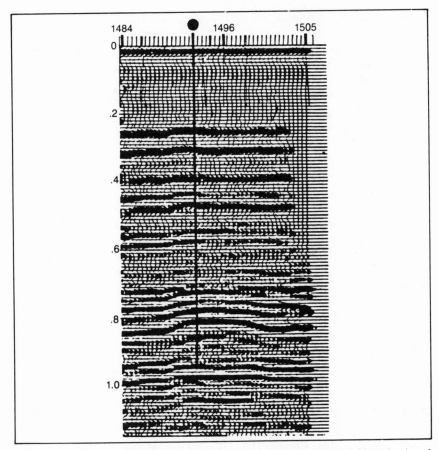

Fig. 17—11 Seismic section across a productive reef (Zama, Alberta). Note draping of reflecting horizon above the reef and the lateral changes of events with it *(after Evans)*

now make possible a good interpretation of the depositional environment of the producing formation. From general geology, the seaward and landward directions of the depositional basin are known. In a vague sort of way, a channel sand will run from the landward toward the seaward, that is, at right angles to the ancient coastline. A beach- (or barrier bar)-type sand will be parallel to the ancient shoreline. One needs only to visualize a modern delta like the Mississippi or the Niger to realize that there may be a maze of channels—some straight and others crooked. Parallel with the coast are whole bands of beach-type sands, some formed on barrier islands and some offshore. Knowing the depositional environment will help, but it certainly will not tell how far to move.

Usually there are other wells in the general area. Careful correlation will identify the intervals in these wells, which are the time equivalents of the producing interval in the discovery well. The depositional environments in those wells will help define the environmental pattern in the discovery well.

Thus, the second well provides an opportunity to obtain essential information that it was not practical to obtain in the first well.

Cores. The second well should be completely cored through the productive intervals. The core will be used to predict the reservoir behavior of the field. Porosity, permeability, relative permeability, and capillary data are used in predictions of ultimate production and rate of production. Especially important to ascertain is the presence of fractures and whether they are open or closed. None of these essential data can be obtained from wire-line logs. However, if they are once determined from cores, wire-line logs can extrapolate them to the wells that will be drilled subsequently to develop the field.

With cores, the geologists can make much better environmental interpretations. When these are extended to the first well and to the other wells in the vicinity, the size and shape of the sand body can be interpreted with more assurance. If the sand body is thick, it is almost certainly multiple. The environmental interpretation will say which intervals of the sand are apt to have considerable continuity of permeability and which are only local. Studies of the earliest wells in the Prudhoe Bay field recognized beach-type sands overlain by braided stream deposits (Eckelman and DeWitt, 1975). Similar identifications were made at Brent (Bowen, 1975, Kingston and Niko, 1975). They are made routinely by Shell-BP in Nigeria (Weber, 1971).

The multiple character of the sand body must be understood to make reservoir predictions. Practically all sand bodies (and most carbonates) contain horizontal permeability barriers. These separate the reservoir into separate subreservoirs. Although nothing has been published as yet, there seems to be some evidence that each subreservoir may have its own initial pressure that differs slightly from the pressure in the other subreservoirs. Certainly after depletion each subreservoir has its own pressure. It would be good to make a detailed pressure survey using the new Schlumberger Repeat Formation Tester.

The transient flow tests will be repeated and reinterpreted in view of the reservoir heterogeneity determined from the core data, extended by the geological interpretations.

It is common practice on the part of some companies to look for the limits of the pool with the second well; that is, the second well is drilled a considerable distance away, off the structure, for the purpose of locating the oil-water contact or other limits of the field. This practice is based on the reasoning that if the pool is small, it is not worth developing. The discovery has been made, and another good well close to the discovery well is not worth drilling. Some companies even drill the first well off the crest of the structure. Unless the field extends some distance down the flanks, it will not be worth developing.

The wisdom of this practice might be questioned. The second well at a good location is really needed to evaluate the field. It provides a chance for a rapid and effective well completion to eliminate, or at least mitigate, well-bore damage. This will make it possible to get a complete core of the reservoir to provide the data that is indispensable to estimate the recoverable oil correctly. It will be possible to subdivide the pay sand with pressure measurements.

Oil occurs in a wide variety of structural and stratigraphic configurations. Some companies have been successful drilling the crest of a diapir, while others have drilled dry holes in diapirs without realizing that the oil might have been on the flanks. In some places the structure is bald-headed; that is, the producing formations are missing over the crest. In Lake Maracaibo there are giant fields on faulted anticlines; only a few kilometers away, the anticline is dry and the oil is in the adjacent syncline. The exploration philosophy of a company is colored by the past experience of the management. It differs from one company to another and changes as the managers change.

In Oklahoma there has been active wildcatting since 1905, but new fields and extensions to old fields are still being found. It has often happened that the first well did not evaluate a new field properly.

3. THIRD AND FOURTH WELLS

After the second well, it is time to look for the pool limits. If the field is a simple structural dome, the third and fourth wells may be located deliberately to drill through the oil-water or gas-water contact. Then the limits of the pool can be drawn with assurance if (1) the sand is regionally permeable, (2) the gas or water contact is level, and (3) the seismic contours are correct. With this data and the core and well-test data, the volume of the reservoir can be calculated with assurance.

However, the first two conditions are not usually true. As an example, the recently discovered supergiant fields Prudhoe Bay and Brent have an oil-water contact on only one side; on the other side, the sand is truncated by an unconformity. In the case of many fields in Venezuela, Nigeria, and the Gulf Coast, one edge of the field is a fault. Faults can be easily recognized on good-quality seismic sections, and under favorable situations unconformities can also be defined. Many great fields (for example Ghawar in Saudi Arabia) have an asphalt layer at the oil-water contact that is higher on one side of the pool than it is on the other.

As soon as the limit of the field is determined on one side, it may be sketched around using the seismic data even without much assurance. It is now time to make preliminary estimates of the reservoir volume and the recoverable reserves. Usually the engineer hesitates to make an estimate until he has what he considers sufficient data. However, as John Arps used to say, the only time we know with real assurance what the recoverable oil reserves of a field are is just after it has produced the last barrel. It is early in the development of a field that the estimates are most needed.

In the case of relatively cheap onshore development in established oil regions, an early estimate is useful but not really essential. Each new well, if it is productive, is tied into the pipeline and the field gradually grows. Normally, development proceeds cautiously with one-location stepouts until finally a dry hole is drilled. Even in relatively inexpensive onshore situations, however, it might be better to make an early estimate of the reserves of the field in order to provide the financial resources for its development. In offshore fields, it is necessary to know early whether the field will be commercial or not. Probably early estimation of the pool size will be too small. The transient test data will, in the case of a large pool, apply only to the subreservoir tested. The best structural part of the pool may coincide with a poor sand development. Cases can be cited where it was not the first or second well on a good structure but the tenth or twentieth, years later, that finally revealed the size of the field. The Ataka field in Indonesia consists of local channel sands. The first few wells were dry or noncommercial, and it was not until another company drilled a different part of the structure that its importance became evident.

REFERENCES

Ausburn, B. E., Nath, A. K., and Wittick, T. R., 1978, Modern seismic methods—an aid for the petroleum engineer: Jour. Petroleum Technology, November, p. 1515–1530.

Brown, A. R., 1979, 3-D seismic survey gives better data: Oil and Gas Journal, Nov. 5, p. 57.

Eckelman, W. R., and Dewitt, P. J., 1975, Prediction of fluvial-deltaic reservoir geometry; Prudhoe Bay Field, Alaska: World Petroleum Congress, Tokyo, Proc., v. 2, p. 223–228.

Evans, H., 1972, Zama, a geophysical case history, *in* R. E. King, ed., Stratigraphic oil and gas fields: AAPG Memoir 16, p. 440–452.

Lyons, P. L., and Dobrin, M. B., 1972, Seismic exploration for stratigraphic traps, *in* R. E. King, ed., Stratigraphic oil and gas fields: AAPG Memoir 16, p. 225–243.

Meckel, L. D. and Nath, A. K., 1977, Geological considerations for stratigraphic modelling and interpretation, *in* C. E. Payton, ed., Seismic Stratigraphy, AAPG Memoir 26, p. 417–449.

Neidell, N. S., and Poggiagliolmi, E., 1977, Stratigraphic modelling and interpretation—geophysical principles and techniques, *in* C. E. Payton ed., Seismic Stratigraphy, AAPG Memoir 26, p. 389–416.

Payton, C. E., Editor, 1977, Seismic stratigraphy: AAPG Memoir 26, 516 p.

Stone, D. G., 1978, Using seismic data to extrapolate well logs: presented to SEG 48th Annual Meeting, San Francisco.

Taner, M. T., and Sheriff, R. E., 1977, Application of amplitude, frequency, and other attributes to stratigraphic and hydrocarbon determination, *in* C. E. Payton, ed., Seismic stratigraphy: AAPG Memoir 26, p. 301–328.

Widess, M. B., 1973, How thin is a thin bed?: Geophysics, v. 38, p. 1176–1180.

Subject Index

Author Index